Decision Making under Deep Uncertainty

Vincent A. W. J. Marchau · Warren E. Walker ·
Pieter J. T. M. Bloemen · Steven W. Popper
Editors

Decision Making under Deep Uncertainty

From Theory to Practice

OPEN

Editors
Vincent A. W. J. Marchau
Nijmegen School of Management
Radboud University
Nijmegen, Gelderland, The Netherlands

Pieter J. T. M. Bloemen
Staff Delta Programme Commissioner
Ministry of Infrastructure and Water Management
The Hague, The Netherlands

Warren E. Walker
Faculty of Technology,
Policy and Management
Delft University of Technology
Delft, Zuid-Holland, The Netherlands

Steven W. Popper
Pardee RAND Graduate School
RAND Corporation
Santa Monica, CA, USA

ISBN 978-3-030-05251-5   ISBN 978-3-030-05252-2 (eBook)
https://doi.org/10.1007/978-3-030-05252-2

Library of Congress Control Number: 2018966116

© The Editor(s) (if applicable) and The Author(s) 2019. This book is an open access publication.
Open Access This book is licensed under the terms of the Creative Commons Attribution 4.0 International License (http://creativecommons.org/licenses/by/4.0/), which permits use, sharing, adaptation, distribution and reproduction in any medium or format, as long as you give appropriate credit to the original author(s) and the source, provide a link to the Creative Commons license and indicate if changes were made.
The images or other third party material in this book are included in the book's Creative Commons license, unless indicated otherwise in a credit line to the material. If material is not included in the book's Creative Commons license and your intended use is not permitted by statutory regulation or exceeds the permitted use, you will need to obtain permission directly from the copyright holder.
The use of general descriptive names, registered names, trademarks, service marks, etc. in this publication does not imply, even in the absence of a specific statement, that such names are exempt from the relevant protective laws and regulations and therefore free for general use.
The publisher, the authors and the editors are safe to assume that the advice and information in this book are believed to be true and accurate at the date of publication. Neither the publisher nor the authors or the editors give a warranty, express or implied, with respect to the material contained herein or for any errors or omissions that may have been made. The publisher remains neutral with regard to jurisdictional claims in published maps and institutional affiliations.

This Springer imprint is published by the registered company Springer Nature Switzerland AG
The registered company address is: Gewerbestrasse 11, 6330 Cham, Switzerland

Preface

Making plans for the future involves anticipating changes, especially when making long-term plans or planning for rare events. When these changes are characterized by a high degree of uncertainty, we consider the resulting situation to be "deeply uncertain"—a situation in which the experts do not know or the parties to a decision cannot agree upon "(1) the appropriate models to describe the interactions among a system's variables, (2) the probability distributions to represent uncertainty about key variables and parameters in the models, and/or (3) how to value the desirability of alternative outcomes."[1]

Several books have been written in the past that deal with different aspects of decisionmaking under uncertainty (in a broad sense). But, there are none that aim to integrate these aspects for the specific subset of decisionmaking under *deep* uncertainty. This book provides a unified and comprehensive treatment of the approaches and tools for developing policies under deep uncertainty, and their application. It elucidates the state of the art in both theory and practice associated with the approaches and tools for decisionmaking under deep uncertainty. It has been produced under the aegis of the Society for Decision Making under Deep Uncertainty (DMDU: http://www.deepuncertainty.org), whose members develop the approaches and tools supporting the design of courses of action or policies under deep uncertainty, and work to apply them in the real world.

The book is intended for use by a broad audience, including students, lecturers, and researchers in the field of decisionmaking under deep uncertainty for various domains, those who commission the application of DMDU approaches and consume the results in the government and the private sectors, and those who carry out such studies. It provides (1) guidance in identifying and applying appropriate approaches and tools to design policies and (2) advice on implementing these policies in the real world. For decisionmakers and practitioners, the book includes realistic examples and practical guidelines that explain what decisionmaking under

---

[1]Lempert R. J., S. W. Popper, and S. C. Bankes (2003) *Shaping the Next One Hundred Years: New Methods for Quantitative, Long-Term Policy Analysis*, MR-1626-RPC, RAND, Santa Monica, California, pp. 3–4.

deep uncertainty is and how it may be of assistance to them. The approaches and tools described and the problems addressed in the book have a cross-sectoral and cross-country applicability.

The book provides the first synthesis of the large body of work on designing policies under deep uncertainty, in both theory and practice. It broadens traditional approaches and tools to include the analysis of actors and networks related to the problem at hand. And it shows how lessons learned in the application process can be used to improve the approaches and tools used in the design process.

Nijmegen, The Netherlands	Vincent A. W. J. Marchau
Delft, The Netherlands	Warren E. Walker
The Hague, The Netherlands	Pieter J. T. M. Bloemen
Santa Monica, USA	Steven W. Popper

Acknowledgements

The publication of this book has been funded by the Radboud University, the RAND Corporation, Delft University of Technology, and Deltares.

Contents

1 **Introduction** 1
Vincent A. W. J. Marchau, Warren E. Walker,
Pieter J. T. M. Bloemen and Steven W. Popper

Part I DMDU Approaches

2 **Robust Decision Making (RDM)** 23
R. J. Lempert

3 **Dynamic Adaptive Planning (DAP)** 53
Warren E. Walker, Vincent A. W. J. Marchau and Jan H. Kwakkel

4 **Dynamic Adaptive Policy Pathways (DAPP)** 71
Marjolijn Haasnoot, Andrew Warren and Jan H. Kwakkel

5 **Info-Gap Decision Theory (IG)** 93
Yakov Ben-Haim

6 **Engineering Options Analysis (EOA)** 117
Richard de Neufville and Kim Smet

Part II DMDU Applications

7 **Robust Decision Making (RDM): Application to Water Planning
and Climate Policy** 135
David G. Groves, Edmundo Molina-Perez, Evan Bloom
and Jordan R. Fischbach

8 **Dynamic Adaptive Planning (DAP): The Case of Intelligent
Speed Adaptation** 165
Vincent A. W. J. Marchau, Warren E. Walker
and Jan-Willem G. M. van der Pas

9 **Dynamic Adaptive Policy Pathways (DAPP): From Theory to Practice** 187
Judy Lawrence, Marjolijn Haasnoot, Laura McKim,
Dayasiri Atapattu, Graeme Campbell and Adolf Stroombergen

10 **Info-Gap (IG): Robust Design of a Mechanical Latch** 201
François M. Hemez and Kendra L. Van Buren

11 **Engineering Options Analysis (EOA): Applications** 223
Richard de Neufville, Kim Smet, Michel-Alexandre Cardin
and Mehdi Ranjbar-Bourani

Part III DMDU-Implementation Processes

12 **Decision Scaling (DS): Decision Support for Climate Change** 255
Casey Brown, Scott Steinschneider, Patrick Ray, Sungwook Wi,
Leon Basdekas and David Yates

13 **A Conceptual Model of Planned Adaptation (PA)** 289
Jesse Sowell

14 **DMDU into Practice: Adaptive Delta Management in The Netherlands** 321
Pieter J. T. M. Bloemen, Floris Hammer, Maarten J. van der Vlist,
Pieter Grinwis and Jos van Alphen

Part IV DMDU-Synthesis

15 **Supporting DMDU: A Taxonomy of Approaches and Tools** 355
Jan H. Kwakkel and Marjolijn Haasnoot

16 **Reflections: DMDU and Public Policy for Uncertain Times** 375
Steven W. Popper

17 **Conclusions and Outlook** 393
Vincent A. W. J. Marchau, Warren E. Walker,
Pieter J. T. M. Bloemen and Steven W. Popper

Glossary 401

About the Editors

Prof. Vincent A. W. J. Marchau (Radboud University (RU), Nijmegen School of Management) holds a chair on Uncertainty and Adaptivity of Societal Systems. This chair is supported by The Netherlands Study Centre for Technology Trends (STT). His research focuses on long-term planning under uncertainty in transportation, logistics, spatial planning, energy, water, and security. Marchau is also Managing Director of the Dutch Research School for Transport, Infrastructure and Logistics (TRAIL) at Delft University of Technology (DUT), with 100 Ph.D. students and 50 staff members across six Dutch universities.

Prof. Warren E. Walker (Emeritus Professor of Policy Analysis, Delft University of Technology) has a Ph.D. in Operations Research from Cornell University, and more than 40 years of experience as an analyst and project leader at the RAND Corporation, applying quantitative analysis to public policy problems. His recent research has focused on methods for dealing with deep uncertainty in making public policies (especially with respect to climate change), improving the freight transport system in the Netherlands, and the design of decision support systems for airport strategic planning. He is the recipient of the 1997 President's Award from the Institute for Operations Research and the Management Sciences (INFORMS) for his "contributions to the welfare of society through quantitative analysis of governmental policy problems."

Dr. Pieter J. T. M. Bloemen (Ministry of Infrastructure and Water Management—Staff Delta Programme Commissioner) is the Chief Strategic Officer of the Dutch Delta Programme. He is responsible for the development and application of Adaptive Delta Management. He led the Strategic Environmental Assessment of the Delta Decisions and preferred regional strategies published in 2014. He presently works on the development of a monitoring and evaluation system that matches the adaptive approach of the Delta Programme and is responsible for the first six-yearly review of the Delta Decisions and preferred regional strategies, planned for 2020.

He is Visiting Researcher at IHE Delft Institute of Water Education (Chair Group Flood Resilience) since January 2015 and works on a Ph.D. thesis on the governance of the adaptive approach.

Dr. Steven W. Popper (RAND Corporation) is a RAND Senior Economist and Professor of science and technology policy in the Pardee RAND Graduate School. His work on macrotransitions led to an invitation by President Vaclav Havel to advise the government of Czechoslovakia, participation in an OECD delegation on the first foreign visit to one of the secret cities of the former Soviet Union, and consultation to the World Bank on issues of industrial restructuring in Hungary and in Mexico. His work on micro-level transition focuses on innovation. From 1996 to 2001, he was the Associate Director of the Science and Technology Policy Institute, providing analytic support to the White House Office of Science and Technology Policy and other executive branch agencies. He has taught planning under deep uncertainty at the Pardee RAND Graduate School, the India School of Business, and the Shanghai Climate Institute. He is co-developer of the Robust Decision-Making (RDM) approach. He is an elected Fellow of the American Association for the Advancement of Science, served as chair of the AAAS Industrial Science and Technology section, and is the founding chair for education and training of the Society for Decision Making under Deep Uncertainty.

List of Contributors

Dayasiri Atapattu Greater Wellington Regional Council, Wellington, New Zealand

Leon Basdekas Black and Veatch, Colorado Springs, CO, USA

Yakov Ben-Haim Technion—Israel Institute of Technology, Haifa, Israel

Pieter J. T. M. Bloemen Staff Delta Programme Commissioner, Ministry of Infrastructure and Water Management, The Hague, The Netherlands

Evan Bloom RAND Corporation, Santa Monica, CA, USA

Casey Brown University of Massachusetts, Amherst, MA, USA

Graeme Campbell Infometrics Consulting, Wellington, New Zealand

Michel-Alexandre Cardin Imperial College London, London, UK

Richard de Neufville Massachusetts Institute of Technology, Cambridge, MA, USA

Jordan R. Fischbach RAND Corporation, Santa Monica, CA, USA

Pieter Grinwis Member of the Municipal Council of The Hague and Policy Officer for the ChristenUnie in the Dutch Senate and the House of Representatives, The Hague, The Netherlands

David G. Groves RAND Corporation, Santa Monica, CA, USA

Marjolijn Haasnoot Water Resources and Delta Management, Deltares, Delft, The Netherlands;
Faculty of Physical Geography, Utrecht University, Utrecht, The Netherlands

Floris Hammer Independent consultant, specializing in executing technical programmes and projects in complex political environments, The Hague, The Netherlands

François M. Hemez Lawrence Livermore National Laboratory, Livermore, CA, USA

Jan H. Kwakkel Faculty of Technology, Policy and Management (TPM), Delft University of Technology, Delft, The Netherlands

Judy Lawrence Victoria University of Wellington, Wellington, New Zealand

R. J. Lempert Rand Corporation, Santa Monica, CA, USA

Vincent A. W. J. Marchau Nijmegen School of Management, Radboud University, Nijmegen, The Netherlands

Laura McKim Greater Wellington Regional Council, Wellington, New Zealand

Edmundo Molina-Perez Tecnológico de Monterrey, Monterrey, Mexico

Steven W. Popper Pardee RAND Graduate School, RAND Corporation, Santa Monica, CA, USA

Mehdi Ranjbar-Bourani University of Science and Technology of Mazandaran, Behshahr, Iran

Patrick Ray University of Cincinnati, Cincinnati, OH, USA

Kim Smet University of Ottawa, Ottawa, Canada

Jesse Sowell Department of International Affairs, The Bush School of Government & Public Service, Texas A&M University, College Station, TX, USA

Scott Steinschneider Cornell University, Ithaca, NY, USA

Adolf Stroombergen Infometrics Consulting, Wellington, New Zealand

Jos van Alphen Staff Delta Programme Commissioner, Ministry of Infrastructure and Water Management, The Hague, The Netherlands

Kendra L. Van Buren Los Alamos National Laboratory, Los Alamos, NM, USA

Jan-Willem G. M. van der Pas Municipality of Eindhoven, Eindhoven, The Netherlands

Maarten J. van der Vlist Rijkswaterstaat, Ministry of Infrastructure and Water Management, The Hague, The Netherlands

Warren E. Walker Faculty of Technology, Policy & Management, Delft University of Technology, Delft, The Netherlands

Andrew Warren Water Resources and Delta Management, Deltares, Delft, The Netherlands

Sungwook Wi University of Massachusetts, Amherst, MA, USA

David Yates National Center for Atmospheric Research, Boulder, CO, USA

Chapter 1
Introduction

Vincent A. W. J. Marchau, Warren E. Walker, Pieter J. T. M. Bloemen and Steven W. Popper

Abstract Decisionmaking for the future depends on anticipating change. And this anticipation is becoming increasingly difficult, thus creating anxiety when we seek to conform short-term decisions to long-term objectives or to prepare for rare events. Decisionmakers, and the analysts upon whom they rely, have had good reason to feel decreasing confidence in their ability to anticipate correctly future technological, economic, and social developments, future changes in the system they are trying to improve, or the multiplicity and time-varying preferences of stakeholders regarding the system's outcomes. Consider, for example, decisionmaking related to the consequences of climate change, the future demand for and means for providing mobility, the planning of mega-scale infrastructure projects, the selection of energy sources to rely on in the future, the role of genomics in health care, or how cities will develop. Or think of rare events like a natural disaster, a financial crisis, or a terrorist attack.

1.1 The Need for Considering Uncertainty in Decisionmaking

Decisionmaking for the future depends on anticipating change. And this anticipation is becoming increasingly difficult, thus creating anxiety when we seek to conform short-term decisions to long-term objectives or to prepare for rare events. Deci-

---

V. A. W. J. Marchau (✉)
Nijmegen School of Management, Radboud University, Nijmegen, The Netherlands
e-mail: v.marchau@fm.ru.nl

W. E. Walker
Faculty of Technology, Policy & Management, Delft University of Technology, Delft, The Netherlands

P. J. T. M. Bloemen
Staff Delta Programme Commissioner, Ministry of Infrastructure and Water Management, The Hague, The Netherlands
e-mail: pieter.bloemen@deltacommissaris.nl

S. W. Popper
Pardee RAND Graduate School, RAND Corporation, Santa Monica, CA, USA

© The Author(s) 2019
V. A. W. J. Marchau et al. (eds.), *Decision Making under Deep Uncertainty*,
https://doi.org/10.1007/978-3-030-05252-2_1

sionmakers, and the analysts upon whom they rely, have had good reason to feel decreasing confidence in their ability to anticipate correctly future technological, economic, and social developments, future changes in the system they are trying to improve, or the multiplicity and time-varying preferences of stakeholders regarding the system's outcomes. Consider, for example, decisionmaking related to the consequences of climate change, the future demand for and means for providing mobility, the planning of mega-scale infrastructure projects, the selection of energy sources to rely on in the future, the role of genomics in healthcare, or how cities will develop. Or think of rare events like a natural disaster, a financial crisis, or a terrorist attack. These topics are all characterized by what can be called "*deep uncertainty.*" In these situations, the experts do not know or the parties to a decision cannot agree upon (i) the external context of the system, (ii) how the system works and its boundaries, and/or (iii) the outcomes of interest from the system and/or their relative importance (Lempert et al. 2003). Deep uncertainty also arises from actions taken over time in response to unpredictable evolving situations (Haasnoot et al. 2013).

In a broad sense, uncertainty (whether deep or not) may be defined simply as *limited knowledge* about future, past, or current events (Walker et al. 2013). With respect to decisionmaking, uncertainty refers to the gap between available knowledge and the knowledge decisionmakers would need in order to make the best policy choice. This uncertainty clearly involves subjectivity, since it relates to satisfaction with existing knowledge, which is colored by the underlying values and perspectives of the decisionmaker (and the various actors involved in the decisionmaking process). But this in itself becomes a trap when implicit assumptions are left unexamined or unquestioned. Uncertainty can be associated with all aspects of a problem of interest (e.g., the system comprising the decision domain, the world outside the system, the outcomes from the system, and the importance stakeholders place on the various outcomes from the system).

The planning for the Channel Tunnel provides an illustration of the danger of ignoring uncertainty. Figure 1.1 shows the forecasts from different studies and the actual number of passengers for the rail tunnel under the English Channel. The competition from low-cost air carriers and the price reactions by operators of ferries, among other factors, were not taken into account in most studies. This resulted in a significant overestimation of the tunnel's revenues and market position (Anguara 2006) with devastating consequences for the project. Twenty years after its opening in 1994, it still did not carry the number of passengers that had been predicted.[1]

Climate change presents a fundamental challenge to bringing analytical insight into policy decisions because of deep uncertainties. Climate change is commonly mentioned as a source of deep uncertainty. The question of assigning probabilities to future scenarios of climate change is particularly controversial. While many argue that scientific uncertainty about emissions simply does not allow us to derive reliable probability distributions for future climate states, others counter by saying that the

---

[1] The projected traffic growth is still far from being achieved, "In 2017, about 21 million passengers, on all services, have travelled through the Channel Tunnel" http://www.eurotunnelgroup.com/uk/eurotunnel-group/operations/traffic-figures/.

1 Introduction

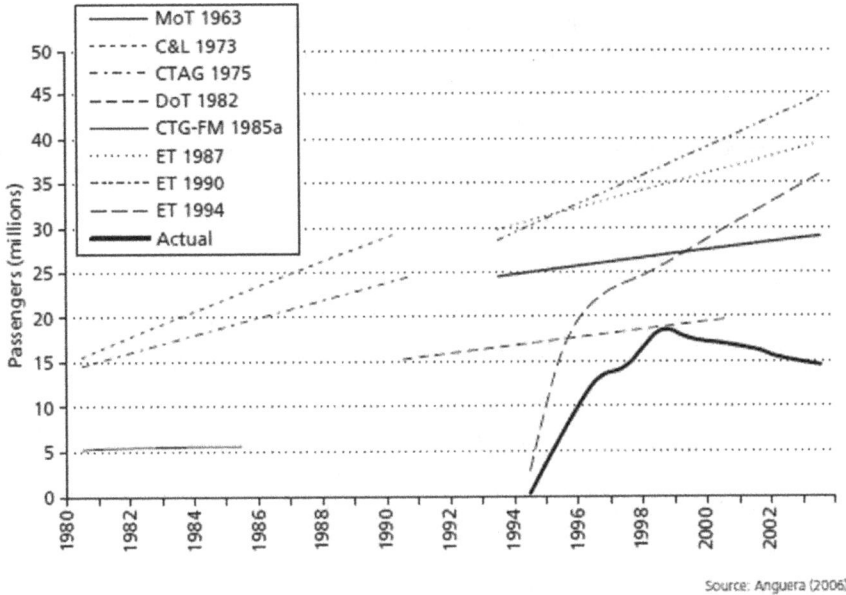

Fig. 1.1 Channel Tunnel passenger traffic forecasts and actual results

lack of assigned probabilities gives non-experts free rein to assign their own, less well-informed probability estimates (see also Sect. 11.5). Climate change research is plagued by imperfect and incomplete understanding about the functioning of natural (environmental) phenomena and processes, about how changes in these phenomena and processes translate into increases in important climate variables, such as precipitation, storm intensities, and global temperatures, and the economic and social consequences of such climatic changes. For a long time, these uncertainties opened the very existence of global climate change to challenge. In recent years, the uncertainty as to whether climate change is taking place has been largely removed (e.g., Cook et al. 2013). There remains, however, considerable uncertainty about (Hallegatte 2009; IPCC 2014):

- The magnitude of climate change (with estimates of increased average temperatures differing greatly across a range of future scenarios);
- The speed of climate change (which determines how quickly policy actions need to be taken);
- The implications for specific areas and regions (the effects of climate change are potentially larger for countries like Bangladesh and the Netherlands than for countries like Mongolia; but even within sub-national regions, such as California, the direction of change is hard to determine);
- The policies that should be implemented to mitigate and/or hedge against the adverse consequences of climate change (because of a lack of knowledge about the costs and benefits of different alternatives for protecting ourselves from the adverse consequences of climate change).

The financial crisis that gripped the world in 2008–2009 is perhaps the most recent acute example of why acknowledging and confronting deep uncertainty are of great importance. The speed and the severity of the decline in world economies were unprecedented in recent times, but it surprised policymakers who were unprepared when it arrived. As Alan Greenspan admitted in October 2008, "I found a flaw in the model that I perceived is the critical functioning structure that defines how the world works.... I was shocked, because I had been going for 40 years or more with very considerable evidence that it was working exceptionally well." (Hearing Before the Committee on Oversight and Government Reform 2008, p. 46).

Hence, for long-term decisionmaking, deep uncertainties are in most cases a given. Whether to take them into consideration in making a decision is a choice. Decisionmaking, even under the most favorable circumstances, is made difficult enough by budget constraints, conflicting stakes, and political turmoil. So, every reduction in uncertainty is more than welcome; ignoring deep uncertainty is attractive. But decisions that ignore deep uncertainty ignore reality. Substituting assumptions for deep uncertainties might simplify choices in the short term but may come at a much higher price in the longer term.

Some would say that it is impossible to make good decisions in the face of deep uncertainty. However, the good news is that recent years have seen the birth and proliferation of ways to include a systematic consideration of deep uncertainty in practical decisionmaking. Many are described in this book. The experience derived from the application of different approaches and tools in a range of policy fields may enable analysts, planners, and the decisionmakers whom they support to come to better recognize, confront, and operate effectively given the deep uncertainties present in their own worlds.

1.2 A Framework for Decision Support

Given a problem, decisionmaking requires an integrated and holistic view of various alternatives, their possible consequences, and conditions (including acceptability, legislation, and institutions) for implementation. Alternatives are the means by which it may be possible to achieve the objectives. Depending on the problem, the alternatives may be policies, strategies, plans, designs, actions, or whatever it appears might attain the objective (Miser and Quade 1985).[2] Walker (2000) views decisionmaking as choosing among alternatives in order to change system outcomes in a desired way (see Fig. 1.2). It involves the specification of policies (P) to influence the behavior of the system to achieve the goals. At the heart of this view is the system that decisionmakers influence directly by their means, distinguishing the system's physical and human elements and their mutual interactions. The results of these interactions (the system outputs) are called *outcomes of interest* (O). They are considered relevant

---

[2]Each chapter of this book uses the term for the alternatives that is most appropriate to its content. In this introductory chapter, we use the terms policy, strategy, and plan interchangeably.

1 Introduction

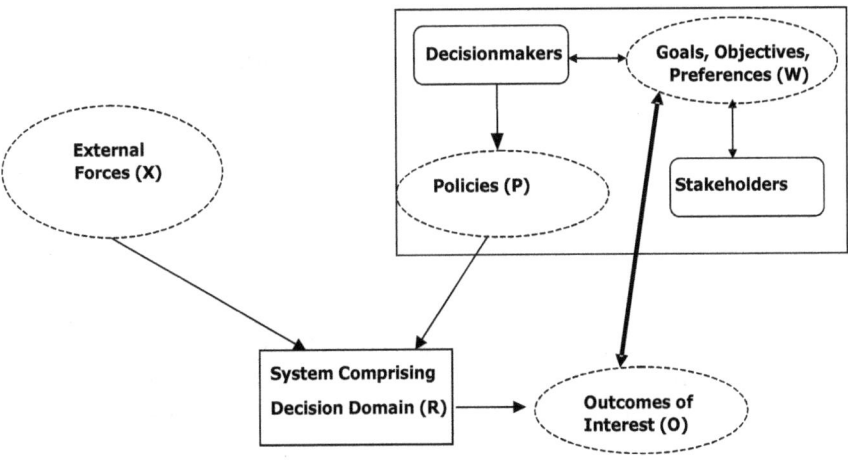

Fig. 1.2 A framework for decision support

criteria for the evaluation of policies. The valuation of outcomes refers to the (relative) weights, not necessarily quantitative, given to the outcomes by crucial stakeholders, including decisionmakers (W). Other external forces (X) act upon the system along with the policies. Both may affect the relationship among elements of the system (R) and hence the structure of the system itself, as well as the outcomes of interest to decisionmakers and other stakeholders. *External forces* refer to forces outside the system that are not controllable by the decisionmaker, but may influence the system significantly (e.g., technological developments, societal developments, economic developments, political developments).[3] Of course, there are uncontrollable autonomous developments inside the system that also may influence the outcomes of interest (i.e., emergent or self-organizing behavior).

Applying the framework shown in Fig. 1.2 to decisionmaking under uncertainty reveals several locations where uncertainties might arise. First, there is likely to be *scenario uncertainty*, since it is difficult to identify which external developments will be relevant for long-term future system performance and, perhaps more important, the size and direction of these changes. Second, even if external developments were certain (that is, we knew how the external world would develop), there might still be *structural uncertainty* about how the system would respond to those external developments (and about the autonomous developments inside the system). The causal mechanisms determining system performance may also be uncertain (i.e., many of the interactions within the system might be insufficiently known) or there may be clashing hypotheses of causation. Finally, the valuation of the various outcomes might be subject to contention. Stakeholders often differ on the importance of future problems and the objectives to be achieved with a plan. This results in different,

---

[3]Clearly, what is external depends on being specific in identifying the decisionmakers. For example, for many decisions, policies of the European Commission would be an external force. This is not the case for analyses in which the European Commission plays the decisionmaking role.

often conflicting, opinions regarding the various policies and the characterization of outcome acceptability. Of course, over time, new stakeholders might emerge, current stakeholders might leave, or preferences might change, all of which will affect the valuation of the outcomes.

1.3 Dealing with Uncertainty in Decisionmaking

In order to manage uncertainty in decisionmaking, one must be aware that an entire spectrum of different levels of knowledge exists, ranging from the unachievable ideal of complete deterministic understanding at one end of the scale to total ignorance at the other. The range of levels of uncertainty, and their challenge to decisionmakers, was acknowledged by Donald
Rumsfeld, who famously said:

> As we know, there are known knowns – these are things we know we know. We also know there are known unknowns – that is to say we know there are some things we do not know; but there are also unknown unknowns – the ones we don't know we don't know.... It is the latter category that tends to be the difficult one.[4]

Uncertainty as inadequacy of knowledge has a very long history, dating back at least to the epistemological questions debated among the ancient Greeks. Its modern history begins around 1921, when Knight made a distinction between risk and uncertainty (Knight 1921). According to Knight, risk denotes the calculable (the probability of an event times the loss if the event occurred) and thus controllable part of all that is unknowable. The remainder is the uncertain—incalculable and uncontrollable. Over the years, this definition has been adopted by different authors to make a distinction between decisionmaking under risk and decisionmaking under uncertainty (Luce and Raiffa 1957; Morgan and Henrion 1990). Note that decisionmaking under uncertainty refers not only to the uncertain future state of the world, but also to the uncertainty resulting from the strategic behavior of actors involved in the decisionmaking (Quade 1989). Hence, risk can be treated as one kind of uncertainty—a low level of uncertainty that can be quantified by using losses and probabilities. Probabilities cannot reliably be associated with the remaining uncertainties. That is, uncertainty is a broader concept than risk.

With application to decisionmaking, a distinction can be made between two extreme levels of uncertainty (determinism and total ignorance) and four intermediate levels (e.g., Courtney 2001; Walker et al. 2003). Walker et al. (2003) define the four levels with respect to the knowledge assumed about four aspects of a problem: (a) the future state of the world, (b) the model of the relevant system for that future world, (c) the outcomes from the system, and (d) the weights that the various stakeholders will put on the outcomes. The levels of uncertainty are briefly discussed below.

---

[4]Donald Rumsfeld, Department of Defense News Briefing, February 12, 2002.

Complete certainty is the situation in which we know everything precisely. This is almost never attainable, but acts as a limiting characteristic at one end of the spectrum.

Level 1 uncertainty represents situations in which one admits that one is not absolutely certain, but one does not see the need for, or is not able, to measure the degree of uncertainty in any explicit way (Hillier and Lieberman 2001, p. 43). These are generally situations involving short-term decisions, in which the system of interest is well defined and it is reasonable to assume that historical data can be used as predictors of the future. Level 1 uncertainty, if acknowledged at all, is generally treated through a simple sensitivity analysis of model parameters, where the impacts of small perturbations of model input parameters on the outcomes of a model are assessed. Several services in our life are predictable, based on the past such as mail delivery and garbage collection. These are examples of this level of uncertainty.

In the case of *Level 2 uncertainties*, it is assumed that the system model or its inputs can be described probabilistically, or that there are a few alternative futures that can be predicted well enough (and to which probabilities can be assigned). The system model includes parameters describing the stochastic—or probabilistic—properties of the underlying system. In this case, the model can be used to estimate the probability distributions of the outcomes of interest for these futures. A preferred policy can be chosen based on the outcomes and the associated probabilities of the futures (i.e., based on "expected outcomes" and levels of acceptable risk). The tools of probability and statistics can be used to solve problems involving Level 2 uncertainties. Deciding on which line to join in a supermarket would be a Level 2 problem.

Level 3 uncertainties involve situations in which there are a limited set of plausible futures, system models, outcomes, or weights, and probabilities cannot be assigned to them—so the tools of neither Level 1 nor Level 2 are appropriate. In these cases, traditional scenario analysis is usually used. The core of this approach is that the future can be predicted well enough to identify policies that will produce favorable outcomes in a few specific, plausible future worlds (Schwartz 1996). The future worlds are called scenarios. Analysts use best-estimate models (based on the most up-to-date scientific knowledge) to examine the consequences that would follow from the implementation of each of several possible policies in each scenario. The "best" policy is the one that produces the most favorable outcomes across the scenarios. (Such a policy is called *robust*.) A scenario does not predict what *will* happen in the future; rather it is a plausible description of what *can* happen. The scenario approach assumes that, although the likelihood of the future worlds is unknown, the range of plausible futures can be specified well enough to identify a (static) policy that will produce acceptable outcomes in most of them. Leaving an umbrella in the trunk of your car in case of rain is an approach to addressing Level 3 uncertainty.

Level 4 uncertainty represents the deepest level of recognized uncertainty. A distinction can be made between situations in which we are still able (or assume) to bound the future around many plausible futures (4a) and situations in which we only know that we do not know (4b). This vacuum can be due to a lack of knowledge or data about the mechanism or functional relationships being studied (4a), but this can also stem from the potential for unpredictable, surprising, events (4b). Taleb (2007) calls these events "black swans." He defines a black swan event as one that lies outside the realm of regular expectations (i.e., "nothing in the past can convincingly point to its possibility"), carries an extreme impact, and is explainable only after the fact (i.e., through retrospective, not prospective, predictability). In these situations, analysts either struggle to (Level 4a) or cannot (Level 4b) specify the appropriate models to describe interactions among the system's variables, select the probability distributions to represent uncertainty about key parameters in the models, and/or value the desirability of alternative outcomes.

Total ignorance is the other extreme from determinism on the scale of uncertainty; it acts as a limiting characteristic at the other end of the spectrum.

Table 1.1 provides a representation of the four intermediate levels of uncertainty for the four locations (X, R, O, and W). Most of the traditional applied scientific work in the engineering, social, and natural sciences assumes that uncertainties result either from a lack of information (i.e., assumes that uncertainties associated with a problem are of Level 1 or 2), which leads to an emphasis on uncertainty reduction through ever-increasing information seeking and processing, or from random variation, which concentrates efforts on stochastic processes and statistical analysis.

The approach to deal with Level 3 uncertainties assumes that a few future worlds can be specified well enough to determine robust policies that will produce favorable outcomes in most of them. These future worlds are described by means of scenarios. The best policy is the policy that produces the most desirable outcomes across the range of scenarios, minimizes the maximum possible regret across the range of scenarios, etc. Although this approach has been successful in the past, the problem is that if the range of assumptions about the future turns out to be wrong, the negative consequences might be large. Also, research points out that scenarios, in general, have rarely been used to address discontinuities in future developments; there has been a natural tendency for scenarios to stay close to evolutionary (discontinuity-averse) business-as-usual situations (van Notten et al. 2005).

However, the important problems faced by decisionmakers are often characterized by a higher level of uncertainty (Level 4a and 4b) that cannot be sufficiently reduced by gathering more information. The uncertainties are unknowable at the present time. Uncertainty permeates some or all aspects of the problem: the external developments, the appropriate (future) system model, the model outcomes, and the valuation of the outcomes by (future) stakeholders. As mentioned at the beginning of this chapter, such situations are defined as *decisionmaking under deep uncertainty*.

1.4 Decisionmaking Under Deep Uncertainty

As individuals, we confront and navigate, more or less successfully, a world filled with many uncertainties. Some characteristic decisions are even classifiable as ones framed by deep uncertainties, e.g., whom to marry; what profession to select; where to live; what jobs to seek. As individuals, the decisionmaking becomes more prone to anxiety as the number of variables increases, thus overwhelming innate abilities

Table 1.1 Progressive transition of levels of uncertainty

	Complete determinism	Level 1	Level 2	Level 3	Level 4 (deep uncertainty)		Total ignorance
					Level 4a	Level 4b	
Context (X)		A clear enough future	Alternate futures (with probabilities)	A few plausible futures	Many plausible futures	Unknown future	
System model (R)		A single (deterministic) system model	A single (stochastic) system model	A few alternative system models	Many alternative system models	Unknown system model; know we don't know	
System outcomes (O)		A point estimate for each outcome	A confidence interval for each outcome	A limited range of outcomes	A wide range of outcomes	Unknown outcomes; know we don't know	
Weights (W)		A single set of weights	Several sets of weights, with a probability attached to each set	A limited range weights	A wide range of weights	Unknown weights; know we don't know	

to think through the combinatorics of potential influences and outcomes. But in the public sector, or for decisionmaking within organizations, the number of actors further increases the practical difficulty of addressing such problems. Hierarchies made up of intelligent, dedicated, and well-disposed individuals may yield collectively poor decisions, because existing mechanisms do not provide appropriate responses in the presence of deep uncertainty.

Integral to DMDU approaches is the use of analytic methods for decision support. Using a common definition from the US National Research Council (2009), decision support represents a "set of processes intended to create the conditions for the production and appropriate use of decision-relevant information." Among its key tenets, decision support emphasizes: (1) that the way in which information is integrated into decisionmaking processes often proves as important as the information products themselves, (2) the coproduction of knowledge among information users and producers, and (3) the need to design the decision process to facilitate learning.

New approaches for decisionmaking under conditions of deep uncertainty are needed because the approaches used for Level 1, Level 2, and Level 3 are rendered inadequate under these conditions (Lempert et al. 2003). Or as Quade (1989, p. 160) phrased this: "Stochastic uncertainties are therefore among the least of our worries; their effects are swamped by uncertainties about the state of the world and human factors for which we know absolutely nothing about probability distributions and little more about the possible outcomes." Goodwin and Wright (2010, p. 355) demonstrate that "all the extant forecasting methods—including the use of expert judgment, statistical forecasting, Delphi and prediction markets—contain fundamental weaknesses." And Popper et al. (2009, p. 50) state that the traditional methods "all founder on the same shoals: an inability to grapple with the long-term's multiplicity of plausible futures." Traditionally, the past has been seen as a reasonably reliable predictor of the future. Such an assumption works well when change is slow and system elements are not too tightly connected. It also works well when there are not too many Black Swans coming toward us. In fact, assuming that the future will be fairly closely related to the past is the basis of classical statistics and probability and the foundation for the way we live our lives. However, if this assumption is not a valid one, plans based on it may lead to very undesirable outcomes.

Box 1.1: When to Use DMDU Approaches

DMDU approaches provide many benefits, but also impose costs. As shown in Fig. 1.3, the benefits are likely to exceed the costs when three conditions are met. Most clearly, these approaches are more useful the more the contextual uncertainties (X) are deep, rather than well characterized. In addition, DMDU approaches are more useful when the set of policies (P) has more rather than fewer degrees of freedom. When uncertainties are well characterized and/or few degrees of decision freedom exist, DMDU approaches yield few benefits over traditional predict-then-act approaches. The third condition, system complexity (R), is a heuristic for how well experts do know and/or disagree on the

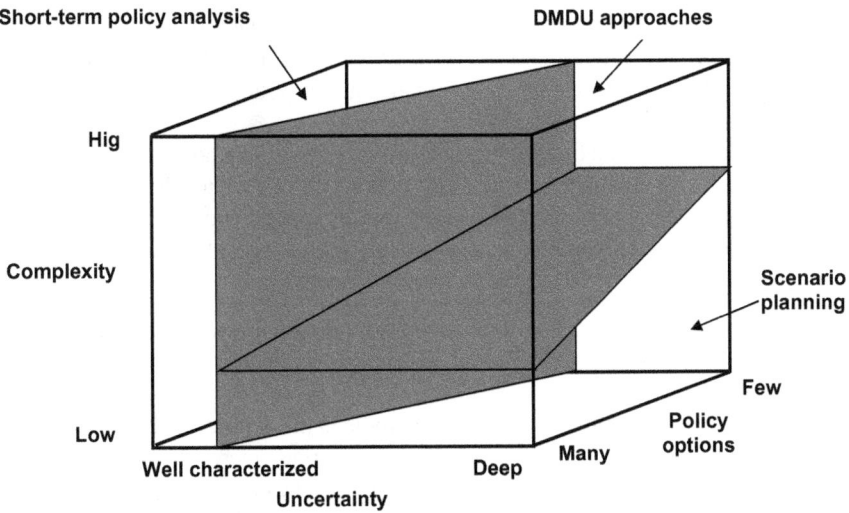

Fig. 1.3 What to do—and when—depends (based on: Dewar 2006)

> proper models, probabilities, and/or system outcomes. When expert intuition is sufficient to link the policies to the relevant outcomes, then scenario planning may suffice. But when the future world, the system, and/or the outcomes have the potential to surprise, a full DMDU analysis may prove valuable.

Decisionmaking in the context of deep uncertainty requires a paradigm that is not based on predictions of the future (known in the literature as the "predict-then-act" paradigm), but that aims to prepare and adapt (in order to prepare for uncertain events) by monitoring how the future evolves and allowing adaptations over time as knowledge is gained (in order to implement long-term strategies). The "monitor and adapt" paradigm explicitly recognizes the deep uncertainty surrounding decision-making for uncertain events and long-term developments and emphasizes the need to take this deep uncertainty into account.

Several books have been written in the past that deal with different aspects of decisionmaking under uncertainty (e.g., Morgan and Henrion 1990; Bedford and Cooke 2001; Lourenco et al. 2014). But, there are practically none that deal with decisionmaking under deep uncertainty, although Lempert et al. (2003) do. In recent years, however, there has been an increasing amount of research devoted to this topic. Some of this research focuses on theoretical approaches and tools for supporting the design of long-term policies or strategies; other research focuses on testing these new approaches and tools in practice (i.e., in the "real world"). With respect to the theory, new analytical tools (e.g., exploratory modeling, scenario discovery) and approaches (e.g., adaptive approaches) have been developed for handling deep uncertainty. And experimentation and application of these tools and approaches in the real world have begun. Up until now, there is no textbook providing a unified and comprehensive treatment of the approaches and tools for designing policies and strategies under deep uncertainty and their application.

This book provides the first synthesis of work on designing policies and strategies under deep uncertainty, both in theory and in practice. It presents the state of the art of approaches and tools supporting the design of policies and strategies that cope with situations of deep uncertainty and recent applications of the approaches and tools. In summary, the book has the following objectives:

- Providing a taxonomy to aid in understanding the relationships among different approaches and tools within the larger context of decisionmaking under deep uncertainty within the problem space;
- Describing the theoretical approaches and tools in the decisionmaking under deep uncertainty "toolkit"—how to choose the appropriate approach and tool(s) for a given situation and how to combine the tools over the course of a single study;
- Describing practical experiences with the approaches and tools;
- Identifying barriers and enablers for the use of the various approaches and tools in practice;
- Drawing insights for improving the tools and approaches, and lessons for practitioners using these tools and approaches.

1.5 Generic Elements of DMDU Approaches—A Framework

Recent years have seen many approaches for decisionmaking under deep uncertainty (DMDU). This section presents a generic set of elements across the approaches presented in the book. Each element includes several steps. Since the application of a DMDU approach is often recursive, involving cycling through iterative loops, there is no strict order in the elements and steps. Moreover, the different DMDU approaches each have their specific focus, which means that not all elements or steps are addressed in all approaches or receive the same emphasis.

In general, each of the DMDU approaches in this book consist of (some or all of) the following elements and steps (Fig. 1.4):

- Frame the analysis:
  - Formulate triggering issue (problem or opportunity)
  - Specify system structure and its boundaries (R)
  - Specify objectives/goals and outcome indicators (O)
  - Specify policies (P)
- Perform exploratory uncertainty analysis:

- Specify uncertainties or disagreements about external forces (X), system structure (R), outcome indicators (O), and valuation of outcomes (W)
- Explore the outcomes of policies and their vulnerabilities (and opportunities), given the uncertainties (using models or expert opinions)

• Choose initial actions and contingent actions:

- Illustrate trade-offs detected by analysis and means for future adjustments as events happen and knowledge is gained
- Select and plan for adoption of the (initial) policy and mechanisms for adjustment
- Plan for communicating, monitoring, and making possible adaptations of the policy

• Iterate and Re-examine

Frame the Analysis:

• Formulate triggering issue (problem or opportunity): This step sets the boundaries for all that follows. It involves examination of the questions or issues involved, establishing the context within which the issues are to be analyzed and the policies will have to function, clarifying constraints on possible actions, identifying the people/organizations that will be involved in making the policy decision, the people who will be affected by the policy decision, and deciding on the analytical approach to be followed. It establishes the trajectory for all that is to follow.

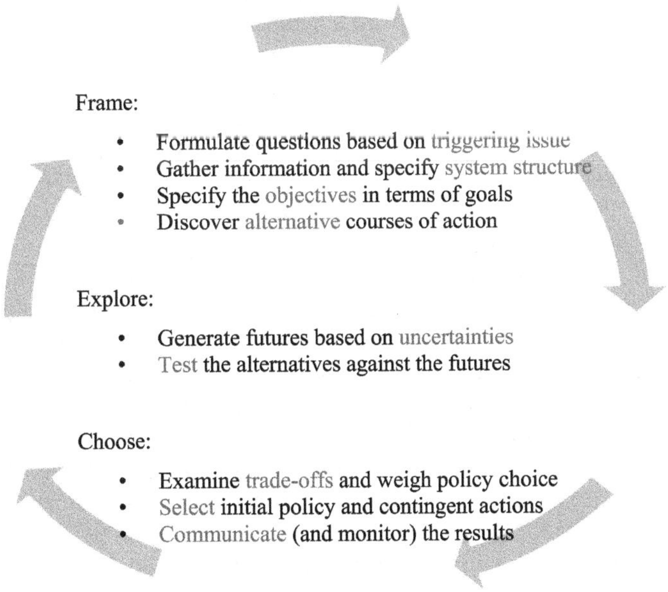

Frame:
- Formulate questions based on triggering issue
- Gather information and specify system structure
- Specify the objectives in terms of goals
- Discover alternative courses of action

Explore:
- Generate futures based on uncertainties
- Test the alternatives against the futures

Choose:
- Examine trade-offs and weigh policy choice
- Select initial policy and contingent actions
- Communicate (and monitor) the results

Fig. 1.4 Elements and steps of DMDU approaches

- Specify system and its boundaries: Different policy actors are able to make changes affecting a (very small) part of the world. The boundaries of this part of the world determine what is inside the "system of interest" and what is outside. A system model (or models) will be designed, built, and used in the remainder of the steps to examine policy alternatives and identify a preferred policy. Within the world of DMDU, a model is intended to be used not as a prediction tool, but as an engine for generating and examining possible futures (i.e., as an exploration tool). Therefore, a process of iterative analysis, rather than an *ex ante* presumption of necessary modeling detail, determines what modeling may be required to inform the decisions surrounding the triggering issue as fully as possible.
- Specify objectives/goals and outcome indicators: The decisionmakers and other stakeholders (the actors in the policy domain) have certain objectives that, if met, would resolve the issue. In this step, the policy objectives or goals are identified. This step also involves identifying policy outcomes directly related to the objectives that can be estimated (qualitatively, or quantitatively using measurable outcome indicators). We are not interested in predicting outcomes, but rather in making better informed choices among alternative policies.
- Identify alternative policies (strategies, plans, courses of action): This step specifies the policies whose outcomes are to be assessed using the system model. In the remainder of the steps, we are interested in understanding the nature of the policies better, discovering additional policies, modifying or combining the policies, possibly arranging them in different timelines, and choosing among them.

Perform Exploratory Uncertainty Analysis:

- Specify uncertainties or disagreements about external forces, system structure, outcome indicators, and valuation of outcomes: In terms of the policy analysis framework in Fig. 1.2, one can identify four primary locations of uncertainty that affect the choice of an appropriate policy (see Table 1.1): uncertainty about the external forces (X); uncertainty about the system response to the external forces and/or policy changes (e.g., uncertainty about the structure of the system model and its parameters (R)); uncertainty about the system outcomes (as propagated by uncertainty about external forces and system response (O)); the relative importance of specific outcomes placed by the actors in the policy domain (W).
- Explore (a) the outcomes of the policies and (b) the vulnerabilities of the policies (and opportunities), given the uncertainties (using, for example, models or expert opinions): This step performs exploratory analysis (often using Exploratory Modeling) to support the process of researching a broad range of assumptions and circumstances. In particular, it involves exploring how a given alternative (A) would perform, in terms of the outcomes (O), under a wide variety of states of the world (X), model structures (R), and alternative value systems (W). The exploration is often carried out using a large number of computational experiments ('cases') under a wide variety of assumptions. This step includes "Scenario Discovery" to identify factors (vulnerabilities or opportunities) that would determine the failure or success of the policies under investigation.

Choose Initial Actions and Contingent Actions:

- Illustrate trade-offs detected by analysis and means for future adjustments as events happen and knowledge is gained: Scenario Discovery and other tools are used to inform further iterations of model or policy architecture. A principal goal of this step is to explore the potential for addressing uncertainty through further changes in policy design and sequencing.
- Select and plan for adoption of an initial policy and mechanisms for future adjustments as events happen and knowledge is gained: Based on the results of earlier steps, a set of initial actions is chosen that should do well given the uncertainties and vulnerabilities. Future contingent actions are also prepared, in order to be ready to respond to uncertain events and developments.
- Plan for monitoring and making possible adaptations to the policy: In addition to the (initial) policy, a "signpost monitoring system" is defined, which specifies what should be watched to know if the underlying assumptions are still valid that implementation is proceeding well and that any needed policy adjustments are taken in a timely and effective manner.

1.6 An Introduction to the DMDU Tools and Approaches

A variety of analytical approaches and tools for decisionmaking under deep uncertainty have been developed. Their underlying paradigm is the need for actions to reduce the vulnerability of a policy or strategy to uncertain future developments. Dewar et al. (1993) called this "Assumption Based Planning" (ABP). Within this paradigm, analysts use "Exploratory Modeling" (EM) and "Scenario Discovery" (SD). EM is a tool to explore a wide variety of scenarios, alternative model structures, and alternative value systems based on computational experiments (Bankes 1993). A computational experiment is a single run with a given model structure and a given parameterization of that structure. It reveals how the real world would behave if the various hypotheses presented by the structure and the parameterization were correct. By exploring a large number of these hypotheses, one can get insights into how the system would behave under a large variety of assumptions (Bankes et al. 2013). SD is a tool to distinguish futures in which proposed strategies meet or miss their goals (Groves & Lempert 2007). It begins with a large database of model runs (e.g., from EM) in which each model run represents the performance of a strategy in one future. The SD algorithms identify those combinations of future conditions that best distinguish the cases in which the policy or strategy does or does not meet its goals (Lempert et al. 2006).

All of the methods discussed in this book are:

- Approaches, offering guidance about the process of creating adaptive strategies; and/or
- Computational decision support tools capable of generating quantitative evidence and insights supporting decisionmaking about different possible courses of action.

ABP was a first step toward an evolving set of analytical approaches for supporting decisionmaking under deep uncertainty. The five approaches presented in the book are:

- *Robust Decisionmaking* (RDM): Robust Decisionmaking (RDM) is a set of concepts, processes, and enabling tools that use computation, not to make better predictions, but to yield better decisions under conditions of deep uncertainty. RDM combines decision analysis, Assumption-Based Planning, scenarios, and Exploratory Modeling to stress test strategies over myriad plausible paths into the future and then to identify policy-relevant scenarios and robust adaptive strategies. RDM analytic tools are often embedded in a decision support process called "deliberation with analysis" that promotes learning and consensus-building among stakeholders (Lempert et al. 2003).
- *Dynamic Adaptive Planning* (DAP): DAP focuses on implementation of an initial plan prior to the resolution of all major uncertainties, with the plan being adapted over time based on new knowledge. DAP specifies the development of a monitoring program and responses when specific trigger values are reached. Hence, DAP makes adaptation over time explicit at the outset of plan formulation. DAP occurs in two phases: (1) the design phase, in which the dynamic adaptive plan, monitoring program, and various pre- and post-implementation actions are designed, and (2) the implementation phase, in which the plan and the monitoring program are implemented and contingent actions are taken, if necessary (Walker et al. 2001).
- *Dynamic Adaptive Policy Pathways* (DAPP): DAPP considers the timing of actions explicitly in its approach. It produces an overview of alternative routes into the future. The alternative routes are based on Adaptation Tipping Points (ATPs). An ATP focuses on "under what conditions will a given plan fail," which is analogous to the question that is asked in ABP or in SD (Haasnoot et al. 2013).
- *Info-Gap Decision Theory* (IG): An information gap is defined as the disparity between what is known and what needs to be known in order to make a reliable and responsible decision. IG is a non-probabilistic decision theory that seeks to optimize robustness to failure (or opportunity for windfall) under deep uncertainty. It starts with a set of alternative actions and evaluates the actions computationally (using a local robustness model). It can, therefore, be considered as a computational support tool, although it could also be categorized as an approach for robust decisionmaking (Ben-Haim 1999).
- *Engineering Options Analysis* (EOA): EOA refers to the process of assigning economic value to technical flexibility. It consists of a set of procedures for calculating the value of an option (i.e., the elements of a system that provide flexibility) and is based on Real Options Analysis (de Neufville and Scholtes 2011).

1.7 Structure of the Book

The book elucidates the state of the art in both the theory and practice associated with the tools and approaches for decisionmaking in the face of deep uncertainty. It has been produced under the aegis of the Society for Decision Making under Deep Uncertainty (DMDU). In 2013–2017, the DMDU Society held workshops involving international scholars and practitioners. The material in the book is a synthesis of, and addition to, the material presented at these workshops, which explored approaches and tools supporting the design of strategic plans under deep uncertainty, and their testing in the real world, including barriers and enablers for their use in practice.

The book consists of the four parts. Part I presents the five approaches for designing policies and strategies under deep uncertainty, which include RDM, DAP, DAPP, IG, and EOA. Each approach is worked out in terms of its theoretical foundations and methodological steps to follow when using the approach, latest methodological insights, and challenges for improvement. In Part II, applications of each of these approaches are presented. Based on recent case studies, the practical implications of applying each approach are discussed in depth.

Part III focuses on using the DMDU approaches and tools in real-world contexts, based on insights from real-world cases. Part III consists of three chapters. The first involves *Decision Scaling* (DS). DS starts with specifying acceptable system performance according to decisionmakers and stakeholders. The vulnerability of the current or best-estimate plan is then assessed by evaluating the system performance for a variety of uncertain developments (possible futures—up until now, mainly climate change) that might occur over management-relevant timescales (Brown 2010). Possible futures of interest can then be investigated further, and probabilities can be assigned if needed. Hence, DS can be seen as extending RDM by assigning probabilities to possible futures. The second chapter in Part III discusses *Planned Adaptation* (PA). PA refers to the design of institutions and processes to regularly review and update policies in light of evolving scientific knowledge and changing technological, economic, social, and political conditions. It is adaptive, but also planned, so it refers not only to the ability of policies to respond to events and information as they arise, but also to a conscious plan to undertake data collection and repeated review over time (McCray et al. 2010). The third chapter in Part III describes *Adaptive Delta Management* (ADM). ADM matches the specific characteristics and context of the Dutch Delta Programme with DMDU approaches and tools. Elements from the DMDU approaches and tools included in Parts I and II of the book are tailored to theme-specific and region-specific needs of the Netherlands.

Part IV of the book contains conclusions and a synthesis of the lessons that can be drawn for designing, applying, and implementing policies and strategies for decisionmaking under deep uncertainty, as well as recommendations for future work.

References

Anguara, R. (2006). The channel tunnel—An ex post economic evaluation. *Transportation Research Part A, 40,* 291–315.

Bankes, S. (1993). Exploratory modeling for policy analysis. *Operations Research, 4*(3), 435–449.

Bankes, S. C., Walker, W. E., & Kwakkel, J. H. (2013). Exploratory modeling and analysis. In S. Gass & M. C. Fu (Eds.), *Encyclopedia of operations research and management science* (3rd ed.). Germany: Springer, Berlin.

Bedford, T. J., & Cooke, R. M. (2001). *Probabilistic risk analysis: Foundations and methods.* Cambridge, UK: Cambridge University Press.

Ben-Haim, Y. (1999). *Info-gap decision theory: Decisions under severe uncertainty* (2nd ed.). London: Academic Press.

Brown, C. (2010). *Decision-scaling for robust planning and policy under climate uncertainty.* World Resources Report 2010-2011 Uncertainty Series, Washington, DC.

Cook, J., Nuccitelli, D., Green, S. A., Richardson, M., Winkler, B., Painting, R., et al. (2013). Quantifying the consensus on anthropogenic global warming in the scientific literature. *Environmental Research Letters, 8*(2). https://doi.org/10.1088/1748-9326/8/2/024024.

Courtney, H. (2001). *20/20 foresight: Crafting strategy in an uncertain world.* Boston: Harvard Business School Press.

de Neufville, R., & Scholtes, S. (2011). *Flexibility in engineering design.* Cambridge, Massachusetts: MIT Press.

Dewar, J. A. (2006). *New practices in long-term planning: Building plans for deep uncertainty.* Presentation given at the Next Generations of Infrastructure (NGI) Workshop (Delft University of Technology, November 16, 2006).

Dewar J. A., Builder, C. H. W., Hix, M., & Levin, M. H. (1993) *Assumption-Based Planning: A planning tool for very uncertain times.* MR-114-A, RAND, Santa Monica, California.

Goodwin, P., & Wright, G. (2010). The limits of forecasting methods in anticipating rare events. *Technological Forecasting and Social Change, 77*(3), 355–368.

Groves, D. G., & Lempert, R. J. (2007). A new analytic method for finding policy-relevant scenarios. *Global Environmental Change, 17,* 73–85.

Haasnoot, M., Kwakkel, J. H., Walker, W. E., & ter Maat, J. (2013). Dynamic adaptive policy pathways: A method for crafting robust decisions for a deeply uncertain world. *Global Environmental Change, 23*(2), 485–498.

Hallegatte, S. (2009). Strategies to adapt to an uncertain climate change. *Global Environmental Change, 19,* 240–247.

Hearing Before the Committee on Oversight and Government Reform. (2008). *The financial crisis and the role of federal regulators,* Serial No. 110-209. U.S. Government Printing Office, Washington, DC.

Hillier, F. S., & Lieberman, G. J. (2001). *Introduction to operations research* (7th ed.). New York: McGraw Hill.

IPCC. (2014). *Climate change 2014: Synthesis report.* In R. K. Pachauri & L. A. Meyer (Eds.), *Contribution of Working Groups I, II and III to the Fifth Assessment Report of the Intergovernmental Panel on Climate Change Core Writing Team.* Geneva, Switzerland: IPCC.

Knight, F. H. (1921). *Risk, uncertainty and profit.* New York: Houghton Mifflin Company (republished in 2006 by Dover Publications, Inc., Mineola, N.Y.).

Lempert, R. J., Groves, D. G., Popper, S. W., & Bankes, S. C. (2006). A general, analytic method for generating robust strategies and narrative scenarios. *Management Science, 52*(4), 514–528. https://doi.org/10.1287/mnsc.1050.0472.

Lempert, R. J., Popper, S. W., & Bankes, S. C. (2003). *Shaping the next one hundred years: New methods for quantitative, long-term policy analysis.* MR-1626-RPC, RAND, Santa Monica, CA.

Lourenco, T. C., Rovisco, A., Groot, A., Nilsson, C., Fussel, H.-M., van Bree, L., et al. (Eds.). (2014). *Adapting to an uncertain climate: Lessons from practice.* Springer.

Luce, R. D., & Raiffa, H. (1957). *Games and decisions: Introduction and critical survey.* Wiley.

McCray, L., Oye, K. A., & Petersen, A. C. (2010). Planned adaptation in risk regulation: An initial survey of U.S. environmental, health, and safety regulation. *Technological Forecasting & Social Change, 77*(6), 951–959.

Miser, H. J., & Quade, E. S. (1985). *Handbook of systems analysis: Overview of uses, procedures, applications, and practice*. New York: Elsevier Science Publishers Co.

Morgan, M. G., & Henrion, M. (1990). *Uncertainty: A guide to dealing with uncertainty in quantitative risk and policy analysis*. Cambridge, UK: Cambridge University Press.

National Research Council. (2009). *Informing decisions in a changing climate*. T. N. A. Press, Washington, DC, Panel on Strategies and Methods for Climate-Related Decision Support, Committee on the Human Dimensions of Climate Change, Division of Behavioral and Social Sciences and Education.

Popper, S. W., Griffin, J., Berrebi, C., Light, T., & Min, E. Y. (2009). *Natural gas and Israel's energy future: A strategic analysis under conditions of deep uncertainty*. TR-747-YSNFF, RAND, Santa Monica, CA.

Quade, E. S. (1989). *Analysis for public decisions* (3rd ed.). Amsterdam: Elsevier Science Publishers B.V.

Schwartz, P. (1996). *The art of the long view: Paths to strategic insight for yourself and your company*. New York: Currency Doubleday.

Taleb, N. N. (2007). *The black swan: The impact of the highly improbable*. New York: Random House.

van Notten, P. W. F., Sleegers, A. M., & van Asselt, M. B. A. (2005). The future shocks: On discontinuity and scenario development. *Technological Forecasting and Social Change, 72*(2), 175–194.

Walker, W. E. (2000). Policy analysis: A systematic approach to supporting policymaking in the public sector. *Journal of Multicriteria Decision Analysis, 9*(1–3), 11–27.

Walker, W. E., Harremoës, P., Rotmans, J., van der Sluijs, J. P., van Asselt, M. B. A., Janssen, P., et al. (2003). Defining uncertainty: A conceptual basis for uncertainty management in model-based decision support. *Integrated Assessment, 4*(1), 5–17.

Walker, W. E., Lempert, R. J., & Kwakkel, J. H. (2013). Deep uncertainty, entry. In S. I. Gass & M. C. Fu (eds.), *Encyclopedia of operations research and management science* (pp. 395–402, 3rd ed.). New York: Springer.

Walker, W. E., Rahman, S. A., & Cave, J. (2001). Adaptive policies, policy analysis, and policymaking. *European Journal of Operational Research, 128*(2), 282–289.

Prof. Vincent A. W. J. Marchau (Radboud University (RU), Nijmegen School of Management) holds a chair on Uncertainty and Adaptivity of Societal Systems. This chair is supported by The Netherlands Study Centre for Technology Trends (STT). His research focuses on long-term planning under uncertainty in transportation, logistics, spatial planning, energy, water, and security. Marchau is also Managing Director of the Dutch Research School for Transport, Infrastructure and Logistics (TRAIL) at Delft University of Technology (DUT), with 100 Ph.D. students and 50 staff members across 6 Dutch universities.

Prof. Warren E. Walker (Emeritus Professor of Policy Analysis, Delft University of Technology). He has a Ph.D. in Operations Research from Cornell University, and more than 40 years of experience as an analyst and project leader at the RAND Corporation, applying quantitative analysis to public policy problems. His recent research has focused on methods for dealing with deep uncertainty in making public policies (especially with respect to climate change), improving the freight transport system in the Netherlands, and the design of decision support systems for airport strategic planning. He is the recipient of the 1997 President's Award from the Institute for Operations Research and the Management Sciences (INFORMS) for his 'contributions to the welfare of society through quantitative analysis of governmental policy problems.'

Drs. Pieter J. T. M. Bloemen (Ministry of Infrastructure and Water management—Staff Delta Programme Commissioner) is the Chief Strategic Officer of the Dutch Delta Programme. He is responsible for the development and application of Adaptive Delta Management. He led the Strategic Environmental Assessment of the Delta Decisions and preferred regional strategies published in 2014. He presently works on the development of a monitoring and evaluation system that matches the adaptive approach of the Delta Programme and is responsible for the first six-yearly review of the Delta Decisions and preferred regional strategies, planned for 2020. Bloemen is Visiting Researcher at IHE Delft Institute of Water Education (Chair Group Flood Resilience) since January 2015 and works on a Ph.D. thesis on the governance of the adaptive approach.

Dr. Steven W. Popper (RAND Corporation) is a RAND Senior Economist and Professor of science and technology policy in the Pardee RAND Graduate School. His work on macrotransitions led to an invitation by President Vaclav Havel to advise the government of Czechoslovakia, participation in an OECD delegation on the first foreign visit to one of the secret cities of the former Soviet Union, and consultation to the World Bank on issues of industrial restructuring in Hungary and in Mexico. His work on microlevel transition focuses on innovation. From 1996 to 2001, he was the Associate Director of the Science and Technology Policy Institute, providing analytic support to the White House Office of Science and Technology Policy and other executive branch agencies. He has taught planning under deep uncertainty at the Pardee RAND Graduate School, the India School of Business, and the Shanghai Climate Institute. He is co-developer of the Robust Decision-Making (RDM) approach. He is an elected Fellow of the American Association for the Advancement of Science, served as chair of the AAAS Industrial Science and Technology section, and is the founding chair for education and training of the Society for Decision Making under Deep Uncertainty.

Open Access This chapter is licensed under the terms of the Creative Commons Attribution 4.0 International License (http://creativecommons.org/licenses/by/4.0/), which permits use, sharing, adaptation, distribution and reproduction in any medium or format, as long as you give appropriate credit to the original author(s) and the source, provide a link to the Creative Commons licence and indicate if changes were made.

The images or other third party material in this chapter are included in the chapter's Creative Commons licence, unless indicated otherwise in a credit line to the material. If material is not included in the chapter's Creative Commons licence and your intended use is not permitted by statutory regulation or exceeds the permitted use, you will need to obtain permission directly from the copyright holder.

Part I
DMDU Approaches

Chapter 2
Robust Decision Making (RDM)

R. J. Lempert

Abstract

- The quest for predictions—and a reliance on the analytical methods that require them—can prove counter-productive and sometimes dangerous in a fast-changing world.
- Robust Decision Making (RDM) is a set of concepts, processes, and enabling tools that use computation, not to make better predictions, but to yield better decisions under conditions of deep uncertainty.
- RDM combines Decision Analysis, Assumption-Based Planning, scenarios, and Exploratory Modeling to stress test strategies over myriad plausible paths into the future, and then to identify policy-relevant scenarios and robust adaptive strategies.
- RDM embeds analytic tools in a decision support process called "deliberation with analysis" that promotes learning and consensus-building among stakeholders.
- The chapter demonstrates an RDM approach to identifying a robust mix of policy instruments—carbon taxes and technology subsidies—for reducing greenhouse gas emissions. The example also highlights RDM's approach to adaptive strategies, agent-based modeling, and complex systems.
- Frontiers for RDM development include expanding the capabilities of multi-objective RDM (MORDM), more extensive evaluation of the impact and effectiveness of RDM-based decision support systems, and using RDM's ability to reflect multiple world views and ethical frameworks to help improve the way organizations use and communicate analytics for wicked problems.

R. J. Lempert (✉)
Rand Corporation, Santa Monica, CA, USA
e-mail: lempert@rand.org

© The Authors(s) 2019
V. A. W. J. Marchau et al. (eds.), *Decision Making under Deep Uncertainty*,
https://doi.org/10.1007/978-3-030-05252-2_2

2.1 Introduction

Toward the end of the Cold War, in the early 1980s, the RAND Corporation invested much effort toward using computer combat simulation models to inform national security decisions regarding weapons procurement and strategy. Designed to predict the course of future military conflicts, these models were obviously imperfect. Everyone recognized the aphorisms about the military's propensity to plan for the last as opposed to the next war. But growing computational capabilities offered new opportunities to simulate complex military campaigns with unprecedented levels of fidelity and realism. The efforts to harness these capabilities provided a focal point for a growing debate on the value of using newly powerful but clearly imperfect analytic tools to inform real and consequential decisions.

The 1982 Falklands/Malvinas war provided a crystalizing moment. Argentine forces landed on the islands in early April and a British task force set sail a few days later for the two-month voyage to the South Atlantic. At that point, all knew the order of battle for the coming conflict. At RAND, those who distrusted the combat simulations used the opportunity to confront the model advocates, challenging them to predict how the conflict would unfold. Two months would provide ample time to conduct the analysis. But the advocates demurred. The models, they claimed, despite all their fidelity and realism, existed to provide "insight" not predictions. The skeptics wondered about the utility of insights gleaned from models whose predictions could not be trusted.

Decisionmakers often seek predictions about the future to inform policy choices. This seems natural because prediction is the bedrock of science, enabling researchers to test their hypotheses and demonstrate understanding of complicated systems. Decisionmakers also find predictions attractive because, when good ones are available, they unquestionably provide valuable input toward better choices. Society's vast enterprise in the physical, biological, and social sciences continually improves its predictive capabilities in order to test and refine scientific hypotheses. Recent analytic innovations, such as big data and machine learning, enhance decision-relevant predictive capabilities. New processes, such as crowd sourcing, prediction markets, and super-forecasting (Tetlock and Gardner 2016), provide new and more reliable means to aggregate human judgments into probabilistic forecasts.

But, as argued throughout this book, the quest for predictions—and a reliance upon analytic methods that require them—can prove counter-productive and sometimes dangerous in a fast-changing, complex world. Prediction-focused analysis risks over-confidence in organizations' decisionmaking and in their internal and external communications (Sarewitz and Pielke 2000). Prediction-focused policy debates can also fall victim to the strategic uses of uncertainty. Opponents may attack a proposed policy by casting doubt on the predictions used to justify it, rather than engaging with the merits of the policy itself, knowing that the policy may be more sound than the predictions (Herrick and Sarewitz 2000; Rayner 2000; Lempert and Popper 2005; Weaver et al. 2013).

A reliance on prediction can also skew the framing of a decision challenge. President Eisenhower (reportedly) advised "if a problem cannot be solved, enlarge it." But science often reduces uncertainty by narrowing its focus, prioritizing questions that can be resolved by prediction, not necessarily on the most decision-relevant inquiries. So-called wicked problems (Rittel and Webber 1973) present this contrast most starkly. In addition to their irreducible uncertainty and nonlinear dynamics, wicked problems are not well-bounded, are framed differently by various stakeholders, and are not well-understood until after formulation of a solution. Using predictions to adjudicate such problems skews attention toward the proverbial lamp post, not the true location of the keys to a policy solution.

This chapter describes Robust Decision Making (RDM), a set of concepts, processes, and enabling tools designed to re-imagine the role of quantitative models and data in informing decisions when prediction is perilous (Lempert et al. 2003, 2006). Rather than regarding models as tools for prediction and the subsequent prescriptive ranking of decision options, models and data become vehicles for systematically exploring the consequences of assumptions; expanding the range of futures considered; crafting promising new responses to dangers and opportunities; and sifting through a multiplicity of scenarios, options, objectives, and problem framings to identify the most important tradeoffs confronting decisionmakers. That is, rather than making better predictions, quantitative models and data can be used to inform better decisions (Popper et al. 2005).

RDM rests on a simple concept (Lempert et al. 2013a). Rather than using computer models and data as predictive tools, the approach runs models myriad times to stress test proposed decisions against a wide range of plausible futures. Analysts then use visualization and statistical analysis of the resulting large database of model runs to help decisionmakers identify the key features that distinguish those futures in which their plans meet and miss their goals. This information helps decisionmakers identify, frame, evaluate, modify, and choose robust strategies—ones that meet multiple objectives over many scenarios.

RDM provides decision support under conditions of deep uncertainty. As described in Sect. 2.2, RDM builds on strong foundations of relevant theory and practice to provide an operational and newly capable synthesis through the use of today's burgeoning information technology. The most commonly used analytic methods for predictive decision and risk analysis have their roots in the 1950s and 1960s, when relative computational poverty made a virtue of analytics recommending a single best answer based on a single best-estimate prediction. Today's ubiquitous and inexpensive computation enables analytics better suited to more complex problems, many of them "wicked" and thus poorly served by the approximation that there exists such an optimal solution.

The British won the Falklands/Malvinas war, but lost a vessel to an unexpectedly effective anti-ship missile attack and sank an Argentine ship under disputed conditions. From an RDM perspective, the model skeptics at RAND asked the wrong question. Rather than judge the combat simulations by how well they could generate an accurate probability density function, the question should have been—can these quantitative models, and the processes for using them, help the British make better decisions regarding what forces were appropriate to bring and how they ought to be employed. While useful, even models that produced a perfect probability density function for precisely selected outcomes would not prove sufficient to answer such questions. Nor are they necessary.[1]

2.2 RDM Foundations

How can quantitative, evidence-based analysis best inform our choices in today's fast-paced and turbulent times? RDM and the other DMDU methods in this book aim to answer this question. In particular, RDM does so by providing a new synthesis of four key concepts: Decision Analysis, Assumption-Based Planning, scenarios, and Exploratory Modeling.

Decision Analysis

A large body of empirical research makes clear that people, acting as individuals or in groups, often make better decisions when using well-structured decision aids. The discipline of Decision Analysis (DA) comprises the theory, methodology, and practice that inform the design and use of such aids. RDM represents one type of quantitative DA method, drawing, for instance, on the field's decision structuring frameworks, a consequentialist orientation in which alternative actions are evaluated in each of several alternative future states of the world, a focus on identifying tradeoffs among alternative decision options, and tools for comparing decision outcomes addressing multiple objectives.

As one key contribution, DA and related fields help answer the crucial question: What constitutes a good decision? No universal criterion exists. Seemingly reasonable decisions can turn out badly, but seemingly unreasonable decisions can turn out well. Good decisions tend to emerge from a process in which people are explicit about their goals, use the best available evidence to understand the potential consequences of their actions, carefully consider the tradeoffs, contemplate the decision

---

[1] As suggested by the Falklands/Malvinas story, RDM had its origins in debates within the RAND Corporation on how best to use models, typically of social interaction or military combat, that could not be validated in the same way as models of less complex decision spaces, or for which probability distribution functions could not be applied with confidence (Hodges 1991; Bankes 1993). The first published examples of the approach that became RDM include: Lempert et al. (1996); Rydell et al. (1997); Brooks et al. (1999). The Third Assessment report of the Intergovernmental Panel on Climate Change (IPCC 2001, Sect. 10.1.4.4) describes such work under the label "computational, multi-scenario approaches." In the 1990s, Evolving Logic (http://www.evolvinglogic.com) developed the first software dedicated to a DMDU methodology—the Computer-Assisted Reasoning® system (CARs™) (Bankes et al. 2001), to support a methodology called Robust Adaptive Planning (RAP). Parallel work at RAND adopted the term Robust Decision Making (RDM) (Lempert et al. 2003). See Light (2005) for a history of approaches to robust and adaptive planning at RAND.

from a wide range of views and vantages, and follow agreed-upon rules and norms that enhance the legitimacy of the process for all those concerned[2] (Jones et al. 2014).

While broad in principle, in practice the DA community often seeks to inform good decisions using an expected utility framework for characterizing uncertainty and comparing decision options (Morgan and Henrion 1990). This expected utility framework characterizes uncertainty with a single joint probability distribution over future states of the world. Such distributions often reflect Bayesian (i.e., subjective) rather than frequentist probability judgments. The framework then uses optimality criteria to rank alternative options. RDM, in contrast, regards uncertainties as deep and thus either eschews probabilities or uses sets of alternative distributions drawing on the concepts of imprecise probabilities (Walley 1991). RDM uses decision criteria based on robustness rather than optimality.

DA based on expected utility can be usefully termed "agree-on-assumptions" (Kalra et al. 2014) or "predict-then-act" (Lempert et al. 2004) approaches, because they begin by seeking agreement regarding the likelihood of future states of the world and then use this agreement to provide a prescriptive ranking of policy alternatives. In contrast, RDM and many of the approaches described in this book follow an "agree-on-decisions" approach, which inverts these steps.[3] They begin with one or more strategies under consideration, use models and data to stress test the strategies over a wide range of plausible paths into the future, and then use the information in the resulting database of runs to characterize vulnerabilities of the proposed strategies and to identify and evaluate potential response to those vulnerabilities. Such approaches seek to expand the range of futures and alternatives considered and, rather than provide a prescriptive ranking of options, often seek to illuminate trade-offs among not-unreasonable choices. As summarized by Helgeson (2018), agree-on-assumptions approaches generally focus on identifying a normative best choice among a fixed menu of decision alternatives, while agree-on-decision approaches focus on supporting the search for an appropriate framing of complex decisions.

Assumption-Based Planning

As part of an "agree-on-decisions" approach, RDM draws on the related concepts of stress testing and red teaming. The former, which derives from engineering and finance (Borio et al. 2014), subjects a system to deliberately intense testing to determine its breaking points. The latter, often associated with best practice in US and other militaries' planning (Zenko 2015), involves forming an independent group to identify means to defeat an organization's plans. Both stress testing and red teaming aim to reduce the deleterious effects of over-confidence in existing systems and plans by improving understanding of how and why they may fail (Lempert 2007).

---

[2] These attributes follow from a broadly consequentialist, as opposed to rule-based (deontological) view of decisionmaking (March 1994).

[3] The DMDU literature often uses different names to describe this inverted analytic process, including "backwards analysis" (Lempert et al. 2013a), "bottom up" (Ghile et al. 2014), "context first" (Ranger et al. 2010), and "assess risk of policy" (Lempert et al. 2004; Carter et al. 2007; Dessai and Hulme 2007).

In particular, RDM draws on a specific form of this concept, a methodology called Assumption-Based Planning (ABP) (Dewar et al. 1993, 2002). Originally developed to help the US Army adjust its plans in the aftermath of the Cold War, ABP begins with a written version of an organization's plans and then identifies load-bearing assumptions—that is, the explicit and implicit assumptions made while developing that plan that, if wrong, would cause the plan to fail. Planners can then judge which of these load-bearing assumptions are also vulnerable—that is, could potentially fail during the lifetime of the plan.

ABP links the identification of vulnerable, load-bearing assumptions to a simple framework for adaptive planning that is often used in RDM analyses. Essential components of an adaptive strategy include a planned sequence of actions, the potential to gain new information that might signal a need to change this planned sequence, and actions to be taken in response to this new information, i.e., contingent actions (Walker et al. 2001). After identifying the vulnerable, load-bearing assumptions, ABP considers *shaping actions* (those designed to make the assumptions less likely to fail), *hedging actions* (those that can be taken if assumptions begin to fail), and *signposts* (trends and events to monitor in order to detect whether any assumptions are failing).

Scenarios

RDM draws from scenario analysis the concept of a multiplicity of plausible futures as a means to characterize and communicate deep uncertainty (Lempert et al. 2003). Scenarios represent internally consistent descriptions of future events that often come in sets of two or more. Most simply, scenarios are projected futures that claim less confidence than probabilistic forecasts. More generally, a set of scenarios often seeks to represent different ways of looking at the world without an explicit ranking of relative likelihood (Wack 1985).

Scenarios are often developed and used in deliberative processes with stakeholders. Deemphasizing probabilistic ranking—focusing on a sense of possibility, rather than probability—helps stakeholders expand the range of futures they consider, allowing them to contemplate their choices from a wider range of views and vantages, thus helping participants consider uncomfortable or unexpected futures (Schoemaker 1993; Gong et al. 2017). The sense of possibility rather than probability can also help scenarios to communicate a wide range of futures to audiences not necessarily eager to have their vantage expanded. By representing different visions of the future without privileging among them, scenarios can offer a comfortable entry into an analysis. Each person can find an initially resonant scenario before contemplating ones that they find more dissonant.

RDM draws from scenario analysis the concept of organizing information about the future into a small number of distinct cases that help people engage with, explore, and communicate deep uncertainty. In particular, the Intuitive Logics school of scenario analysis (Schwartz 1996) uses qualitative methods to craft a small number of scenarios, distinguished by a small number of key uncertain determinants that differentiate alternative decision-relevant paths into the future. As described below, RDM uses quantitative "Scenario Discovery" algorithms to pursue the same ends.

The resulting scenarios summarize the results of the ABP-style stress tests and can link to the development of adaptive strategies (Groves and Lempert 2007; Groves et al. 2014). Note that while the scenario literature traditionally distinguishes between probabilistic and non-probabilistic treatments, RDM often employs a third alternative—entertaining multiple views about the likelihood of the scenarios.

The scenario literature also describes a process for seeking robust strategies that include choosing a set of scenarios that include the most important uncertainties facing the users and then identifying strategies that perform well across all of them (van der Heijden 1996). This process provides an animating idea for RDM.

Exploratory Modeling

RDM integrates these concepts—DA, ABP, and scenarios—through Exploratory Modeling (EM). Bankes (1993) encapsulated the 1980s RAND debates on useful and predictive models by dividing computer simulations into two types: (1) consolidative models, which gather all known facts together into a single package that, once validated, can serve as a surrogate for the real world, and (2) exploratory models, which map a wide range of assumptions onto their consequences without privileging one set of assumptions over another. Exploratory models are useful when no single model can be validated because of missing data, inadequate or competing theories, or an irreducibly uncertain future.

Running a model many times is not profound. But as perhaps its key insight, EM notes that when used with an appropriate experimental design—that is, appropriate questions and a well-chosen set of cases designed to address those questions—the large database of results generated from non-predictive, exploratory models can prove surprisingly useful toward informing policy choices. Bankes describes several types of questions one may address with EM, including hypothesis generation, reasoning from special cases, and assessing properties of the entire ensemble (Weaver et al. 2013 and Box 2.1). RDM uses them all, but focuses in particular on robustness as a property of the entire ensemble. That is, identifying and evaluating robust strategies become key questions one can inform with EM.

EM provides RDM with a quantitative framework for stress testing and scenario analysis. While consolidative models most usefully support deductive reasoning, exploratory models serve best to support inductive reasoning—an iterative cycle of question and response. As described in the process section below (Sect. 2.3), RDM also aims to support a decision-analytic human-machine collaboration that draws upon what each partner does best.

RDM also exploits another EM advantage: the focus on the simple computational task of running models numerous times in the forward direction. This facilitates exploration of futures and strategies by reducing the requirements for analytic tractability on the models used in the analysis, relative to approaches that rely on optimization or Dynamic Programming. In addition, EM enables truly global sensitivity explorations, since it privileges no base case or single future as an anchor point.

> **Box 2.1: Key Elements of RDM**
>
> RDM meets its goals by proceeding in multiple iterations, as humans and computers alternatively test each other's conclusions about futures and strategies. Four key elements govern these interactions (Lempert et al. 2003):
> - Consider a multiplicity of plausible futures. The ensemble of futures should be as diverse as possible to adequately stress test proposed policies. The ensemble can also facilitate group processes by including futures that correspond to different groups' worldviews.
> - Seek robust, rather than optimal strategies. Robust strategies perform well, compared to the alternatives, over a wide range of plausible futures.
> - Employ adaptive strategies to achieve robustness. Adaptive strategies are designed to evolve over time in response to new information. Generally, such strategies reflect decisionmaking rules and in practice are often organized around near-term actions, signposts to monitor, and contingency actions to take in response to those signposts.
> - Use the computer to facilitate human deliberation over explorations, options, and tradeoffs, not as a device for recommending a particular ordering of strategies.

2.3 RDM Process

RDM explicitly follows a learning process called "deliberation with analysis" in which parties to a decision deliberate on their objectives and options; analysts generate decision-relevant information using system models; and the parties to the decision revisit their objectives, options, and problem framing influenced by this quantitative information (NRC 2009). Among learning processes, deliberation with analysis proves most appropriate for situations with diverse decisionmakers who face a changing decision environment and whose goals can evolve as they collaborate with others. Deliberation with analysis also supports continuous learning based on indicators and monitoring (NRC 2009, p. 74), a process important to the literature on adaptive policymaking (Swanson and Bhadwal 2009; Walker et al. 2010).

Step 1 As shown in Fig. 2.1, the RDM process starts with a decision framing exercise in which stakeholders define the key factors in the analysis: the decisionmakers' objectives and criteria; the alternative actions they can take to pursue those objectives; the uncertainties that may affect the connection between actions and consequences; and the relationships, often instantiated in computer simulation models, between actions, uncertainties, and objectives. This information is often organized

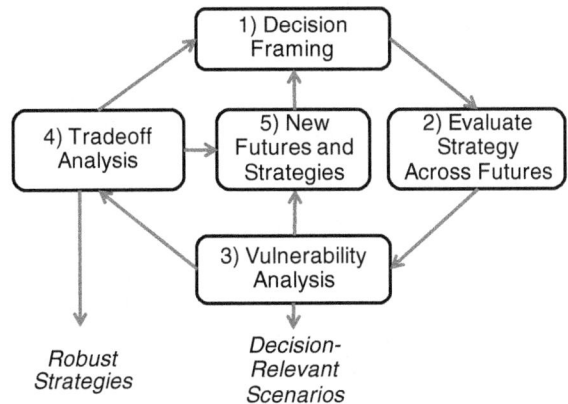

Fig. 2.1 Steps in an RDM analysis. *Source* Lempert et al. (2013a), with earlier versions in Lempert and Groves (2010); Lempert and Kalra (2011)

in a 2 × 2 matrix called "XLRM" (Lempert et al. 2003), for exogenous uncertainties (X), policy levers (L), relationships (R), and measures of performance (M).[4]

Step 2 As an "agree-on-decision" approach, RDM next uses simulation models to evaluate proposed strategies in each of many plausible paths into the future, which generate a large database of simulation model results. The proposed strategies can derive from a variety of sources. In some cases, an RDM analysis might start with one or more specific strategies drawn from the relevant public debate. For instance, an RDM analysis for a water agency might begin with that agency's proposed plan for meeting its supply requirements (Groves et al. 2014) or water quality requirements (Fischbach et al. 2015). In other cases, optimization routines for one or more expected futures or decision criteria might yield the initial proposed strategies (Hall et al. 2012). Additionally, an analysis might begin with a wide span of simple strategies covering the logical spectrum and then refine, select, and modify to yield a small group of more sophisticated alternatives (Popper et al. 2009). Often, such as in the example in Sect. 2.5, an application uses a combination of these approaches.

Step 3 Analysts and decisionmakers next use visualization and data analytics on these databases to explore for and characterize vulnerabilities. Commonly, RDM analyses use statistical Scenario Discovery (SD) algorithms (see below) to identify and display for users the key factors that best distinguish futures in which proposed strategies meet or miss their goals. These clusters of futures are usefully considered policy-relevant scenarios that illuminate the vulnerabilities of the proposed policies. Because these scenarios are clearly, reproducibly, and unambiguously linked to a policy stress test, they can avoid the problems of bias and arbitrariness that sometimes afflict more qualitative scenario exercises (Lempert 2013; Parker et al. 2015).

Step 4 Analysts and decisionmakers may use these scenarios to display and evaluate the tradeoffs among strategies. For instance, one can plot the performance of one or more strategies as a function of the likelihood of the policy-relevant scenarios

---

[4]Chapter 1 (see Fig. 1.2) uses the acronym XPROW for external forces, policies, relationships, outcomes, and weights.

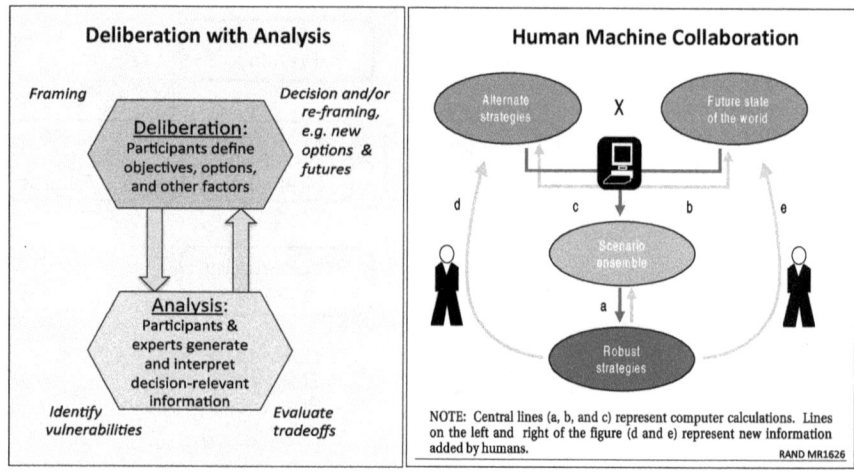

Fig. 2.2 Two views of RDM as a deliberative process. *Source* for right panel, Lempert et al. (2003)

(e.g., see Fig. 2.2) to suggest the judgments about the future implied by choosing one strategy over another. Other analyses plot multi-objective tradeoff curves—for instance comparing reliability and cost (Groves et al. 2012)—for each of the policy-relevant scenarios to help decisionmakers decide how to best balance among their competing objectives.

Step 5 Analysts and decisionmakers could then use the scenarios and tradeoff analyses to identify and evaluate potentially more robust strategies—ones that provide better tradeoffs than the existing alternatives. These new alternatives generally incorporate additional policy levers, often the components of adaptive decision strategies: short-term actions, signposts, and contingent actions to be taken if the pre-designated signpost signals are observed. In some analyses, such adaptive strategies are crafted using expert judgment (e.g., see Lempert et al. 1996, 2000, 2003; Popper et al. 2009; Lempert and Groves 2010). In other analyses, optimization algorithms may help suggest the best combination of near-term actions, contingent actions, and signposts for the new adaptive strategies (Lempert et al. 2006; Lempert and Collins 2007; Kasprzyk et al. 2013; Herman et al. 2014). (This RDM step provides a natural point of connection with the Dynamic Adaptive Policy Pathways (DAPP) methods in Chaps. 4 and 9.)

RDM uses both absolute and relative performance measures to compare strategies in the vulnerability and tradeoff analyses. Absolute performance measures are useful when decisionmakers are focused on one or more outcomes, such as profit, energy produced, or lives saved. Absolute performance measures are also useful when decisionmakers are focused on some invariant standard—for instance, a regulatory requirement on reliability or environmental quality, a required threshold for an economic rate of return, or a requirement that benefits exceed costs. Relative performance measures are often useful when uncertainties create a wide range of outcomes, so decisionmakers seek strategies that perform well compared to alternatives over a wide range of futures. RDM often uses regret to represent relative performance.

At each of the RDM steps, information produced may suggest a re-framing of the decision challenge. The process produces key deliverables including (1) the scenarios that illuminate the vulnerabilities of the strategies, and (2) potential robust strategies and the tradeoffs among them.

The left panel of Fig. 2.2 shows the RDM steps in support of a process of deliberation with analysis. Stakeholders begin by deliberating over the initial decision framing. In the vulnerability and tradeoff analysis steps, stakeholders and analysts produce decision-relevant information products. Using these products, stakeholders deliberate over the choice of a robust strategy or return to problem framing, for instance seeking new alternatives or stress testing a proposed strategy over a wider range of futures. In practice, the process often moves back and forth between problem framing, generating the scenarios that illuminate vulnerabilities, identifying new alternatives based on those scenarios, and conducting a tradeoff analysis among the alternatives.

People teaming with computers—each doing what they do best—are more capable than computers or people alone.[5] RDM uses EM to support deliberation with analysis in a process of human/machine collaboration (Lempert et al. 2003). As shown in the right panel of Fig. 2.2, people use their creativity and understanding to pose questions or suggest solutions—for instance candidate robust strategies. Computers consider numerous combinations of strategies and futures to help users address their questions, search for initially unwelcome counter-examples to proposed solutions, and help people find new candidate robust strategies to propose (Lempert et al. 2002; Lempert and Popper 2005).

The RDM steps and deliberative processes are consistent with others in the DMDU literature. For instance, Multi-Objective RDM (MORDM) offers a similar iterative process but with the major advance of more articulation of the step of generating alternative strategies (Kasprzyk et al. 2013). The DAPP process (see Chaps. 4 and 9), also similarly, emphasizes monitoring and policy adjustment (Haasnoot et al. 2013). Among related literatures, many-objective visual analytics uses interactive visualizations to support problem framing and re-framing (Kollat and Reed 2007; Woodruff and Reed 2013), often with a posteriori elicitation of preferences (Maass et al. 1962; Cohon and Marks 1975).

Overall, the RDM process aims to provide quantitative decision support that helps meet the criteria for good decisions even in the presence of deep uncertainty and the other attributes of wicked problems. The process encourages participants to be explicit about their goals and consider the most important tradeoffs. The process uses scenario concepts linked to the idea of policy stress tests, along with computer-assisted exploration, to encourage and facilitate consideration of the decision from a wide range of views. It helps recognize the legitimacy of different interests, values, and expectations about the future by using models as exploratory, rather than

---

[5]For example, Thompson (2013) notes that competent chess players teamed with computers can defeat both grand masters without computers and computers without human assistants.

predictive, tools within an evidence-based, multi-scenario, multi-objective decision-making process.

> **Box 2.2: What is a Robust Strategy?**
>
> A robust strategy is one that performs well, compared to the alternatives, over a wide range of plausible futures (Rosenhead et al. 1972; Lempert et al. 2003). Other definitions exist, including trading some optimal performance for less sensitivity to assumptions (Lempert and Collins 2007) and keeping options open (Rosenhead 1990). Choosing a specific quantitative criterion to judge robustness can, however, prove complicated because many robustness criteria exist, and in some cases they can yield a different ordering of strategies (Giuliani and Castelletti 2016). No robustness criterion is best in all circumstances and, as befits a decision support methodology designed to facilitate problem framing, RDM often includes the choice of a robustness criterion as part of its problem-framing step.
>
> RDM and other deliberative "agree-on-decision" DMDU methods generally draw robustness criteria from the normative, "agree-on-assumptions" decision-analytic literature. This literature identifies four traditional criteria —called Wald, Hurwicz, Savage, and Laplace—for ranking choices without well-defined probability distributions over future states of the world (Luce and Raiffa 1957; Schneller and Sphicas 1983). These criteria envision a set of future states of the world f_j and a set of strategies s_i, each with a known utility, u_{ij}, in each state of the world. If the probability of each future, p_j, were known, the best strategy would be that which yielded the maximum expected utility, $\text{Max}_i \sum_j p_j u_{ij}$. Lacking such probabilities, Wald selects the strategy that gives the best worse case, $\text{Max}_i \text{Min}_j (u_{ij})$, and Savage's mini-max regret selects the strategy with the least regret, that is, which deviates least from the best one could choose with perfect information, $\text{Min}_i \text{Max}_j [\text{Max}_i (u_{ij}) - u_{ij}]$. Both Wald and Savage are conservative in that they attempt to avoid worst cases. In contrast, Hurwicz interpolates between selecting the strategy with the best case and the best worst case, $\text{Max}_i [\alpha \text{Max}_j (u_{ij}) + (1 - \alpha) \text{Min}_j (u_{ij})]$, where α represents a level of confidence. Laplace's criterion of insufficient reason assumes equal weighting over all the futures and then selects the strategy that maximizes expected utility. Starr (1962) subsequently proposed the domain criterion, which selects the strategy that has highest utility in the most futures.
>
> The domain criterion comes closest to the idea of performing well over a wide range of futures and can also facilitate tradeoff analyses using scenario-themed visualizations. RDM analyses commonly use some form of the domain criterion, although RDM (see Table 7.2 in Cervigni et al. 2015) and other robustness approaches (e.g., Hansen and Sargent 2008) have also employed Savage's mini-max regret. Traditionally, the domain criterion uses an absolute

performance measure (e.g., the utility u_{ij}, or a benefit–cost ratio (Lempert 2014)), as its argument. RDM often extends the criterion by also using as arguments relative performance measures, e.g., the regret $\text{Max}_i(u_{ij}) - u_{ij}$, or satisficing criteria for both absolute and relative performance measures. For instance, employed with a regret measure and satisficing criterion, the domain criterion might select the strategy that has a regret below some threshold value over the widest range of futures (Lempert and Collins 2007). In contrast to a strict application of Savage's mini-max regret, such a criterion would retain the comparative quality of the regret measure without undue attention to worst cases.

RDM also uses the domain criterion over the space of probabilities rather than the space of futures; that is, the analysis considers a set of plausible probability distributions over the futures (Walley 1991), and the domain criterion selects the strategy with the highest expected utility or expected regret over the largest number of distributions in the set (see, e.g., Fig. 2.3 in this chapter). RDM often uses the domain criterion, or other robustness decision criteria, to focus on tradeoffs rather than on a normative ranking of strategies. For instance, Lempert and Collins (2007) employ a definition of robustness as trading some optimal performance for less sensitivity to assumptions, which it implements with a domain criterion using expected regret as its argument. The study displays the choice of a robust strategy as a function of the confidence one places in the best-estimate distribution. The *Combined* Strategy discussed in Sect. 2.5.3 is robust by both a domain criterion and by mini-max regret.

2.4 Tools

The concepts underlying RDM—scenario thinking, robustness decision criteria, stress testing proposed plans, and the use of exploratory models—have long pedigrees. But over the last decade, new computer and analytic capabilities have made it possible to combine them in practical decision analyses. In particular, RDM often relies on Scenario Discovery and visualization, robust multi-objective optimization, integrated packages for EM, and high-performance computing.

SD algorithms often implement the RDM vulnerability analysis step (Step 3) in Fig. 2.1. SD begins with a large database of runs in which each model run represents the performance of a strategy in one future. The SD cluster-finding algorithms then offer concise descriptions of those combinations of future conditions that best distinguish the cases in which the implementation plan does or does not meet its goals. The requisite classification algorithms—often Patient Rule Induction Method (PRIM) (Friedman and Fisher 1999) or Classification and Regression Tree (CART) (Breiman et al. 1984), combined with a principal component analysis (Dalal et al.

2013)—seek to balance between the competing goals of simplicity and accuracy in order to describe sets of strategy-stressing futures as concise, understandable, and decision-relevant scenarios (Groves and Lempert 2007; Bryant and Lempert 2010). Software to implement the PRIM algorithm is available, both in stand-alone routines[6] and embedded in the EM software packages described below. Overall, SD replicates analytically the ideas of qualitative Intuitive Logics scenario analysis and provides information products that can prove compelling in stakeholder deliberations (Lempert 2013). RDM analyses also use computer visualization of the database of runs to support the vulnerability and tradeoff analyses. Tableau, a commercially available platform, has proven particularly useful in much RDM work (Groves et al. 2013; Cervigni et al. 2015).

The "new futures and strategies" step (Step 5) can employ a variety of methods. While some RDM analyses use only expert judgment to craft responses to potential vulnerabilities (e.g., Lempert and Groves 2010), many applications use some multi-objective robust optimization tool. Some such applications have used constrained optimization to trace out a range of potentially robust solutions, both in single objective (Lempert and Collins 2007) and multi-objective (Groves et al. 2012) cases. The latter instance involved a planning tool that allowed analysts to trace the Pareto optimal tradeoff curves in each scenario for any two objectives using constrained optimizations over the other objectives (Groves et al. 2012), and has been widely used. Other applications, such as the Colorado Basin Supply and Demand Study (Groves et al. 2013; also see Chap. 7), run large portfolio optimizations for many futures; identify the individual actions that occur in the optimal set for most, some, and few of the futures; and use this information to craft adaptive strategies that begin with the actions that occur in most of the futures' optimal sets, and implement the others depending on which future comes to pass (Groves et al. 2013; Bloom 2015). MORDM tools provide a more general solution, using evolutionary algorithms to identify the regions of a Pareto surface over many objectives that are most robust to uncertainty (Kasprzyk et al. 2013). MORDM has been used to identify adaptive strategies modeled as controllers in the control theory sense (Quinn et al. 2017), through a process called direct policy search.

While many RDM analyses use stand-alone software to generate and analyze large numbers of model runs, several integrated EM packages exist that can greatly facilitate such analysis. These packages generally consider a simulation model as a mapping of inputs to outputs and then provide connectivity between the model and various analysis and visualization tools. For instance, EM packages allow the user to specify an experimental design over the inputs; run the resulting cases on one or more processors; store the results in a database; use a variety of visualization routines to examine the data; run SD and other statistical algorithms on the data; and use robust optimization routines to identify new, potentially more robust strategies. While they require study and practice to use, EM software can make it much easier to move seamlessly through the steps of an RDM analysis. Computer-Assisted Reasoning system (CARs) was the earliest such software package (Bankes et al.

---

[6]https://cran.r-project.org/web/packages/sdtoolkit/index.html.

2001). More recently, new open-source platforms have become available, such as the Exploratory Modeling and Analysis Workbench,[7] Open MORDM (Hadka et al. 2015), and Rhodium.[8]

While many RDM analyses may be conveniently run on a laptop or desktop computer (e.g., Lempert et al. 2013b), recent studies have also used high-performance computation—either large-scale cluster computing (Zeff et al. 2014; Groves et al. 2016) or cloud-based computer services (Isley 2014; Cervigni et al. 2015)—to quickly and inexpensively conduct a very large number of runs.

2.5 Example: Carrots and Sticks for New Technology

An early RDM application provides an ideal example of the approach, the types of tools used, and how EM can draw together concepts from decision analysis, scenarios, and ABP. The example, called "carrots and sticks for new technology," focused on determining the most robust combination of two policy instruments—carbon prices and technology subsidies—to reduce climate-altering greenhouse gas emissions (Robalino and Lempert 2000; Lempert 2002). We describe this study using the generic DMDU framework presented in Sect. 1.5 (the major subheadings below) linked to the relevant RDM steps shown in Fig. 2.1.

2.5.1 Frame the Analysis

Formulate Question (RDM Step 1)

It is well-understood that an economically ideal greenhouse gas emission reduction policy should include an economy-wide carbon price implemented through mechanisms such as a carbon tax or a cap-and-trade system. This early RDM study addressed the question of whether and under what conditions technology incentives, such as technology-specific tax credits or subsidies, also prove necessary and important as part of a greenhouse gas reduction strategy. Many national and regional jurisdictions worldwide employ such incentives in their climate policies because they prove politically popular and have a compelling logic, if for no other reason than significant technology innovation will prove crucial to limiting climate change. But technology incentives have a mixed record of success (Cohen and Noll 2002), and sometimes link to larger debates about the appropriate role of government (Wolf 1993). Standard economic analysis proves a poor platform to adjudicate such questions, because the extent to which technology incentives prove economically

---

[7] http://simulation.tbm.tudelft.nl/ema-workbench/contents.html.
[8] https://github.com/Project-Platypus/Rhodium/wiki/Philosophy.

important may depend on coordination failures that occur in the presence of increasing returns to scale, imperfect information, and heterogeneous preferences—deeply uncertain factors not well-represented in standard economic models. Addressing the technology incentive question with RDM thus proved useful due to this deep uncertainty and because people's views on technology incentives can be strongly affected by their worldviews.

Identify Alternatives (RDM Step 1)

The study was organized as a head-to-head comparison between strategies incorporating two types of "policy instruments" or "actions": an economy-wide carbon tax and a technology price subsidy for low-carbon-emitting technologies. We considered four combinations: (1) neither taxes nor subsidies *(No Action)*, (2) tax only *(Taxes Only)*, (3) subsidies only *(Subsidies Only)*, and (4) a combination of both taxes and subsidies *(Combined)*. As described below, both the tax and subsidy were configured as adaptive strategies designed to evolve over time in response to new information.

Specify Objectives (RDM Step 1)

This study compared the strategies using two output measures: the present value of global economic output (pvGDP) and the mid-twenty-first-century level of greenhouse gas emissions. We focused on pvGDP to facilitate comparison of this work with other analyses in the climate change policy literature using more standard economic formulations. As described below, we calculated the regret for each strategy in each future and used a domain criterion for robustness (see Box 2.2), thus looking for strategies with low regret over a wide range of plausible futures.

Specify System Structure (RDM Step 1)

To compare these adaptive strategies, we employed an agent-based model of technology diffusion, linked to a simple macromodel of economic growth that focused on the social and economic factors that influence how economic actors choose to adopt, or not to adopt, new emission-reducing technologies. The agent-based representation proved useful, because it conveniently represents key factors potentially important to technology diffusion, such as the heterogeneity of technology preferences among economic actors and the flows of imperfect information that influence their decisions. Considering this tool as an exploratory, rather than predictive, model proved useful, because it allowed the study to make concrete and specific comparisons of price- and subsidy-based strategies even though, as described below, available theory and data allowed the model's key outputs to vary by over an order of magnitude.

The model was rooted in the microeconomic understanding of the process of technology diffusion (Davies 1979). Each agent in our model represents a producer of a composite good, aggregated as total GDP, using energy as one key input. Each period the agents may switch their choice of energy-generation technology, choosing among high-, medium-, or low-emitting options. Agents choose technology to maximize their economic utility. The agents estimate utility based on their expectations regarding each technology's cost and performance. Costs may or may not decline significantly due to increasing returns to scale as more agents choose to adopt. The agents have imperfect information and current and future technology cost and performance, but can gain information based on their own experience, if any, and by querying other agents who have used the technology. Thus, the model's technology diffusion rates depend reflexively on themselves, since by adopting a technology each agent generates new information that may influence the adoption decisions of other potential users. The model also used simple, standard, but deeply uncertain relationships from the literature on the connections between greenhouse gas emissions and the economic impacts due to climate change (Nordhaus 1994).

2.5.2 Perform Exploratory Uncertainty Analysis

Specify Uncertainties or Disagreements (RDM Step 2)

The agent-based model had thirty input parameters representing the deeply uncertain factors, including the macroeconomic effects of potentially distortionary taxes and subsidies on economic growth, the microeconomic preferences that agents use to make technology adoption decisions, the future cost and performance of high- (e.g., coal), low- (e.g., gas), and non-emitting (e.g., solar) technologies, the way information about new technologies flows through agent networks, and the impacts of climate change.

We employed three sources of information to constrain our EM. First, the agent-based model embodied the theoretical economic understanding of technology diffusion. Second, we drew plausible ranges for each individual parameter using estimates from the microeconomics literature. Third, we required the model to reproduce macroeconomic data regarding the last fifty years of economic growth and market shares for different types of energy technology. We also constrained future technology diffusion rates in the model to be no faster than the fastest such rates observed in the past.

Consistent with these constraints, the model nonetheless was able to generate a vast range of plausible futures. To choose a representative sample of futures from this vast set, the study launched a genetic algorithm over the model inputs, searching for the most diverse set of model inputs consistent with the theoretical, macroeconomic, and microeconomic constraints (Miller 1998). This process yielded an ensemble of 1,611 plausible futures, with each future characterized by a specific set of values for each of the thirty uncertain model input parameters. Each member of the ensemble reproduced the observed history from 1950 to 2000, but differed by up to an order of magnitude in projected mid-twenty-first-century emissions.

Evaluate Strategies Against Futures (RDM Step 2)

The study represented its carbon tax and technology subsidies as adaptive strategies using a single set of parameters for each to describe their initial conditions, and how they would evolve over time in response to new information.

As with many carbon price proposals, the study's carbon tax would start with an initial price per ton of CO_2 at time zero and rise at a fixed annual rate, subject to two conditions meant to reflect political constraints. First, we assumed that the government would not let the tax rise faster than the increase in the observed social cost of carbon. Second, we assumed that if the economy dropped into recession (defined as a global growth rate below some threshold), the government would drop the carbon tax back to its original level. Three parameters defined the tax strategy: the initial tax rate, the annual rate of increase, and the minimum threshold for economic growth required not to repeal the tax.

The study's technology subsidy reduces the cost to users of a technology to a fixed percentage of its unsubsidized cost. The subsidy stays in effect until the policy either succeeds in launching the technology (defined as its market share rising above some threshold value) or the policy fails (defined as the market share failing to reach a minimum threshold after a certain number of years). Meeting either of these conditions permanently terminates the subsidy. Four parameters defined the subsidy strategy: the subsidy level, the market share defining success, the market share defining failure, and the number of years before the subsidy can be judged a failure.

The study chose a single set of parameter values to define the tax and subsidy policies, each set chosen to optimize the pvGDP for the future represented by the average value for each of the thirty model parameters. The results of the study were relatively insensitive to this simplification.

We used the agent-based model to calculate the pvGDP and mid-century greenhouse gas emissions for each strategy in each of the 1,611 plausible futures. This ensemble of runs made it immediately clear that the *Taxes Only* and *Combined* strategies consistently perform better than the *No Action* and *Subsidies Only* strategies. The remainder of the analysis thus focused on the first two.

2.5.3 Choose Initial Actions and Contingent Actions

Illustrate Tradeoffs (RDM Steps 3 and 4)

Lacking (not-yet-developed) SD algorithms and faced with too many dimensions of uncertainty for an exhaustive search, the study used importance sampling to find the five uncertain input parameters most strongly correlated with mid-century GHG emissions. Four of these key uncertainties related to the potential for coordination failures to slow technology diffusion—the rate of cost reductions for non-emitting technologies caused by increasing returns to scale, the rate at which agents learn from one another about the performance of new technologies, the agents' risk aversion, and the agents' price-performance preferences for new technologies—and one related to the damages from climate change.

We then examined the regret in pvGDP across the cases for the *Taxes Only* and *Combined* strategies as a function of all ten two-dimensional combinations of the five key uncertainties.[9] Each visualization told a similar story—that the *Combined* strategy cases had lower mean and variability in pvGDP regret than those for *Taxes Only* except in the corner of the uncertainty space with low potential for coordination failures and/or low impacts from climate change.

This analysis provides the study's basic comparison among the strategies: *Taxes Only* performs best when the potential for coordination failures and the impacts from climate change are small, and *Combined* performs best otherwise. Both strategies perform better than *No Action* or *Subsidies Only*.

Select and Plan for Adaptation (RDM Step 4)

To help decisionmakers understand the conditions under which the *Taxes Only* and *Combined* strategies would each be favored over the other, we first collapsed the ten visualizations into a single two-dimensional graph by combining all four key uncertain parameters relating to potential coordination failures into a single variable. Figure 2.3 shows the sets of expectations in which pvGDP for the *Taxes Only* strategy exceeds that of *Combined*, and vice versa, as a function of the probability assigned to high rather than low values of the four uncertain parameters related to potential coordination failures, which we labeled "probability of a non-classical world," and "probability of high damages" due to climate change. The region dominated by *Combined* is larger than that of *Taxes Only* for two reasons. First, *Combined* dominates *Taxes Only* over larger regions of the state space. Second, in the regions of the uncertainty space where *Taxes Only* is the better strategy, *Combined*'s regret is relatively small, while in the regions where *Combined* is better, *Taxes Only*'s regret is relatively large. Existing scientific understanding proves insufficient to define with certainty where the future lies in Fig. 2.3. Different parties to the decision may have different views. But the boundary between the two regions is consistent with increasing returns to scale much smaller than those observed for some energy technologies—such as natural gas turbines, wind, and solar—and the middle of the figure is consistent with relatively small levels of risk aversion, learning rates, and heterogeneity of preferences compared to those seen in various literatures. These results suggest that a combination of price instruments and technology subsidies may prove the most robust strategy over a wide range of plausible futures.

Implementation, Monitoring, and Communication

This study addressed a high-level question of policy architecture—the best mix of policy instruments for decarbonization. While the study did not provide detailed implementation plans, it does suggest how a national or state/provincial government might pursue the study's recommendations. The study envisions policymakers choosing a strategy which includes the rules by which the initial actions will be adapted over time (Swanson et al. 2007). The carbon price, presumably set by the legislature, would follow the social cost of carbon as periodically updated by executive agencies (National Academies of Sciences 2016) whenever the economy was not in recession.

---

[9]Ten combinations because $\binom{5}{2} = 10$.

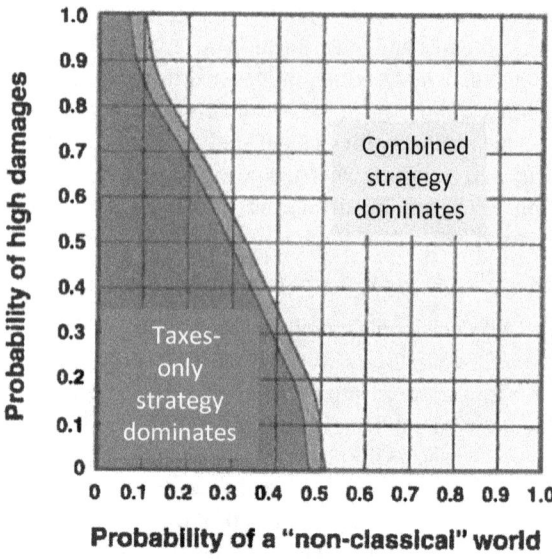

Fig. 2.3 Scenario map comparing "Taxes Only" and "Combined" strategies

The legislature would also set the technology subsidy and terminate it when the subsidized technologies either succeeded or failed based on market share data gathered by executive agencies. The study did not examine pre-commitment issues—that is, how the current legislature could ensure that future legislatures would in fact follow the adaptive strategy. These interesting issues of political economy have, however, been recently explored using RDM methods (Isley et al. 2015).

2.5.4 Iterate and Re-Examine (RDM Steps 2, 3, and 5)

The study's results are based on an examination of only six of the 30 dimensions of uncertainty in the model, representing a small subset of the full range of plausible futures. As a key final step, we tested the policy recommendations by launching a genetic search algorithm across the previously unexamined dimensions looking for additional futures that would provide counter-examples to our conclusions. This process represents the computer feedback loop in the right panel of Fig. 2.2. The genetic algorithm ran for most of the time the authors spent writing their manuscript and found no plausible counter-examples.

Overall, this study suggests that if decisionmakers hold even modest expectations that market imperfections are likely to inhibit the diffusion of new, emissions-reducing technologies or that the impacts of climate change will turn out to be serious, then strategies combining both carbon taxes and technology incentives may be a promising component of a robust strategy for reducing greenhouse gas emissions.

2.6 Recent Advances and Future Challenges

The "carrots and sticks for new technology" example of Sect. 2.5 includes all the steps of an RDM analysis shown in Fig. 2.1. It used optimization algorithms to define its alternative, adaptive strategies, and generated its futures using genetic algorithms to perform what has more recently been called scenario diversity analysis (Carlsen et al. 2016a, b). The study employed the process of human-machine collaboration shown in Fig. 2.2, in particular in the computer search for counter-examples to the human-derived patterns that constitute its policy conclusions.

Since this early example, the methods and tools for MORDM analyses have approached maturity, now reaching the point at which one can describe with some specificity how to conduct multi-scenario, MORDM for many wicked problems. For instance, a recent study, described in detail in Chap. 7, used RDM on a topic similar to the "carrots and sticks" example—examining how international finance institutions such as the Green Climate Fund (GCF) can best craft long-term investment strategies to speed decarbonization in the face of deep technological and climate uncertainty (Molina Perez 2016). This more recent study was made possible by powerful new SD algorithms and visualization tools.

Recent work for four North Carolina cities illustrates the power of MORDM, a combination of RDM with new evolutionary algorithms for multi-objective robust optimization (Herman et al. 2014, 2016; Zeff et al. 2014, 2016). The study helped the neighboring cities of Raleigh, Durham, Chapel Hill, and Cary link their short-term operational and long-term investment water plans by shifting the former from rule-based procedures to new dynamic risk-of-failure triggers, and the latter from static to adaptive policy pathways. The study also helped the four independent cities coordinate their plans in the presence of different objectives and deep uncertainty. Linking such MORDM with the Dynamic Adaptive Policy Pathways (DAPP) approach of Chaps. 4 and 9 represents an exciting direction for the DMDU community.

Recent work for the US Bureau of Reclamation and the parties to the Colorado Compact, also described in Chap. 7, showcases RDM's ability to facilitate deliberation with analysis, helping contesting parties to agree on the vulnerabilities they face and adaptive strategies for addressing them (Groves et al. 2013; Bloom 2015).

Some important technical hurdles still remain before these capabilities fully mature. First, an approach is needed that provides full Pareto satisficing surfaces. Current MORDM analyses identify Pareto *optimal* surfaces for best-estimate cases and measure the robustness of alternative strategies, represented by different regions on the Pareto surface, to the deep uncertainties (Kasprzyk et al. 2013). In the future, MORDM could produce sets of strategies chosen specifically because their performance across multiple objectives was largely insensitive to the deep uncertainties. Recent work has taken steps toward providing such Pareto satisficing surfaces (Watson and Kasprzyk 2017), but more needs to be done. Furthermore, despite the availability of ubiquitous computation on the cloud, and through high-performance computation facilities, it still remains difficult in many cases to conduct a full MORDM analysis using realistic system models, which would require running many thousands of cases to perform the multi-objective robust optimization over each of many thousands of scenarios. Research is needed on what we might call adaptive sampling approaches to help navigate more efficiently through the set of needed runs. In addition, research could usefully provide guidance on when to use alternative robustness criteria, as well as the conditions under which RDM's iterative analytic process is guaranteed to converge independent of the initial problem framing, or when path dependence may lead analyses to different answers (Kwakkel et al. 2016).

Finally, the cost of developing the needed system models often puts RDM analyses out of reach for many decisionmakers. Research on "RDM-lite"—means to quickly develop such models through approaches such as expert elicitation and participatory modeling—could greatly increase the use of these methods (see, e.g., O'Mahony et. al. 2018).

Evaluation plays a crucial role in the design and use of any successful decision support system (NRC 2009; Pidgeon and Fischhoff 2011; Wong-Parodi et al. 2016). Some evaluations of RDM tools, visualizations, and processes exist, both in the laboratory (Budescu et al. 2013; Parker et al. 2015; Gong et al. 2017) and through field experiments (Groves et al. 2008). Recent work has proposed frameworks for evaluating the impacts of RDM-based decision support in urban environments (Knopman and Lempert 2016). But much more such evaluation work is required to improve the practical application of RDM decision support (Bartels et. al., forthcoming).

More broadly, as DMDU methods reach technical maturity, they offer the opportunity to reshape the relationship between quantitative decision analytics and the way in which organizations use this information with their internal and external audiences and processes. The potential for such reshaping presents a rich menu of research needs to understand the organizational, anthropological, political, and ethical implications.

As one example, the concept of *risk governance* embeds risk management, which often has a narrow, more technocratic perspective, in a broader context that considers institutions, rules conventions, processes, and mechanisms through which humans acting as individuals and groups make choices affecting risk (Renn 2008). Recent work has explored how to embed RDM methods and tools in a risk governance framework (Knopman and Lempert 2016). For instance, RDM can help decisionmakers working within a multi-agent and multi-jurisdictional system organize their strategies into "tiers of transformation," which derives from the ideas of triple-loop learning. Lower tiers represent actions the decisionmakers can address on their own, while the outer tiers represent large-scale, transformative system changes that only the decisionmakers can help to catalyze. Any understanding of how to implement and use such capabilities, and the extent to which they would prove useful, remains nascent.

Future work can also usefully situate the types of moral reasoning and social choice embodied in alternative approaches to decision support. In his treatise, the *Idea of Justice*, Amartya Sen (2009) describes two classes of moral reasoning—the transcendental and the relational. The former, represented by Sen's teacher John Rawls (1971), seeks to inform ethical societal choices by first envisioning a common vision of a perfectly just world. People can then use that vision to inform their near-term choices. The latter, Sen's preferred alternative, rests on the assumption that irreducible uncertainty about the consequences of our actions, and a diversity of priorities, goals, and values, are fundamental attributes of our world. Thus, no such transcendental vision of the type envisioned by Rawls is possible because the level of agreement and commonality of values it presupposes does not, and should not, exist in a diverse society in which people are free to pursue their lives according to their own, often very different, visions of what is good. In addition, Sen argues, even if a common transcendental vision were possible, it would prove insufficient to inform near-term choices, because human knowledge is too fallible and the uncertainties too deep to chart an unambiguous path to the ideal. But humans can obtain sufficient knowledge to craft near-term options and differentiate the better from the worse. Relational reasoning thus involves an iterative process of debating, choosing, learning, and revisiting choices, always trying to move in the direction of more justice in the face of imperfect knowledge and conflicting goals.

"Agree-on-assumptions" approaches to decision support reflect transcendental reasoning, while "agree-on-decisions" approaches reflect relational reasoning (Lempert et al. 2013c). Sen emphasizes the importance of deliberation in a relational process of social choice. His framework provides attributes for judging what constitutes an ethical process of deliberation with analysis. In particular, such deliberations work best when they recognize the inescapable plurality of competing views; facilitate re-examination and iterative assessments; demand clear explication of reasoning and logic; and recognize an "open impartiality" that accepts the legitimacy and importance of the views of others, both inside and outside the community of interest to the immediate policy discussion.

Recent work has pioneered methods for conducting ethical-epistemological analysis on the extent to which decision support products, methods, and systems meet such ethical criteria (Tuana 2013; Bessette et al. 2017; Mayer et al. 2017), but much more remains to be done. Today's world presents numerous, complex decision challenges—from sustainability to national security—that require quantitative decision support to successfully address. But "agree-on-assumptions" methods often lure decisionmakers toward over-confidence and can make it difficult to engage and promote consensus among participants with diverse expectations and interests. Such methods—built on the assumption that the decision analytics aim to provide a normative ranking of decision options—have their foundations in a time of computational poverty and rest on a narrow understanding of how quantitative information can best inform decisions. Recent years have seen an explosion of computational capabilities and a much richer understanding of effective decision support products and processes. RDM—a multi-objective, multi-scenario "agree-on-decision" approach—exploits these new capabilities and understanding to facilitate deliberative processes in which decisionmakers explore, frame, and reach consensus on the "wicked" problems that today's decisionmakers increasingly face.

References

Bankes, S. C. (1993). Exploratory modeling for policy analysis. *Operations Research, 41*(3), 435–449.

Bankes, S. C., Lempert, R. J., & Popper, S. W. (2001). Computer-assisted reasoning. *Computing in Science & Engineering, 3*(2), 71–77.

Bartels, E., Mikolic-Torreira, I, Popper, S. W., & Predd, J. (forthcoming). *What is the value proposition of analysis for decisionmaking?* Santa Monica, CA: RAND Corporation, PR-3485-RC.

Bessette, D. L., Mayer, L. A., Cwik, B., Vezer, M., Keller, K., Lempert, R., et al. (2017). Building a values-informed mental model for New Orleans climate risk management. *Risk Analysis, 37*(10), 1993–2004.

Bloom, E. (2015). *Changing midstream: Providing decision support for adaptive strategies using robust decision making.* Santa Monica, CA: RAND Corporation, RGSD-348.

Borio, C., Drehmann, M., & Tsatsaronis, K. (2014). Stress-testing macro stress testing: Does it live up to expectations? *Journal of Financial Stability, 12*(20), 3–15.

Breiman, L., Friedman, J. H., Olshen, R. A., & Stone, C. J. (1984). *Classification and regression trees.* Wadsworth Statistics/Probability Series. Monterey, CA: Wadsworth.

Brooks, A., Bennett, B., & Bankes, S. C. (1999). An application of exploratory analysis: The weapon mix problem. *Military Operations Research, 4*(1), 67–80.

Bryant, B. P., & Lempert, R. J. (2010). Thinking inside the box: A participatory, computer-assisted approach to scenario discovery. *Technological Forecasting and Social Change, 77,* 34–49.

Budescu, D. V., Lempert, R. J., Broomell, S., & Keller, K. (2013). Aided and unaided decisions with imprecise probabilities. *European Journal of Operational Research, 2*(1–2), 31–62.

Carlsen, H., Eriksson, E. A., Dreborg, K. H., Johansson, B., & Bodin, Ö. (2016a). Systematic exploration of scenario spaces. *Foresight, 18*(1), 59–75.

Carlsen, H., Lempert, R. J., Wikman-Svahn, P., & Schweizer, V. (2016b). Choosing small sets of policy-relevant scenarios by combining vulnerability and diversity approaches. *Environmental Modelling and Software, 84,* 155–164.

Carter, T. R., Jones, R. N., Lu, S. B. X., Conde, C., Mearns, L. O., O'Neill, B. C., et al. (2007). New assessment methods and the characterisation of future conditions. In M. L. Parry, O. F. Canziani, J. P. Palutikof, P. J. V. D. Linden, & C. E. Hanson, (Eds.), *Climate change 2007: Impacts, adaptation and vulnerability. Contribution of working group II to the fourth assessment report of the intergovernmental panel on climate change* (Vol. 1, pp. 33–171). Cambridge, UK: Cambridge University Press.

Cervigni, R., Liden, R., Neumann, J. E., & Strzepek, K. M. (Eds.). (2015). *Enhancing the climate resilience of Africa's infrastructure: The water and power sectors.* Africa Development Forum Series. Washington, DC: World Bank.

Cohen, L. R., & Noll, R. G. (2002). *Technology pork barrel.* Washington, DC: Brookings Institution Press.

Cohon, J., & Marks, D. (1975). A review and evaluation of multiobjective programing techniques. *Water Resources Research,* 11(2).

Dalal, S., Han, B., Lempert, R. J., Jaycocks, A., & Hackbarth, A. (2013). Improving scenario discovery using orthogonol rotations. *Environmental Modeling and Software, 48,* 1–16.

Davies, S. (1979). *The diffusion of process innovations.* Cambridge, MA: Cambridge University Press.

Dessai, S., & Hulme, M. (2007). Assessing the robustness of adaptation decisions to climate change uncertainties: A case study on water resources management in the East of England. *Global Environmental Change, 17*(1), 59–72.

Dewar, J. A., Builder, C. H., Hix, W. M., & Levin, M. H. (1993). Assumption-based planning—A planning tool for very uncertain times. Santa Monica, CA, RAND Corporation. https://www.rand.org/pubs/monograph_reports/MR114.html. Retrieved July 20, 2018.

Dewar, J. A. (2002). *Assumption-based planning—A tool for reducing avoidable surprises.* Cambridge: Cambridge University Press.

Fischbach, J. R., Lempert, R. J., Molina-Perez, E., Tariq, A., Finucane, M. L., & Hoss, F. (2015). *Managing water quality in the face of uncertainty: A robust decision-making demonstration for EPA's National Water Program*. Santa Monica, CA: RAND Corporation, PR-1148-EPA.

Friedman, J. H., & Fisher, N. I. (1999). Bump hunting in high-dimensional data. *Statistics and Computing, 9,* 123–143.

Ghile, Y. B., Taner, M. Ü., Brown, C., Grijsen, J. G., & Talbi, A. (2014). Bottom-up climate risk assessment of infrastructure investment in the Niger River Basin. *Climatic Change, 122,* 97–110.

Giuliani, M., & Castelletti, A. (2016). Is robustness really robust? How different definitions of robustness impact decision-making under climate change. *Climatic Change, 135*(3–4), 409–424.

Gong, M., Lempert, R. J., Parker, A. M., Mayer, L. A., Fischbach, J., Sisco, M., et al. (2017). Testing the scenario hypothesis: An experimental comparison of scenarios and forecasts for decision support in a complex decision environment. *Environmental Modeling and Software, 91,* 135–155.

Groves, D. G., Bloom, E. W., Lempert, R. J., Fischbach, J. R., Nevills, J., & Goshi, B. (2014). Developing key indicators for adaptive water planning. *Journal of Water Resources Planning Management, 141*(7).

Groves, D. G., Fischbach, J. R., Bloom, E., Knopman, D., & Keefe, R. (2013). *Adapting to a changing Colorado river: Making future water deliveries more reliable through robust management strategies*. Santa Monica, CA: RAND Corporation, RR-242-BOR.

Groves, D. G., Knopman, D., Lempert, R. J., Berry, S., & Wainfan, L. (2008). *Presenting uncertainty about climate change to water resource managers—Summary of workshops with the Inland Empire Utilities Agency*. Santa Monica, CA: RAND Corporation, TR-505-NSF.

Groves, D. G., & Lempert, R. J. (2007). A new analytic method for finding policy-relevant scenarios. *Global Environmental Change, 17,* 73–85.

Groves, D. G., Lempert, R. J., May, D. W., Leek, J. R., & Syme, J. (2016). *Using high-performance computing to support water resource planning, a workshop demonstration of real-time analytic facilitation for the Colorado River Basin*. Santa Monica, CA: RAND Corporation and LLNL, CF-339-RC.

Groves, D. G., Sharon, C., & Knopman, D. (2012). *Planning tool to support Louisiana's decision-making on coastal protection and restoration*. Santa Monica, CA: RAND Corporation, TR-1266-CPRA.

Haasnoot, M., Kwakkel, J. H., Walker, W. E., & ter Maat, J. (2013). Dynamic adaptive policy pathways: A new method for crafting robust decisions for a deeply uncertain world. *Global Environmental Change, 23*(2), 485–498.

Hadka, D., Herman, J., Reed, P., & Keller, K. (2015). An open source framework for many-objective robust decision making. *Environmental Modelling and Software, 74,* 129–144.

Hall, J. M., Lempert, R. J., Keller, K., Hackbarth, A., Mijere, C., & McInerney, D. (2012). Robust Climate Policies under uncertainty: A comparison of Info-Gap and RDM methods. *Risk Analysis, 32*(10), 1657–1672.

Hansen, L. P., & Sargent, T. J. (2008). *Robustness*. Princeton, NJ: Princeton University Press.

Helgeson, C. (2018). Structuring decisions under deep uncertainty. *Topoi,* pp. 1–13.

Herman, J., Zeff, H., Lamontagne, J., Reed, P., & Characklis, G. (2016). Synthetic drought scenario generation to support bottom-up water supply vulnerability assessments. *Journal of Water Resources Planning and Management,* 142(11).

Herman, J., Zeff, H., Reed, P., & Characklis, G. (2014). Beyond optimality: Multistakeholder robustness tradeoffs for regional water portfolio planning under deep uncertainty. *Water Resources Research, 50*(10), 7692–7713.

Herrick, C., & Sarewitz, D. (2000). Ex post evaluation: A more effective role for scientific assessments in evnironmental policy. *Science, Technology and Human Values, 25*(3), 309–331.

Hodges, J. (1991). Six (or so) things you can do with a bad model. *Operations Research, 39*(3), 355–365.

IPCC. (2001). *Climate change 2001: Mitigation, intergovernmental panel on climate change*.

Isley, S. (2014). *The political sustainability of carbon control policies in an evolutionary economics setting*. Santa Monica, CA: RAND Corporation, RGSD-331.

Isley, S. C., Lempert, R. J., Popper, S. W., & Vardavas, R. (2015). The effect of near-term policy choices on long-term greenhouse gas transformation pathways. *Global Environmental Change, 34,* 147–158.

Jones, R. N., Patwardhan, A., Cohen, S., Dessai, S., Lammel, A., Lempert, R. J., et al. (2014). Foundations for decision making. In *Climate change 2014: Impacts, adaptation, and vulnerability*. Intergovernmental Panel on Climate Change (IPCC).

Kalra, N., Hallegatte, S., Lempert, R. J., Brown, C., Fozzard, A., Gill, S., et al. (2014). Agreeing on robust decisions: A new process of decision making under deep uncertainty. *Policy Research Working Paper*. World Bank, WPS-6906.

Kasprzyk, J. R., Nataraj, S., Reed, P. M., & Lempert, R. J. (2013). Many-objective robust decision making for complex environmental systems undergoing change. *Environmental Modeling and Software, 42,* 55–71.

Knopman, D., & Lempert R. J. (2016). *Urban responses to climate change: Framework for decision-making and supporting indicators* (156). Santa Monica, CA: RAND Corporation, RR-1144-MCF.

Kollat, J., & Reed, P. (2007). A framework for visually interactive decision-making and design using evolutionary multi-objective optimization (VIDEO). *Environmental Modeling and Software, 22*(12), 1691–1704.

Kwakkel, J. H., Haasnoot, M., & Walker, W. E. (2016). Comparing robust decision-making and dynamic adaptive policy pathways for model-based decision support under deep uncertainty. *Environmental Modelling and Software, 86,* 168–183.

Lempert, R. J. (2013). Scenarios that illuminate vulnerabilities and robust responses. *Climatic Change, 117,* 627–646.

Lempert, R. J., & Groves, D. G. (2010). Identifying and evaluating robust adaptive policy responses to climate change for water management agencies in the American West. *Technological Forecasting and Social Change, 77,* 960–974.

Lempert, R. J., Popper, S. W., Groves, D. G., Kalra, N., Fischbach, J. R., Bankes, S. C. et al. (2013a). *Making Good Decisions Without Predictions: Robust Decision Making for Planning Under Deep Uncertainty*. Santa Monica, CA: RAND, RB-9701.

Lempert, R. J., Kalra, N., Peyraud, S., Mao, Z., Tan, S. B., Cira, D., & Lotsch, A. (2013b). Ensuring Robust Flood Risk Management in Ho Chi Minh City: A robust decision-making demonstration. World Bank, WPS-6465.

Lempert, R. J., Groves, D. G., & Fischbach, J. (2013c). *Is it ethical to use a single probability density function?* Santa Monica, CA: RAND Corporation, WR-992.

Lempert, R. J., & Kalra, N. (2011). *Managing climate risks in developing countries with robust decision making*. Washington, DC: World Resources Institute. https://www.wri.org/our-work/project/world-resources-report/managing-uncertainty. Retrieved October 12, 2018.

Lempert, R. J., Nakicenovic, N., Sarewitz, D., & Schlesinger, M. (2004). Characterizing climate-change uncertainties for decision-makers—An editorial essay. *Climatic Change, 65*(1–2), 1–9.

Lempert, R. J. (2002). A new decision sciences for complex systems. *Proceedings of the National Academy of Sciences, 99*(3), 7309–7313.

Lempert, R. J. (2007). Can scenarios help policymakers be both bold and careful? In F. Fukuyama (Ed.), *Blindside: How to anticipate forcing events and wild cards in global politics*. Washington, DC: Brookings Institute Press.

Lempert, R. J. (2014). Embedding (some) benefit-cost concepts into decision support processes with deep uncertainty. *Journal of Benefit Cost Analysis, 5*(3), 487–514.

Lempert, R. J., & Collins, M. (2007). Managing the risk of uncertain threshold responses: Comparison of robust, optimum, and precautionary approaches. *Risk Analysis, 27*(4), 1009–1026.

Lempert, R. J., Groves, D. G., Popper, S. W., & Bankes, S. C. (2006). A general, analytic method for generating robust strategies and narrative scenarios. *Management Science, 52*(4), 514–528.

Lempert, R. J., & Popper, S. W. (2005). High-Performance Government in an Uncertain World. In R. Klitgaard & P. Light (Eds.), *High Performance Government: Structure, Leadership, and Incentives*. Santa Monica, CA: RAND Corporation.

Lempert, R. J., Popper, S. W., & Bankes, S. C. (2002). Confronting Surprise. *Social Science Computer Review, 20*(4), 420–440.

Lempert, R. J., Popper, S. W., & Bankes, S. C. (2003). *Shaping the Next One Hundred Years: New Methods for Quantitative, Long-term Policy Analysis*. Santa Monica, CA, RAND Corporation, MR-1626-RPC.

Lempert, R. J., Schlesinger, M. E., & Bankes, S. C. (1996). When we don't know the costs or the benefits: Adaptive strategies for abating climate change. *Climatic Change, 33*(2), 235–274.

Lempert, R. J., Schlesinger, M. E., Bankes, S. C., & Andronova, N. G. (2000). The impact of variability on near-term climate-change policy choices. *Climatic Change, 45*(1), 129–161.

Light, P. C. (2005). *The four pillars of high performance: how robust organizations achieve extraordinary results*. New York, NY: McGraw-Hill.

Luce, R. D., & Raiffa, H. (1957). *Games and decisions*. New York: Wiley.

Maass, A., Hufschmidt, M., Dorfman, R., Thomas, H., Marglin, S. Fair, G. et al. (1962). *Design of water resources systems; new techniques for relating economic objectives, engineering analysis, and governmental planning*. Cambridge, Harvard University Press.

March, J. (1994). *A primer on decision making: how decisions happen*. Toronto, Canada: The Free Press.

Mayer, L. A., Loa, K., Cwik, B., Tuana, N., Keller, K., Gonnerman, C., et al. (2017). Understanding scientists' computational modeling decisions about climate risk management strategies using values-informed mental models. *Global Environmental Change, 42*, 107–116.

Miller, J. H. (1998). Active Nonlinear Tests (ANTs) of complex simulations models. *Management Science, 44*(6), 820–830.

Molina Perez, E. (2016). *Directed international technological change and climate policy: New methods for identifying robust policies under conditions of deep uncertainty*, Pardee RAND Graduate School, RGSD-369.

Morgan, M. G., & Henrion, M. (1990). *Uncertainty: A guide to dealing with uncertainty in quantitative risk and policy analysis*. Cambridge, UK: Cambridge University Press.

National Academies of Sciences, E., and Medicine. 2016 (NAS) (2016). *Assessment of Approaches to Updating the Social Cost of Carbon: Phase 1 Report on a Near-Term Update*. Washington, DC, The National Academies Press.

National Research Council (NRC) (2009), *Informing decisions in a changing climate*. National Academies Press.

Nordhaus, W. D. (1994). *Managing the global commons: the economics of climate change*. Cambridge, MA: MIT Press.

O'Mahony, A., Blum, I., Armenta, G., Burger, N., Mendelsohn, J., McNerney, M. et al. (2018). *Assessing, Monitoring and Evaluating Army Security Cooperation: A Framework for Implementation*. Santa Monica, CA: RAND Corporation, RR-2165-A.

Parker, A. M., Srinivasan, S., Lempert, R. J., & Berry, C. (2015). Evaluating simulation-derived scenarios for effective decision support. *Technological Forecasting and Social Change, 91*, 64–77.

Pidgeon, N., & Fischhoff, B. (2011). The role of social and decision sciences in communicating uncertain climate risks. *Nature Climate Change, 1*, 35–41.

Popper, S. W., Berrebi, C., Griffin, J., Light, T., Min, E. Y., & Crane, K. (2009). *Natural gas and Israel's energy future: Near-term decisions from a strategic perspective*. Santa Monica, CA, RAND Corporation, MG-927.

Popper, S. W., Lempert, R. J., & Bankes, S. C. (2005). Shaping the future. *Scientific American, 292*(4), 66–71.

Quinn, J. D., Reed, P. M., & Keller, K. (2017). Direct policy search for robust multi-objective management of deeply uncertain socio-ecological tipping points. *Environmental Modelling and Software, 92*, 125–141.

Ranger, N., Millner, A., Dietz, S., Fankhauser, S., Lopez, A., & Ruta, G. (2010). *Adaptation in the UK: A decision-making process*. London: Granthan/CCEP Policy Brief.

Rawls, J. (1971). *A theory of justice*. Harward University Press

Rayner, S. (2000). Prediction and other approaches to climate change policy. In D. Sarewitz (Ed.), *Prediction: Science, decision making, and the future of nature* (pp. 269–296). Washington, DC: Island Press.

Renn, O. (2008). *Risk governance: Coping with uncertainty in a complex world*. London Earthscan.

Rittel, H., & Webber, M. (1973). Dilemmas in a general theory of planning. *Policy Sciences, 4*, 155–169.

Robalino, D. A., & Lempert, R. J. (2000). Carrots and sticks for new technology: Abating greenhouse gas emissions in a heterogeneous and uncertain world. *Integrated Assessment, 1*(1), 1–19.

Rosenhead, J. (1990). Rational analysis: Keeping your options open. In: J. Rosenhead & J. Mingers (Eds.), *Rational analysis for a problematic world: Problem structuring methods for complexity, uncertainty and conflict*. Chichester, England, Wiley.

Rosenhead, M. J., Elton, M., & Gupta, S. K. (1972). Robustness and optimality as criteria for strategic decisions. *Operational Research Quarterly, 23*(4), 413–430.

Rydell, C., Peter, J. P. Caulkins, & S. S. Everingham. (1997). *Enforcement or treatment? Modeling the relative efficacy of alternatives for controlling cocaine*. Santa Monica, CA, RAND Corporation, RP-614.

Sarewitz, D., & Pielke, R. A. (2000). *Science, prediction: Decisionmaking, and the future of nature*. Washington, DC: Island Press.

Schneller, G. O., & Sphicas, G. P. (1983). Decision making under uncertainty: Starr's Domain Criterion. *Theory and Decision, 15*, 321–336.

Schoemaker, P. J. H. (1993). Multiple scenario development: Its conceptual and behavioral foundation. *Strategic Management Journal, 14*(3), 193–213.

Schwartz, P. (1996). *The art of the long view—planning for the future in an uncertain world*. New York, NY: Currency-Doubleday.

Sen, A. (2009). *The idea of justice*. Cambridge, Massachusetts: Belknap Press.

Starr, M. K. (1962). *Product design and decision theory*. Englewood Cliffs, NJ: Prentice-Hall.

Swanson, D., & Bhadwal, S. (2009). *Creating adaptive policies: A guide for policy-making in an uncertain world*. Sage Publications.

Swanson, D., Venema, H., Barg, S., Tyler, S., Drexage, J., Bhandari P., & Kelkar, U. (2007). *Initial conceptual framework and literature review for understanding adaptive policies*.

Tetlock, P. E., & Gardner, D. (2016). *Superforecasting: The art and science of prediction*. Random House.

Thompson, C. (2013). *Smarter than you think: How technology is changing our minds for the better*. New York: Penguin.

Tuana, N. (2013). Embedding philosophers in the practices of science: Bringing humanities to the sciences. *Synthese, 190*(11), 1955–1973.

Van der Heijden, K. (1996). *Scenarios: The art of strategic conversation*. New York: Wiley.

Wack, P. (1985). The gentle art of reperceiving—scenarios: Uncharted waters ahead (part 1 of a two-part article). *Harvard Business Review (September–October)*: 73–89.

Walker, W., Marchau, V., & Swanson, D. (2010). Addressing deep uncertainty using adaptive policies. *Technology Forecasting and Social Change, 77*, 917–923.

Walker, W. E., Rahman, S. A., & Cave, J. (2001). Adaptive policies, policy analysis, and policymaking. *European Journal of Operational Research, 128*, 282–289.

Walley, P. (1991). *Statistical reasoning with imprecise probabilities*. London: Chapman and Hall.

Watson, A. A., & Kasprzyk, J. R. (2017). Incorporating deeply uncertain factors into the many objective search process. *Environmental Modeling and Software, 89*, 159–171.

Weaver, C. P., Lempert, R. J., Brown, C., Hall, J. A., Revell, D., & Sarewitz, D. (2013). Improving the contribution of climate model information to decision making: The value and demands of robust decision frameworks. *WIREs Climate Change, 4*, 39–60.

Wolf, C. (1993). *Markets or governments: Choosing between imperfect alternatives*. MIT Press.

Wong-Parodi, G., Krishnamurti, T., Davis, A., Schwartz, D., & Fischhoff, B. (2016). A decision science approach for integrating social science in climate and energy solutions. *Nature Climate Change, 6,* 563–569.

Woodruff, M., & Reed, P. (2013). Many objective visual analytics: Rethinking the design of complex engineered systems. *Structural and Multidiciplinary Optimization, 48*(1), 201–219.

Zeff, H., Herman, J., Reed, P., & Characklis, G. (2016). Cooperative drought adaptation: integrating infrastructure development, conservation, and water transfers into adaptive policy pathways. *Water Resources Research, 52*(9), 7327–7346.

Zeff, H. B., Kasprzyk, J. R., Herman, J. D., Reed, P. M., & Characklis, G. W. (2014). Navigating financial and supply reliability tradeoffs in regional drought management portfolios. *Water Resources Research, 50*(6), 4906–4923.

Zenko, M. (2015). *Red Team: How to succeed by thinking like the enemy.* Basic Books.

Robert Lempert is a principal researcher at the RAND Corporation and Director of the Frederick S. Pardee Center for Longer Range Global Policy and the Future Human Condition. His research focuses on risk management and decision-making under conditions of deep uncertainty. Dr. Lempert is a Fellow of the American Physical Society, a member of the Council on Foreign Relations, a chapter lead for the Fourth US National Climate Assessment, and a convening lead author for Working Group II of the United Nation's Intergovernmental Panel on Climate Change (IPCC) Sixth Assessment Report. Dr. Lempert was the Inaugural EADS Distinguished Visitor in Energy and Environment at the American Academy in Berlin and serves as the inaugural president of the Society for Decision Making Under Deep Uncertainty. A Professor of Policy Analysis in the Pardee RAND Graduate School, Dr. Lempert is an author of the book *Shaping the Next One Hundred Years: New Methods for Quantitative, Longer-Term Policy Analysis.*

Open Access This chapter is licensed under the terms of the Creative Commons Attribution 4.0 International License (http://creativecommons.org/licenses/by/4.0/), which permits use, sharing, adaptation, distribution and reproduction in any medium or format, as long as you give appropriate credit to the original author(s) and the source, provide a link to the Creative Commons licence and indicate if changes were made.

The images or other third party material in this chapter are included in the chapter's Creative Commons licence, unless indicated otherwise in a credit line to the material. If material is not included in the chapter's Creative Commons licence and your intended use is not permitted by statutory regulation or exceeds the permitted use, you will need to obtain permission directly from the copyright holder.

Chapter 3
Dynamic Adaptive Planning (DAP)

Warren E. Walker, Vincent A. W. J. Marchau and Jan H. Kwakkel

Abstract

- DAP is a DMDU approach for designing a plan that explicitly includes provisions for adaptation as conditions change and knowledge is gained.
- The resulting plan combines actions to be taken right away with those that make important commitments to shape the future and those that preserve needed flexibility for the future.
- The approach includes the specification of a monitoring system, together with the specification of actions to be taken when specific trigger values are reached.
- This chapter describes the DAP approach and illustrates it with a (more or less) fictitious case. A real-life application is given in Chap. 8.

3.1 Introduction

Most strategic plans implicitly assume that the future can be predicted. A static plan is developed using a single future, often based on the extrapolation of trends, or a static 'robust' plan is developed that will produce acceptable outcomes in a small set of plausible future worlds. However, if the future turns out to be different from the hypothesized future(s), the plan might fail. Furthermore, not only is the future highly uncertain, the conditions planners need to deal with are changing over time (the economic situation, annual rainfall, etc.). This chapter describes an approach for planning under conditions of deep uncertainty called Dynamic Adaptive Planning (DAP). This approach is based on specifying a set of objectives and constraints, designing an initial plan consisting of short-term actions, and establishing a framework to guide future (contingent) actions. A plan that embodies these is explicitly designed to be adapted over time to meet changing circumstances.

W. E. Walker (✉) · J. H. Kwakkel
Delft University of Technology, Delft, The Netherlands
e-mail: w.e.walker@tudelft.nl

V. A. W. J. Marchau
Radboud University (RU), Nijmegen, The Netherlands

© The Author(s) 2019
V. A. W. J. Marchau et al. (eds.), *Decision Making under Deep Uncertainty*,
https://doi.org/10.1007/978-3-030-05252-2_3

DAP was first outlined by Walker et al. (2001) and made more concrete by Kwakkel et al. (2010). DAP has been explored in various applications, including flood risk management in the Netherlands in light of climate change (Rahman et al. 2008) and policies with respect to the implementation of innovative urban transport infrastructures (Marchau et al. 2008), congestion road pricing (Marchau et al. 2010), intelligent speed adaptation (Agusdinata et al. 2007), and 'magnetically levitated' (Maglev) rail transport (Marchau et al. 2010). Central to DAP is the acknowledgment of uncertainty, that 'in a rapidly changing world, fixed static policies are likely to fail' (Kwakkel et al. 2010). As new information becomes known, the plan should incorporate the ability to adapt dynamically through learning mechanisms (Kwakkel et al. 2010; Walker et al. 2001).

DAP is based in part on concepts related to Assumption-Based Planning (ABP) (Dewar et al. 1993). In ABP, an assumption is an assertion about some characteristic of the world that underlies a plan. A critical (load-bearing) assumption is an assumption whose failure would mean that the plan would not meet its objectives (i.e., would not be successful). An assumption is vulnerable if plausible events could cause it to fail within the expected lifetime of the plan. In brief, DAP involves specifying goals and objectives, developing an initial plan to meet these goals and objectives, identifying the vulnerabilities of the plan (i.e., how it might fail), adding to the plan a set of initial actions to be taken immediately upon implementation to protect it against some of these vulnerabilities, and establishing signposts to monitor the remaining uncertain vulnerabilities. During implementation, if the monitoring program indicates that one or more of the signposts reaches a predetermined critical level, predetermined contingent actions are taken ('triggered') to ensure that the plan stays on track to meet its goals and objectives. The plan, monitoring program, and contingent adaptations remain in place unless monitoring indicates that the intended outcomes can no longer be achieved, or if the goals and objectives change. In these instances, the entire plan is then reassessed, and a new plan is designed. The elements of flexibility, adaptability, and learning enable the plan to adjust to new information as it becomes available and therefore to deal with deep uncertainty. (The new information might reveal developments that can make the plan more successful, or succeed sooner; the adaptive plan should also be designed to take advantage of such opportunities.)

The DAP approach is carried out in two phases: (1) the design phase, in which the plan, monitoring program, and various pre- and post-implementation actions are designed, and (2) the implementation phase, in which the plan and the monitoring program are implemented and contingent actions are taken, if triggered. The five steps of the design phase are shown in Fig. 3.1. Once the plan is established through the five design steps shown, the plan is implemented and monitoring commences.

3 Dynamic Adaptive Planning (DAP)

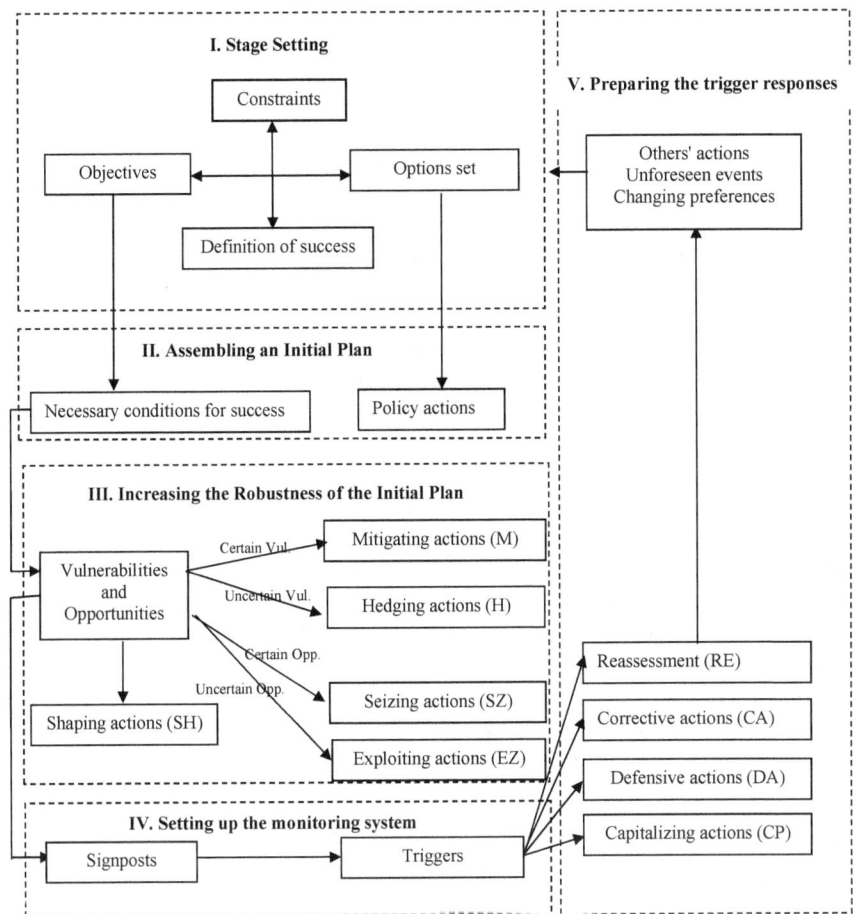

Fig. 3.1 Steps for DAP (Walker et al. 2013, p. 344)

3.2 The DAP Approach

Step I (Stage Setting) and Step II (Assembling an Initial Plan)

As a foundation for the plan, the goals and objectives that are important to the planners and stakeholders are defined—i.e., what constitutes a successful outcome. Constraints on the plan are identified, and a set of alternative actions to achieve the objectives are analyzed. In Step II, an initial plan that meets the goals and objectives is assembled from the alternatives that have been identified (as might be done, for example, in a traditional policy analysis study (Walker 2000)). The necessary conditions for success are outlined (e.g., social, technological, physical, political, economic, or other conditions necessary for the plan to succeed). It is very important in this step to identify the range of necessary conditions for success, as these are used in later steps to identify vulnerabilities, signposts, and triggers. For this reason, it is important to involve managing agencies, as well as other stakeholders.

Step III (Increasing the Robustness of the Initial Plan)

The static robustness of the initial plan is increased through a series of anticipatory actions taken in direct response to vulnerabilities and opportunities. Vulnerabilities that can diminish the success of the initial plan, and opportunities that can increase the success of the initial plan, are first identified. Analytical tools, such as Exploratory Modeling and Analysis (EMA) (Bankes et al. 2013), and scenario analysis (van der Heijden 1996), or expert opinions using Strengths, Weaknesses, Opportunities, and Threats (SWOT) analysis (Osita et al. 2014), may be used to investigate plausible future conditions to ensure that relevant vulnerabilities are identified. An approach based on EMA, called Scenario Discovery (Bryant and Lempert 2010; Kwakkel et al. 2012), can be used to identify the scenarios in which a plan would perform poorly. These scenarios highlight the vulnerabilities of the plan. Then, actions can be specified to protect the plan from failing if any one of these scenarios occurs. A way to specify protective actions is to use threats, opportunities, weaknesses, and strengths (TOWS) analysis (Weihrich 1982), which uses a SWOT analysis as input, and translates the outcomes of the SWOT analysis into actions. Other possible techniques include Delphi (Rowe et al. 1991), ABP (Dewar et al. 1993), and scenario analysis (van der Heijden 1996).

Five types of anticipatory actions can be taken immediately upon implementation of the plan to address these vulnerabilities (and opportunities), thus increasing the robustness of the initial plan. These five types of actions are (Walker et al. 2013):

- Mitigating actions (M)—Actions that reduce adverse impacts on a plan stemming from *certain* (or very likely) vulnerabilities.
- Hedging actions (H)—Actions that reduce adverse impacts on a plan, or spread or reduce risks that stem from *uncertain* vulnerabilities (much like buying car insurance).
- Seizing actions (SZ)—Actions that take advantage of certain (or very *likely*) opportunities that may prove beneficial to the plan.
- Exploiting actions (E)—Actions that take advantage of (*uncertain*) new developments that can make the plan more successful, or succeed sooner.
- Shaping actions (SH)—Actions taken proactively to affect external events or conditions that could either reduce the plan's chance of failure or increase its chance of success.

Mitigating actions and hedging actions prepare the initial plan for potential adverse effects and in this way try to make the plan more robust. Seizing actions and exploiting actions are actions taken now to change the policy in order to take advantage of available opportunities, which can also make the plan more robust. In contrast, shaping actions are proactive and aim at affecting external forces in order to reduce the chances of negative outcomes or to increase the chances of positive outcomes. As such, shaping actions aim not so much at making the plan more robust, but at changing the external situation in order to change the nature of the vulnerability or opportunity. For example, marketing is an attempt to increase the demand for a given product. In this way, one tries to prevent insufficient demand for the product.

Step IV (Setting up the Monitoring System)

A monitoring system is developed that will inform decisionmakers about actions that can be taken in response to new conditions. This constitutes the learning component that gives DAP the flexibility to adapt to new conditions over time. This introduces the element of adaptive robustness, which makes DAP able to deal with deep uncertainty, and distinguishes it from scenario-based approaches that are based on responding to a single or small set of hypothesized futures to achieve static robustness. The monitoring program consists of signposts and triggers. *Signposts* specify the types of information and variables that should be monitored to show (1) whether the initial plan is currently achieving its goals and/or (2) whether the vulnerabilities and opportunities identified in Step 3 are impeding the plan from achieving its goals in the future. *Triggers* are the critical signpost levels or events that, when they occur, signify that (contingent) actions should be taken to ensure the initial plan remains on course in order to continue to achieve its specified goals.

Step V (Preparing the Trigger Responses)

A series of trigger events and associated responsive actions are developed prior to implementation to allow the plan to adapt to new conditions if a trigger event occurs over the life of the plan. Preparation of these contingent actions may include carrying out studies, engineering design work, or developing supporting political and financial plans. The results of these efforts are then saved for use if trigger events occur after the actions in Steps II and III have been implemented. Walker et al. (2013) describe the four types of contingent actions that can be taken:

- Defensive actions (DA)—Responsive actions taken *after implementation of the initial plan* to clarify the plan, preserve its benefits, or meet outside challenges in response to specific triggers, but that leave the initial plan unchanged.
- Corrective actions (CR)—Adjustments to the initial plan in response to specific triggers.
- Capitalizing actions (CP)—Responsive actions taken *after implementation of the initial plan* to take advantage of opportunities that further improve its performance.
- Reassessment (RE)—A process initiated when the analysis and assumptions critical to the plan's success have lost validity (i.e., when unforeseen events cause a shift in the fundamental goals, objectives, and assumptions underlying the initial plan).

The dynamic adaptive plan is then implemented. This involves the implementation of:

- the initial plan identified in Step II;
- the mitigating, hedging, seizing, and shaping actions developed in Step III;
- and the monitoring program developed in Step IV.

If one of the signposts' trigger events occurs after implementation of the initial plan, one or more of the contingent actions developed in Step V is executed. If the original objectives of the plan and constraints on it remain in place upon occurrence of the trigger event, then defensive or corrective actions will be taken. If the monitoring program encounters an opportunity, then capitalizing actions will be taken. If the monitoring program indicates a change that invalidates the initial plan's goals, objectives, or intended outcomes (e.g., vulnerabilities exist or evolve beyond those considered during Step III—for example, the occurrence of a 'Black Swan' event (Taleb 2007)), then the complete plan is reassessed. However, reassessment does not mean completely starting over, as the knowledge of outcomes, objectives, measures, etc., learned during the initial DAP process would accelerate the new planning process.

3.3 A DAP Illustration: Strategic Planning for Schiphol Airport[1]

Amsterdam Airport Schiphol's position as a hub within Europe is under pressure. The merger of Air France and KLM has resulted in the threat that KLM, Schiphol's hub carrier, which is responsible for more than half of the scheduled aircraft movements at the airport, might move a significant portion of its operations to Charles de Gaulle Airport. The other major airports in Europe are planning on expanding their capacity or are developing dual airport systems, while Schiphol's capacity is under threat of being reduced due to climate change induced changes in wind conditions and lack of societal support for capacity expansion. Together, this makes the long-term planning for Schiphol both urgent and problematic.

In the remainder of this section, we illustrate how each of the steps of DAP might be applied to the case of the long-term development of Schiphol. Its purpose is to illustrate DAP and how it could be applied in practice. To make the approach clear and understandable, the example simplifies some of the key challenges Schiphol faced in the recent past, when this study was performed. Therefore, the case should not be understood as presenting a realistic plan for the long-term development of Schiphol. It is merely an example loosely based on real planning issues and debates that planners are currently facing with respect to the long-term development of an airport.

Step I: Stage Setting

As outlined in its long-term vision (Schiphol Group and LVNL 2007), the main goals of the Schiphol Group are (1) to create room for the further development of the network of KLM and its Sky team partners, and (2) to minimize (and, where possible, reduce) the negative effects of aviation in the region. Underlying the first goal is the implicit assumption that aviation will continue to grow. However, in light of recent

---

[1] Based on Kwakkel et al. (2010).

developments such as climate change and the financial crisis, this assumption is questionable. Therefore, as part of our 'thought experiment,' we rephrased this first goal more neutrally as 'retain market share.' If aviation in Europe grows, Schiphol will have to accommodate more demand in order to retain its market share, while if aviation declines, Schiphol could still reach its goal of retaining market share.

There are several types of changes that can be made at Schiphol in order to achieve its goals of retaining market share and minimizing the negative effects of aviation. Schiphol can expand its capacity by using its existing capacity more efficiently and/or building new capacity. It can also expand its capacity or use its existing capacity in a way that mitigates the negative effects of aviation. More explicitly, among the alternatives that Schiphol might consider are:

1. Add a new runway
2. Add a new terminal
3. Use the existing runway system in a more efficient way, in order to improve capacity
4. Use the existing runway system in a way that minimizes noise impacts
5. Move charter operations out of Schiphol (e.g., to Lelystad)
6. Move Schiphol operations to a new airport (e.g., in the North Sea)
7. Invest in noise insulation of surrounding buildings and houses.

Some of these actions can be implemented immediately (e.g., using the existing runway system in a more efficient way). For others, an adaptive approach would be to begin to prepare plans and designs (e.g., for a new runway), but to begin actual building only when conditions show it to be necessary (i.e., when it is triggered). The various alternative actions can, of course, be combined. The actions are constrained by costs, spatial and legal restrictions, public acceptance, and the landside accessibility of Schiphol. The definition of success includes that Schiphol maintains its market share and that living conditions improve compared to some reference situation (e.g., number of people affected by noise within a specified area).

Step II: Assembling an Initial Plan

An initial plan might be to immediately implement existing plans for using the runways more efficiently (alternative 3) and in a way that reduces noise impacts (alternative 4). It might also include all alternatives that focus on planning capacity expansions, without beginning to build any of them (i.e., alternatives 1, 2, and 5). A final element of the initial plan would be alternative 7: invest in noise insulation. The choice for only planning capacity expansions but not yet building them is motivated by the fact that Schiphol is currently constrained by the environmental rules and regulations, not by its physical capacity. This also motivates the choice for alternatives 3 and 4, which together can reduce the negative externalities of aviation.

In addition to the capacity expansions, Schiphol can develop plans to move charter operations to nearby Lelystad Airport, which would reduce noise around Schiphol and increase Schiphol's capacity for regular flight operations. In order to realize such a move, Lelystad Airport would need to be expanded considerably, so planning should be started right away. Charter operations should then be moved there as soon as possible. In the short run, this would create additional capacity and reduce noise at the edges of the night, which is favorable for Schiphol, because the current noise regulation system heavily penalizes flights in the evening (19.00–23.00) and during the night (23.00–07.00). Note that Schiphol is currently implementing a plan for moving some operations to Lelystad.

In summary, the initial plan involves using the existing runway system in a more efficient and noise-reducing way; investing in noise insulation; and initiating plans for capacity expansion.

In light of Schiphol's goals (retaining market share and minimizing the negative effects of aviation (Schiphol Group and LVNL 2007)), several necessary conditions for the success of the initial plan can be specified, including

- Support from crucial stakeholders
- Sufficient landside accessibility
- 'Self-hubbing' grows as expected
- 'Airport city' development increases
- Schiphol retains its current market share
- The population affected by noise and the number of noise complaints does not increase
- Schiphol's competitive position in terms of available capacity in Europe does not decrease.

Step III: Increasing the Robustness of the Initial Plan

The long-term development of Schiphol is complicated by the many and diverse trends and developments that can affect Schiphol. These developments and trends present both opportunities and vulnerabilities for the initial plan. Some of these vulnerabilities are fairly certain. These are given in Table 3.1. Two fairly certain vulnerabilities of the initial plan are resistance from stakeholders and a reduction in landside accessibility. The mitigating actions for addressing these vulnerabilities are very similar to actions under discussion by the Dutch government (V&W 2007). A shaping action for the vulnerability of landside accessibility is investment in research. In addition to vulnerabilities, there are also some opportunities available to Schiphol. First, there is research showing the potential for 'self-hubbing' (Burghouwt 2007; Malighetti et al. 2008). Self-hubbing means that passengers arrange their own flights and routes, using low-cost carriers or a variety of alliances, in order to minimize costs and/or travel time. Schiphol has a great potential for attracting such self-hubbing passengers because it has direct links to about 300 destinations. Schiphol can seize

3 Dynamic Adaptive Planning (DAP)

Table 3.1 Certain (very likely) vulnerabilities and opportunities, and responses to them

Vulnerabilities and opportunities	Mitigating (M), Shaping (SH), and Seizing (SZ) actions
Reduction of the landside accessibility of the airport	M: Develop a system for early check-in and handling of baggage at rail stations SH: Invest in R&D into the landside accessibility of the Randstad area
Resistance from Schiphol stakeholders (e.g., environmental groups, people living around Schiphol)	M: Develop plans for green areas to compensate for environmental losses M: Offer financial compensation to residents in the high noise zone
Rise of self-hubbing	SZ: Design and implement a plan for supporting self-hubbing passengers with finding connection flights, transferring baggage, and acquiring boarding passes
Rise of the airport city	SZ: Diversify revenues by developing non-aeronautical landside real estate

this opportunity by developing and implementing services tailored to self-hubbing passengers, such as services for baggage transfer and help with acquiring boarding passes. Furthermore, Schiphol could take into account walking distances between connecting European flights when allocating aircraft to gates. A second opportunity is presented by the fact that airports in general, and Schiphol in particular, are evolving into 'airport cities.' Given the good transport connections available, an airport is a prime location for office buildings. Schiphol can seize this opportunity by investing in non-aeronautical landside real estate development.

Not all vulnerabilities and opportunities are very likely. The real challenge for the long-term development of Schiphol is presented by uncertain vulnerabilities and opportunities. Table 3.2 presents some of the uncertain vulnerabilities, together with possible hedging (H) and shaping actions (SH) to take right away to handle them. The vulnerabilities and opportunities can be directly related and categorized according to the success conditions specified in the previous step. With respect to the success condition of retaining market share, air transport demand might develop significantly different from what is hoped and anticipated. Schiphol can respond to this development by making Lelystad Airport suitable for handling non-hub-essential flights. Another vulnerability is that KLM might decide to move a significant part of its operations to Charles de Gaulle Airport in Paris. This will leave Schiphol without its hub carrier, significantly reducing demand, and changing the demand to origin–destination (O/D) demand. Schiphol could prepare for this vulnerability by making plans for adapting the terminal to the requirements of an O/D airport, and by diversifying the carriers that serve Schiphol. Schiphol can also try to directly affect KLM by investing in a good working relationship, reducing the chance that KLM will leave. There is also uncertainty about the future of the hub-and-spoke network structure. Due to Open Sky agreements and the development of the Boeing 787,

Table 3.2 Uncertain vulnerabilities and opportunities, and responses to them

Vulnerabilities and opportunities	Hedging (H) and Shaping (SH) actions
Necessary condition for success: retain market share	
Demand for air traffic grows faster than forecast	H: Prepare Lelystad Airport to receive charter flights
Demand for air traffic grows slower than forecast	SH: Advertise for flying from Schiphol
Collapse or departure of the hub carrier (KLM) from Schiphol	H: Prepare to adapt Schiphol to be an O/D airport H: Diversify the carriers serving Schiphol SH: Develop a close working relation with KLM
Rise of long-haul low-cost carriers	H: Design existing and new LCC terminal to allow for rapid customization to airline wishes.
Rise of self-hubbing, resulting in increasing transfers among LCC operations	H: Design a good connection between the existing terminal and the new LCC terminal, first with buses, but leave room for replacing it with a people mover
Necessary condition for success: population affected by noise and the number of noise complaints should not increase	
Maintain current trend of decrease of environmental impact of aircraft	SH: Negotiate with air traffic control on investments in new air traffic control equipment that can enable noise abatement procedures, such as the continuous descent approach SH: Invest in R&D, such as noise abatement procedures
Increase in the population density in area affected by noise	H: Test existing noise abatement procedures, such as the continuous descent approach, outside the peak periods (e.g., at the edges of the night) SH: Negotiate with surrounding communities to change their land use planning SH: Invest in R&D, such as noise abatement procedures
Change in the valuation of externalities by the public	SH: Invest in marketing of the airport to brand it as an environmentally friendly organization SH: Join efforts to establish an emission trading scheme
Necessary condition for success: Schiphol's competitive position in terms of available capacity in Europe does not decrease	
Other major airports in Europe increase capacity	No immediate action required
Development of wind conditions due to climate change	H: Have plans ready to quickly build the sixth runway, but do not build it yet. If wind conditions deteriorate even further, start construction

long-haul low-cost, hub bypassing, and self-hubbing become plausible, resulting in the emergence of long-haul low-cost carriers (LCC) and increasing transfer between short-haul low-cost, and long-haul carriers (both LCC and legacy carriers). Schiphol can prepare for this by developing a plan to change its current terminal to serve a different type of demand, and by taking these plausible developments into consideration when designing the new LCC terminal and its connection with the existing terminal. If a transformation to international O/D traffic and/or a no-frills airport is needed, this plan can be implemented, making sure that the transformation can be achieved quickly.

The second success condition is that the population affected by noise, and the number of noise complaints, should not increase. Vulnerabilities and opportunities associated with this condition are that the environmental impacts of aircraft decreases, the population density in the area affected by noise increases, and the valuation of externalities (predominantly noise) by the large public changes. If the current trend of decreasing environmental impact slows down, the area affected by noise might increase. If demand increases, it is also possible that the area affected by noise will increase. On the other hand, the trend could also accelerate, giving Schiphol the opportunity to expand the number of flights that is handled. Given the potential impact of this trend, Schiphol should try and shape its development by investing in R&D and negotiate with air traffic control about testing noise abatement procedures, such as continuous descent approaches. If the population density changes, the situation is similar. If it increases, the number of people affected by noise will increase, while if it decreases, the number of people affected by noise will decrease. Schiphol can try and shape this development by negotiating with surrounding communities about their land use planning, and invest in research (or even implement measures) that can make the area affected by noise smaller. It can also hedge against a growing population density by starting to test noise abatement procedures outside peak hours. This will make the area affected by noise smaller. Thus, even if the population density increases, the total number of people affected will not increase. A third uncertainty is how the valuation of noise will change in the future. If noise will be considered more of a nuisance, complaints are likely to go up and vice versa. Schiphol could try to affect this valuation by branding the airport as environmentally friendly and support the development of an emission trading scheme that also includes aviation.

The third success condition is that Schiphol's competitive position in terms of available capacity in Europe does not decrease. Schiphol is vulnerable to capacity developments at other airports in Europe. The major hubs in Europe are all working on expanding their capacities, either by adding runways and expanding terminals, or by moving non-hub-essential flights to alternative airports in the region. Schiphol should monitor these developments closely and, if necessary, speed up its capacity investments. A second vulnerability is the robustness of Schiphol's peak hour capacity across weather conditions. Under southwesterly wind conditions, Schiphol's hourly capacity is almost halved, resulting in delays and cancellations. If (e.g., due to climate change) these wind conditions were to become more frequent, Schiphol would no longer be able to guarantee its capacity. Schiphol should hedge against this by having plans ready for building the sixth runway.

Step IV: Setting up the Monitoring System and Step V: Preparing the Trigger Responses

Step IV sets up the monitoring system and identifies the actions to be taken when trigger levels of the signposts are reached. The vulnerabilities and opportunities are those presented in Table 3.2. Table 3.3 shows the signpost to be set up for each vulnerability and each opportunity, and the possible responsive actions in case of a trigger event. The numbers used as triggers are for illustrative purposes only. For example, if demand increases twice as fast as expected, this presents an opportunity (extra business) and a vulnerability (increased noise) and triggers capitalizing and defensive actions. Suppose that demand grows, e.g., 25% slower than anticipated. This presents a threat to the plan. In reaction, investments in capacity are delayed or even canceled. If demand either fully breaks down or explodes, then the plan should be reassessed.

3.4 Implementation and Adaptation

After Step V has been completed, the dynamic adaptive plan has been completely designed. That plan is then implemented. It consists of the initial plan specified in Step II, the actions specified in Tables 3.1 and 3.2, and the system of monitoring, triggers, and actions specified in Table 3.3. Note that the new runway being planned in the initial plan is not built yet, but can be built easily when necessary in light of demand increases or capacity decreases at other major European airports. As such, it is what is known as a 'real option.' Planning should also be started for the new terminal (including its connections to the highway system, the rail system, and required utilities). However, construction of the terminal itself should begin only if triggered by demand developments or capacity developments at other airports.

During implementation, Schiphol monitors the development. Schiphol might experience faster growth than anticipated in the plan. The signposts might indicate that Schiphol is maintaining its position as a major airport for the Sky Team alliance and its partners; however, the boundaries set for safety, the environment, and quality of life, and spatial integration with its surroundings might be violated. If so, construction of the new terminal can start. In addition, actions need to be taken to defend the plan with respect to the negative external effects. The noise insulation program can be expanded, and more investment can be made in branding and marketing that aim at explaining the plan. If these actions prove to be insufficient, the noise insulation program can be expanded; Schiphol should start to buy out residents that are heavily affected by noise and increase landing fees for environmentally unfriendly planes. If this still is insufficient, Schiphol should consider limiting the number of available slots, especially during the night and edges of the night. If these actions are still insufficient, either because demand grows very fast or because the environmental impact grows too fast, the plan should be reassessed. If this alternative is chosen, the decisionmakers would reiterate through the adaptive planning steps in order to develop a new (adaptive) plan.

Table 3.3 Monitoring, triggers, and actions

Vulnerabilities and opportunities	Monitoring and trigger system	Actions (Reassessment (RE), Corrective (CR), Defensive (DA), Capitalizing (CP))
Necessary condition for success: retain market share		
Demand for air traffic grows faster than forecast	Monitor the growth of Schiphol in terms of passenger movements, aircraft movements (and related noise and emissions). If double demand (trigger), take CP-action; if demand explodes, take RE-action	CP & DA: Begin to implement the plan for the new terminal and the new runway RE: Reassess entire plan
Demand for air traffic grows slower than forecast	Monitor types of demand. If overall demand is decreasing by half of forecast, take DA-actions. If demand fully breaks down, take RE-action. If transfer rate decreases below 30%, take CR-action	DA: Delay investments and reduce landing fees RE: Reassess entire plan CR: Cancel terminal capacity expansions
Collapse or departure of the hub carrier (KLM) from Schiphol	Monitor the network of KLM–Air France. If 25% of flights are moved, take DA-action; if 50%, take CR-action; if 80% or more, take RE-action	DA: Diversify the carriers that fly from Schiphol CR: Switch airport to an O/D airport by changing terminal RE: Reassess entire plan
Rise of long-haul low-cost carriers	Monitor development of the business model of low-cost carriers. If long-haul LCC carriers make profit for 2 years, take CP-action	CP: Attract long-haul LCC by offering good transfer between LCC terminal and existing terminal and/or by offering wide-body aircraft stands at the LCC terminal
Rise of self-hubbing, resulting in increasing transfers between LCC operations	Monitor transfer rate among LCC flights and between LCC and legacy carriers. If transfer rate becomes more than 20%, take CP-action	CP: Expand transfer capabilities between the new LCC terminal and the existing terminal
Necessary condition for success: population affected by noise and the number of noise complaints should not increase		
Maintain current trend of decrease of environmental impact of aircraft	Monitor noise footprint and emissions of the fleet mix serving Schiphol and of the new aircraft entering service. If there is an increase of noise or emissions of 10%, take CR-action	CR: Change landing fees for environmentally unfriendly planes

(continued)

Table 3.3 (continued)

Vulnerabilities and opportunities	Monitoring and trigger system	Actions (Reassessment (RE), Corrective (CR), Defensive (DA), Capitalizing (CP))
Increase in the population density in area affected by noise	Monitor population affected by noise. If population affected by noise increases by 2%, take DA-action; by 5%, take CR-action; by 7.5%, take RA-action. If population density decreases by 2%, take CP-action	DA: Expand insulation program and explain initial plan again CR: Slow down growth by limiting available slots RE: Reassess entire plan CP: Make new slots available
Change in the valuation of externalities by the public	Monitor the complaints about Schiphol. If complaints increase by an average of 5% over 2 years, take DA-action; if complaints increase by an average of 10% or more over 2 years, take CR-action	DA: Increase investments in marketing and branding CR: Slow down the growth of Schiphol by limiting the available slots

Necessary condition for success: Schiphol's competitive position in terms of available capacity in Europe does not decrease

Other major airports in Europe increase capacity	Monitor declared capacity for the major airports in Europe. If declared capacity is up by 25%, take DA-action	DA: Speed up expansions
Development of wind conditions due to climate change	Monitor the prevailing wind conditions throughout the year. If for 2 years in a row the number of days with crosswind conditions exceeds 50, take DA-action	DA: Begin to implement the plan for the new runway

3.5 Conclusions

Long-term plans must be devised in spite of profound uncertainties about the future. When there are many plausible scenarios for the future, it may be impossible to construct any single static plan that will perform well in all of them. It is likely, however, that over the course of time new information will become available. Thus, plans should be adaptive—devised not to be optimal for the best estimate future, but robust across a range of plausible futures.

Such plans combine actions that are time urgent with those that make important commitments to shape the future and those that preserve flexibility needed for the future. DAP is an approach to plan design and implementation that explicitly confronts the pragmatic reality that traditional (static) plans need to be adjusted as the world changes and as new information becomes available. The approach allows planners to cope with the uncertainties that confront them by creating plans that respond to changes over time and that make explicit provision for learning. The approach makes adaptation explicit at the outset of plan design. Thus, the inevitable changes in the plan become part of a larger, recognized process and would not be forced to be made repeatedly on an ad hoc basis, which would be the case if the plan were static.

The DAP approach has several strengths. First, it is relatively easy to understand and explain. Second, it encourages planners to think about 'what if' situations and their outcomes, and to make decisions over time to adapt while maintaining flexibility with respect to making future changes; this also helps in foreseeing undesirable lock-ins or other path dependencies so that they can be avoided. Third, it makes explicit that adaptation is a dynamic process that takes place over time; it forces planners to consider changes continuously over time, rather than at one or a few points in time as most scenario approaches do. On the other hand, the resulting plan might end up costing more if no responsive actions are needed. Also, setting up the monitoring system may be complicated, the monitoring itself may be expensive over a long period, and policymakers and politicians may resist the idea of committing to adaptive actions in advance.

References

Agusdinata, D. B., Marchau, V. A. W. J., & Walker, W. E. (2007). Adaptive policy approach to implementing intelligent speed adaptation. *IET Intelligent Transport Systems (ITS), 1*(3), 186–198.

Bankes, S., Walker, W. E., & Kwakkel, J. H. (2013). Exploratory modeling and analysis. In S. Gass & M. Fu (Eds.), *Encyclopedia of operations research and management science* (3rd ed.). New York: Springer.

Bryant, B. P., & Lempert, R. J. (2010). Thinking inside the box: A participatory computer-assisted approach to scenario discovery. *Technological Forecasting and Social Change, 77,* 34–49.

Burghouwt, G. (2007). *Airline network development in Europe and its implications for airport planning.* Burlington, USA: Ashgate Publishing Company.

Dewar, J. A., Builder, C. H., Hix, W. M., & Levin, M. (1993). *Assumption-based planning: A planning tool for very uncertain times.* Santa Monica, CA: MR114-A, RAND.

Kwakkel, J. H., Walker, W. E., & Marchau, V. A. W. J. (2010). Adaptive airport strategic planning. *European Journal of Transport and Infrastructure Research, 10*(3), 249–273.

Kwakkel, J. H., Auping, W., & Pruyt, E. (2012). Dynamic scenario discovery under deep uncertainty: The future of copper. *Technological Forecasting and Social Change, 80*(4), 789–800. https://doi.org/10.1016/j.techfore.2012.09.012.

Malighetti, P., Paleari, S., & Redondi, R. (2008). Connectivity of the European airport network: 'SelfHelp Hubbing' and business implicatons. *Journal of Air Transport Management, 14*(2), 53–65.

Marchau, V. A. W. J., Walker, W. E., & van Duin, R. (2008). An adaptive roach to implementing innovative urban transport solutions. *Transport Policy, 15*(6), 405–412.

Marchau, V. A. W. J., Walker, W. E., & van Wee, G. P. (2010). Dynamic adaptive transport policies for handling deep uncertainty. *Technological Forecasting and Social Change, 77,* 940–950.

Osita, H. C., Idoko O. R., & Nzwkwe, J. (2014). Organizations stability and productivity: The role of SWOT analysis. *International Journal of Innovative and Applied Research, 2*(9), 23–32.

Rahman, S. A., Walker, W. E., & Marchau, V. A. W. J. (2008). *Coping with uncertainties about climate change in infrastructure planning: An adaptive policymaking approach.* Rotterdam: Ecorys.

Rowe, G., Wright, G., & Bolger, F. (1991). Delphi, a re-evaluation of research and theory. *Technological Forecasting and Social Change, 53,* 235–251.
Schiphol Group, & LVNL. (2007). *Verder Werken aan de toekomst van Schiphol en de regio.* Schiphol Group, Air Traffic Control the Netherlands.
Taleb, N. N. (2007). *The black swan.* New York: Random House.
van der Heijden, K. (1996). *Scenarios: The art of strategic conversation.* Chichester, UK: Wiley.
V&W. (2007). *Lange Termijn Verkenning Schiphol: probleemanalyse.* Ministerie van Verkeer en Waterstaat.
Walker, W. E., Rahman, S. A., & Cave, J. (2001). Adaptive policies, policy analysis, and policymaking. *European Journal of Operations Research, 128,* 282–289.
Walker, W. E., Marchau, V. A. W. J., & Kwakkel, J. H. (2013). Uncertainty in the framework of policy analysis. In W. E. Walker & W. A. H. Thissen (Eds.), *Public policy analysis: New developments.* New York: Springer.
Walker, W. E. (2000). Policy analysis: A systematic approach to supporting policymaking in the public sector. *Journal of Multicriteria Decision Analysis, 9*(1–3), 11–27.
Weihrich, H. (1982). The TOWS matrix: A tool for situational analysis. *Long Range Planning, 15*(2), 54–66.

Prof. Warren E. Walker (Emeritus Professor of Policy Analysis, Delft University of Technology). He has a Ph.D. in Operations Research from Cornell University, and more than 40 years of experience as an analyst and project leader at the RAND Corporation, applying quantitative analysis to public policy problems. His recent research has focused on methods for dealing with deep uncertainty in making public policies (especially with respect to climate change), improving the freight transport system in the Netherlands, and the design of decision support systems for airport strategic planning. He is the recipient of the 1997 President's Award from the Institute for Operations Research and the Management Sciences (INFORMS) for his 'contributions to the welfare of society through quantitative analysis of governmental policy problems.'

Prof. Vincent A. W. J. Marchau (Radboud University (RU), Nijmegen School of Management) holds a chair on Uncertainty and Adaptivity of Societal Systems. This chair is supported by The Netherlands Study Centre for Technology Trends (STT). His research focuses on long-term planning under uncertainty in transportation, logistics, spatial planning, energy, water, and security. Marchau is also Managing Director of the Dutch Research School for Transport, Infrastructure and Logistics (TRAIL) at Delft University of Technology (DUT), with 100 Ph.D. students and 50 staff members across 6 Dutch universities.

Dr. Jan H. Kwakkel is Associate Professor at the Faculty of Technology, Policy and Management (TPM) of Delft University of Technology. He is also the Vice President of the Society for Decision Making under Deep Uncertainty, and a member of the editorial board of *Environmental Modeling and Software, Futures and Foresight.* His research focuses on model-based support for decision making under uncertainty. He is involved in research projects on smart energy systems, high-tech supply chains, transport, and climate adaptation. Jan is the developer of the Exploratory Modeling Workbench, an open-source software implementing a wide variety of state-of-the-art techniques for decision making under uncertainty.

Open Access This chapter is licensed under the terms of the Creative Commons Attribution 4.0 International License (http://creativecommons.org/licenses/by/4.0/), which permits use, sharing, adaptation, distribution and reproduction in any medium or format, as long as you give appropriate credit to the original author(s) and the source, provide a link to the Creative Commons licence and indicate if changes were made.

The images or other third party material in this chapter are included in the chapter's Creative Commons licence, unless indicated otherwise in a credit line to the material. If material is not included in the chapter's Creative Commons licence and your intended use is not permitted by statutory regulation or exceeds the permitted use, you will need to obtain permission directly from the copyright holder.

Chapter 4
Dynamic Adaptive Policy Pathways (DAPP)

Marjolijn Haasnoot, Andrew Warren and Jan H. Kwakkel

Abstract

- Dynamic Adaptive Policy Pathways (DAPP) is a DMDU approach that explicitly includes decision making over time. The essence is proactive and dynamic planning in response to how the future actually unfolds.
- DAPP explores alternative sequences of decisions (adaptation pathways) for multiple futures and illuminates the path dependency of alternative strategies. It opens the decision space and helps to overcome policy paralysis due to deep uncertainty. There are different routes that can achieve the objectives under changing conditions (like 'different roads leading to Rome').
- Policy actions have an uncertain design life and might fail sooner or later to continue achieving objectives as the operating conditions change (i.e. they reach an adaptation tipping point (ATP)). Similarly, opportunity tipping points may occur.
- Multiple pathways are typically visualized in a metro map or decision tree, with time or changing conditions on one of the axes.
- DAPP supports the design of a dynamic adaptive strategy that includes initial actions, long-term options, and adaptation signals to identify when to implement the long-term options or revisit decisions.

4.1 Introduction

Nowadays, decision-makers face deep uncertainties about a myriad of external factors, such as climate change, population growth, new technologies, economic developments, and their impacts. Moreover, not only environmental conditions, but also societal perspectives and preferences may change over time, including stakeholders' interests and their evaluation of plans (Offermans 2011; van der Brugge et al. 2005). For investments in infrastructure, where capital expenditures are high and asset lifespans long, decision-makers need to be confident that the decisions they take today will continue to apply, and that the actions are designed to be able to cope with the changing conditions. Traditional planning approaches are ill-equipped to take into account these uncertainties, resting as they do on approaches for handling Level 1, 2, and 3 uncertainties (see Table 1.1). This chapter describes an approach to planning under conditions of deep uncertainty called Dynamic Adaptive Policy Pathways (DAPP) (Haasnoot et al. 2013), which recognizes that decisions are made over time in dynamic interaction with the system of concern, and thus cannot be considered independently of each other. It explicitly considers the sequencing and path dependencies of decisions over time.

DAPP integrates two partially overlapping and complementary adaptive planning approaches—Dynamic Adaptive Planning (Walker et al. 2001, Chap. 3) and adaptation pathways (Haasnoot et al. 2012). Also central to the approach is the concept of Adaptation Tipping Points (Kwadijk et al. 2010). The Adaptation Tipping Point (ATP) approach was developed in the Netherlands in response to a desire of the national government for a planning approach less dependent on any particular set of scenarios. The publication of a new generation of national climate scenarios in 2006 (van den Hurk et al. 2007; Katsman et al. 2008) meant that regulations had to be changed, and that existing strategic plans based on the old scenarios all had to be updated to bring them into line with the new scenarios. ATP, and therefore also DAPP, address the need to be less scenario-dependent by refocusing the policy analysis process away from an assessment of what may happen when and towards identifying the general conditions under which a policy will fail (referred to as 'adaptation tipping point conditions' (Kwadijk et al. 2010)). Scenarios are then used to assess when this may happen. After these ATPs, additional actions are needed to prevent missing the objectives, and pathways start to emerge. As alternatives are possible, multiple pathways can be explored. The DAPP approach has been adopted by, and further co-developed with, the Dutch Delta Programme (Chap. 14). The chair of the Delta Programme summarized the main challenge of the Programme as follows: "One of the biggest challenges is dealing with uncertainties in the future climate, but also in population, economy and society. This requires a new way of planning, which we call adaptive delta planning. It seeks to maximize flexibility, keeping options open and avoiding 'lock-in'" (Kuijken 2010). In the UK, similar ideas were developed and applied in the Thames Estuary 2100 study (Reeder and Ranger 2010; Ranger et al. 2013) and applied in New Zealand in a flood risk management setting (see Chap. 9).

The DAPP approach aims at building flexibility into an overall plan by sequencing the implementation of actions over time in such a way that the system can be adapted to changing conditions, with alternative sequences specified to deal with a range of plausible future conditions. Although originally developed for implementing climate-resilient pathways for water management, it is a generic approach that can be applied to other long-term strategic planning problems under uncertain changing conditions.

4.2 The DAPP Approach

Within the DAPP approach (Haasnoot et al. 2013), a plan is conceptualized as a series of actions over time (pathways), including initial actions and long-term options. The essence is the proactive planning for flexible adaptation over time, in response to how the future actually unfolds. The approach starts from the premise that policies/decisions have a design life and might fail when operating conditions change (Kwadijk et al. 2010). Once actions fail, additional or other actions are needed to ensure that the original objectives are still achieved, and a set of potential pathways emerges. Depending on how the future unfolds, the course can be changed when predetermined conditions occur, in order to ensure that the objectives are still achieved. The preference for specific pathways over others is actor specific and will depend on the trade-offs, such as the costs (including negative externalities) and benefits of the different pathways. Based on an evaluation of the various possible pathways, an adaptive plan that includes initial actions and long-term options can be designed. The plan is monitored for signals that indicate when the next step of a pathway should be implemented, or whether an overall reassessment of the plan is needed.

Figure 4.1 shows the overall DAPP approach (Haasnoot et al. 2013). Based on a problem analysis for the current and future situations, policy actions are identified to address vulnerabilities and seize opportunities. The conditions and timing of ATPs are assessed based on their efficacy in achieving the desired outcomes over changing conditions or time. Once the set of policy actions is deemed adequate, pathways can be designed and evaluated. A (policy) pathway consists of a sequence of policy actions, where a new action is activated once its predecessor is no longer able to meet the specified objectives. Pathways can focus on adapting to changing conditions (adaptation pathways), enabling socio-economic developments (development pathways), or transitioning to a desired future (transition pathways). Based on the evaluation of the pathways, a plan can be made, which describes the initial actions, long-term options, the developments to monitor, and under what conditions the next actions on a pathway should be taken to stay on track. Initial actions include actions aimed at reaching the policy objectives and actions aimed at keeping long-term options open. In practice, Steps 1–5 are often first carried out qualitatively based on expert judgement, followed by a more detailed model-based assessment of these steps.

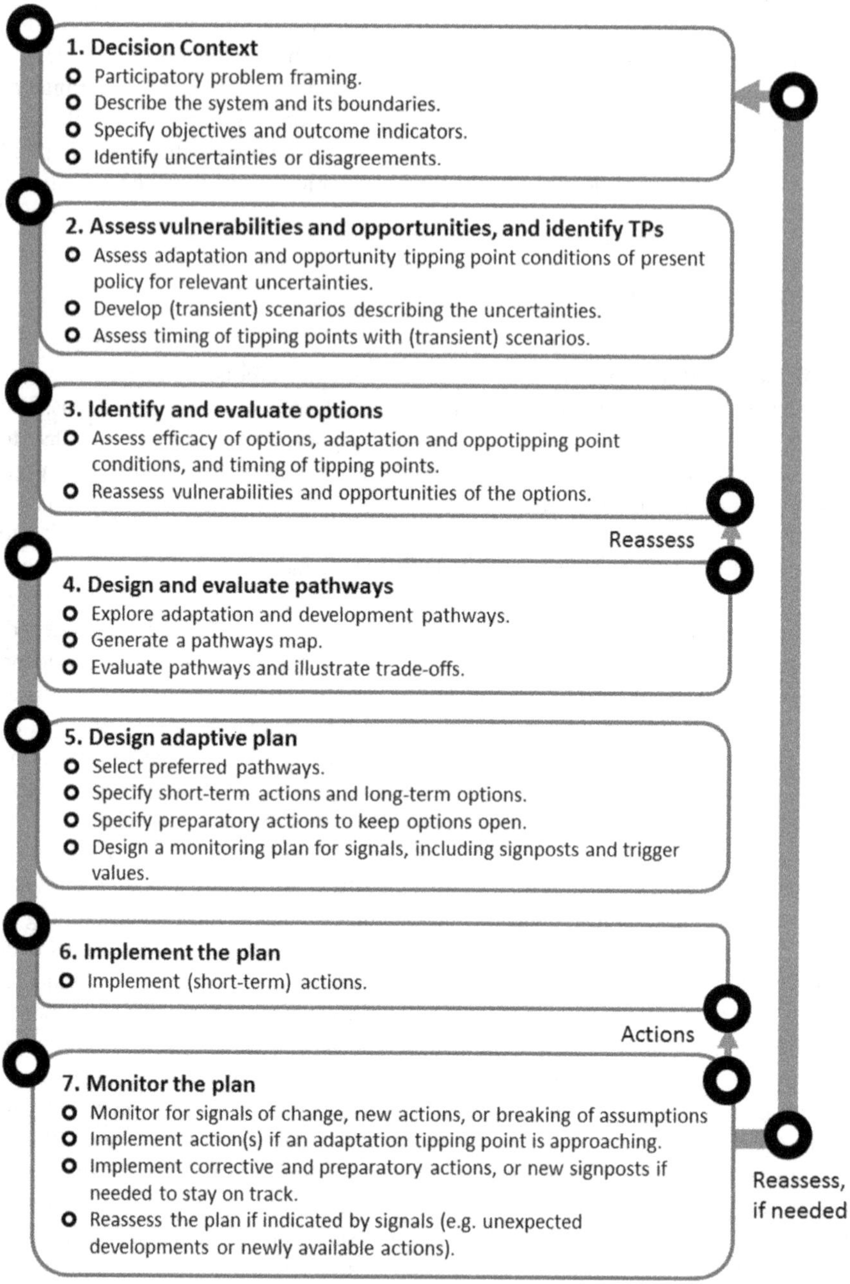

1. Decision Context
- Participatory problem framing.
- Describe the system and its boundaries.
- Specify objectives and outcome indicators.
- Identify uncertainties or disagreements.

2. Assess vulnerabilities and opportunities, and identify TPs
- Assess adaptation and opportunity tipping point conditions of present policy for relevant uncertainties.
- Develop (transient) scenarios describing the uncertainties.
- Assess timing of tipping points with (transient) scenarios.

3. Identify and evaluate options
- Assess efficacy of options, adaptation and oppotipping point conditions, and timing of tipping points.
- Reassess vulnerabilities and opportunities of the options.

Reassess

4. Design and evaluate pathways
- Explore adaptation and development pathways.
- Generate a pathways map.
- Evaluate pathways and illustrate trade-offs.

5. Design adaptive plan
- Select preferred pathways.
- Specify short-term actions and long-term options.
- Specify preparatory actions to keep options open.
- Design a monitoring plan for signals, including signposts and trigger values.

6. Implement the plan
- Implement (short-term) actions.

Actions

7. Monitor the plan
- Monitor for signals of change, new actions, or breaking of assumptions
- Implement action(s) if an adaptation tipping point is approaching.
- Implement corrective and preparatory actions, or new signposts if needed to stay on track.
- Reassess the plan if indicated by signals (e.g. unexpected developments or newly available actions).

Reassess, if needed

Fig. 4.1 DAPP approach. Adapted from Haasnoot et al. (2013)

Step 1: Participatory problem framing, describe the system, specify objectives, and identify uncertainties

The first step is to describe the setting, including the system characteristics, objectives, constraints in the current situation, and potential constraints in future situations. The result is a definition of success, which is a specification of the desired outcomes in terms of indicators and targets. These are used in subsequent steps to evaluate the performance of actions and pathways, and to assess the conditions and timing of ATPs. This step also includes a specification of the major uncertainties or disagreements that play a role in the decision making, such as (changes in) external forces, system structure, and valuation of outcomes. The specified uncertainties are used to generate an ensemble of plausible futures in the form of scenarios. Such scenarios can be static scenarios (describing an end point into the future) or time-dependent transient scenarios (describing developments over time) (Haasnoot et al. 2015).

Step 2: Assess system vulnerabilities and opportunities, and identify adaptation tipping points

The second step is to assess the current situation against the ensemble of plausible futures, using the specified indicators and targets, in order to identify the conditions under which the system starts to perform unacceptably (ATPs)[1]. These are those conditions under which a specified indicator fails to meet its target. Each plausible future is treated as a 'reference case' assuming that no new policies are implemented. They consist of (transient) scenarios that span the uncertainties identified in Step 1. Both opportunities and vulnerabilities should be considered, with the former consisting of those developments that can help achieve the defined objectives, and the latter consisting of those developments that can prevent the objectives from being reached.

Various approaches can be used to identify ATPs (Fig. 4.2). These include 'bottom-up' vulnerability assessments that establish unacceptable outcome thresholds before assessing the timing of tipping points using scenarios, as well as 'top-down' approaches that use traditional scenario analyses to determine the range and timing of these points. 'Bottom-up' approaches can rely on model-based assessments (e.g. stress tests, sensitivity analyses, Scenario Discovery) to establish the failure conditions (thresholds), or these can be specified via expert judgment or stakeholder consultation. Stakeholders could assess tipping point values in terms of absolute or relative values (Brown et al. 2016). The latter is especially useful when objectives cannot be translated into clear target indicators and values (e.g. see Chap. 9 on defining ATPs in practice). For example: absolute: Action A is sufficient to avoid flood damage at a sea level rise of 1 m, while Action B can do the same for a sea level rise of 2 m; relative: Action B can accommodate more change than Action A. Scenarios are then used to assess the timing of the tipping point conditions.

'Top-down' approaches are largely dependent on model-based assessments and can use either multiple static scenarios (e.g. Bouwer et al. under review) or transient

---

[1] This is essentially performing Scenario Discovery (see Chap. 2).

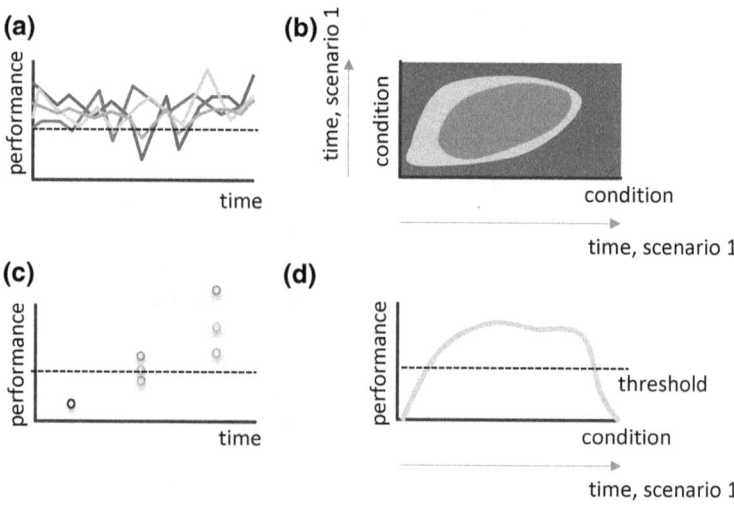

Fig. 4.2 Approaches for identifying tipping points

scenario inputs (e.g. Haasnoot et al. 2012) to represent the variety of relevant uncertainties and their development over time. The results of this analysis reveal if and when policy actions are needed to achieve the desired outcomes. Such an analysis can be useful in case of multiple and correlated uncertainties that can be represented in scenarios. The pathway maps then only show different time axes for the considered scenarios (and not the axes with changing conditions).

Step 3: Identify contingent actions and assess their ATP conditions and timing

Based on the problem analysis in Steps 1 and 2, alternative policy actions are identified to address vulnerabilities and seize opportunities. As in Step 2, the conditions and timing of the ATPs for each of the identified actions are assessed based on their efficacy in achieving the desired outcomes over changing conditions or time, using the same approach as before. Ineffective actions are screened out (Walker 1988), and only the promising actions are used in the next steps as the building blocks for the assembly of adaptation pathways.

Step 4: Design and evaluate pathways

Once the set of policy actions is deemed adequate, pathways can be designed and evaluated. Note that the alternatives not only may be single actions, but can also include portfolios of actions that are enacted simultaneously. The result is a pathway map (Fig. 4.3), which summarizes all policy actions and the logical potential pathways in which the specified objectives are reached under changing conditions. The map shows different axes—for example, for (multiple) changing conditions and for the assessed timing of these conditions for different scenarios. With the map, it is possible to identify opportunities, no-regret actions, lock-ins, and the timing of actions,

4 Dynamic Adaptive Policy Pathways (DAPP)

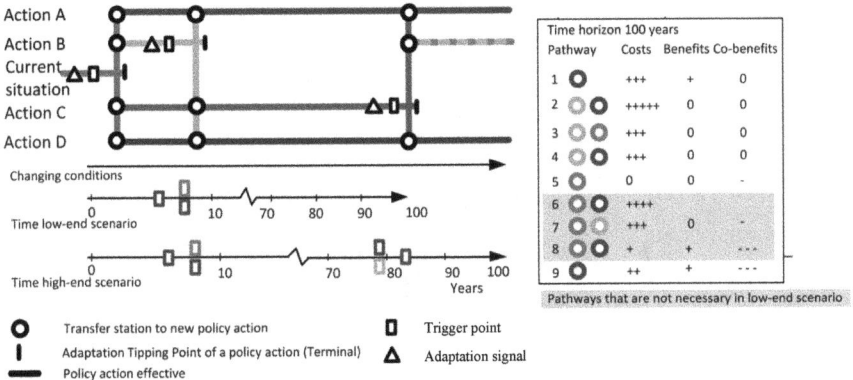

Fig. 4.3 A pathways map and a scorecard presenting the costs and benefits of the pathways presented in the map. Adapted from Haasnoot et al. (2018)

in order to support decision making in a changing environment. Some sequences of actions may be impossible, undesirable, or are less likely (e.g. when they exclude a transfer to other actions or when it is very costly to add or shift actions), while other sequences will match well and enable an adaptive response to changing conditions.

Adaptation pathways can be assembled in different ways. For example, the ATPs established in Step 3 can be used to explore all the possible routes for all the alternative actions. The resulting pathways can then be drawn either manually or using an application specifically designed for this purpose (e.g. the Pathways Generator, which is available from http://pathways.deltares.nl).

Alternatively, pathways can be developed and explored directly using models (e.g. agent-based or multi-objective robust optimization models (Kwakkel et al. 2015)), serious games (Lawrence and Haasnoot 2017, and Chap. 9), or more qualitatively during stakeholder focus group discussions (e.g. Campos et al. 2016) or a combination of model-based and group discussions. In group discussions, stakeholders could identify initial actions and long-term options based on their expert judgment, possibly supported with modelling results, and then draw pathways (see Chap. 9). Developing 'storylines' that take into consideration the socio-economic and environmental conditions and specify the relevant adaptation triggers is another way of developing pathways directly together with stakeholders. Based on multiple storylines, a pathway map can be drawn (Haasnoot 2013). One must recognize that modelling the performance of multiple adaptation pathways across many futures can be computationally intensive. This has consequences for the design of models that are fit for this purpose (see also Haasnoot et al. (2014) for a more in-depth discussion on this topic). In practice, tipping points and pathways are often first defined and developed qualitatively according to expert judgment before more intensive model-based quantitative investigations are carried out.

Each pathway is then evaluated (e.g. using scorecards, cost–benefit analysis, multi-criteria analysis, or engineering options analysis) on its performance, as well as

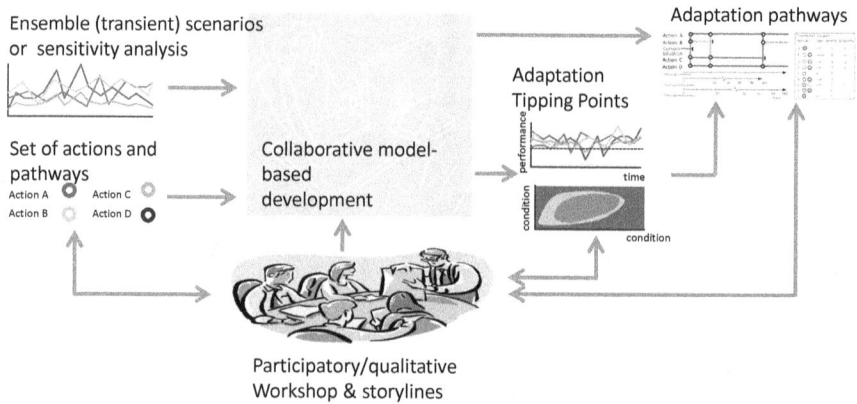

Fig. 4.4 Various approaches for generating adaptation pathways

on other fundamental criteria, such as the urgency of actions, the severity of impacts, the uncertainty involved, path dependency, and the desire to keep future options open. Note that costs and benefits can change over time, which can be considered by using separate scorecards for different periods. Based on this evaluation that illustrates the trade-offs among pathways, a manageable number of promising and preferred pathways can be identified and subjected to further economic analysis and evaluation. Such analyses need not only take into account the initial and recurrent costs for any actions, but also the costs associated with transferring from one action to another (e.g. for modifications or impacts due to the relocation, removal, or adjustments of the previous action (Haasnoot et al. under review).

Figure 4.4 summarizes the process of generating pathways directly or via the identification of ATPs, either using expert judgment/group discussions, computational models, or both.

Step 5: Design the adaptive strategy

The fifth step is to identify initial actions and long-term options for a set of preferred pathways that are selected based on the trade-offs. The robustness of the preferred pathways is improved through contingency planning. This requires the specification of enabling corrective, defensive, and capitalizing actions (Kwakkel et al. 2010, Chap 3) to stay on track, and a set of preparatory actions to keep the preferred longer-term options open for as long as possible. Such an analysis can involve the identification of institutional and socio-cultural conditions that can enable preferred pathways, as was done by van der Brugge and Roosjen (2016).

The adaptive plan also requires an associated monitoring system, describing signposts to monitor and related trigger points that indicate the necessity to implement the next actions. Ideally, a monitoring system gives signals before a decision needs to be taken to implement actions (thus before a decision point). The specification of the right signpost variables and how to analyze the information is important for ensuring that pathways do not evolve into maladaptation or leave preparations for the next actions until too late. The signal may be different from the related ATP, which is related to objectives (impacts). Signals could be related not only to impacts but also to driving forces, such as trends and events in the physical environment, human-driven impacts on the system, technological developments, or changes in societal values and perspectives. An example of how to design signposts for timely and reliable signals is given by Haasnoot et al. (2015), and a more extensive description is given by Haasnoot et al. (2018). 'Timely' means that there is enough time to have the actions performing effectively before an ATP is reached, thus taking into account a lead time for preparation, implementation, and activation of the action. Time needed for implementation of follow-up actions depends on both the existing situation and the action itself, and may thus vary over time and between actions (see also example in Fig. 4.3). 'Reliable' means no missed signal or false alarm. From a policy perspective, it may seem logical to select indicators and trigger points that are related to norm values, objectives, or acceptability values, as these are the values the policies are evaluated upon. However, Haasnoot et al. (2015) argue that these indicators may not give timely and reliable signals, since they are often linked to infrequently occurring extreme conditions, which makes them unsuitable for detecting (systematic) trends. Their results show that other trigger indicators—not necessarily policy related—can be used instead as signals for change. How to design the governance of such a monitoring effort (e.g. who should monitor what, when, and for whom) is described by Hermans et al. (2017).

After producing the final pathway map, a plan for action is drawn up, specifying the actions to be taken immediately (initial actions), as well as an overview of longer-term options, the enabling measures needed to keep them open, the developments to monitor, and the conditions under which the next actions should be taken in order to remain on track to follow the preferred pathway(s). The plan should essentially summarize the results from all the previous steps, including all targets, problems, potential and preferred pathways, enabling actions, and the monitoring system.

Step 6: Implement the strategy, and Step 7: Monitor the strategy

Finally, the initial actions and the enabling actions needed to keep the long-term options open are implemented and the monitoring system is established. Then, time starts running, signpost information related to the triggers is collected, and actions are started, altered, stopped, or expanded in response to this information. After implementation of the initial actions, activation of other actions is suspended until a signal is given. A signal could trigger more research, preparation, or implementation of actions, further enabling actions, or reassessment of the plan.

4.3 A DAPP Illustration: Navigation along the Waas River

The Waas River is a fictitious river flowing through a typical lowland area (Fig. 4.5). Although the river and floodplain area have been highly schematized, they retain

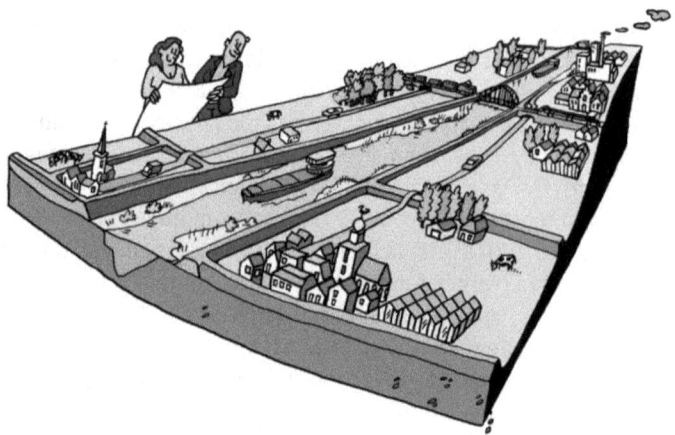

Fig. 4.5 Waas River area

realistic characteristics. The river is bounded by embankments, and the floodplain is separated into five dike rings. A large city is situated on higher ground in the top left of the area. Smaller villages exist in the remaining area, which also includes greenhouses, industry, conservation areas, and pastures. The river serves important economic functions in terms of providing critical navigation services, as it is the principal mode of bulk transport from the upstream hinterland to the downstream coastal port.

The Waas River area has in the past had to contend with both high- and low-flow events. While flooding poses a greater risk to the general population, low flows present significant economic risks in terms of river navigation. Water levels regularly reach levels too low to permit navigation during dry spells. When this occurs, agricultural produce and other industrial goods cannot be transported into or out of the area. Climate change poses a significant risk to the navigation function of the River Waas. Although uncertain, the growing scientific consensus is that warmer temperatures will lead to reduced precipitation during summer months. This will likely further increase dry spells and reduce flows during these times.

In the remainder of this section, we illustrate how each of the steps of DAPP might be applied to the Waas case, following a top-down analytical approach using transient scenarios.

Step 1: Describe the system, specify objectives, and identify uncertainties

Based on the preceding description, the Waas area is identified as a riverine environment vulnerable to both high and low flows (only low flows will be considered for the remainder of this example). Local industries are particularly exposed to these climate-related risks, as they are dependent on river navigation for goods transport. 'Historical records' (based on real-world Dutch data) indicate that over the past 25 years the average annual non-navigable duration was 29 days. Given this situation, the objective for the adaptive planning exercise could be to minimize the number of days that boats are unable to navigate the river channel each year. This could be expressed in terms of an annual non-navigable time indicator (%), with a target being less than 2% (equivalent to approximately 7 days). Policies could be assumed to fail when this target is exceeded for a certain number of years in a row, for example.

The major uncertainty in this example relates to climate change and the extent to which this will affect future precipitation patterns and river discharges. An ensemble of plausible futures can then be set up to encompass this uncertainty for a time horizon of 2000–2100. For this example, three climate scenarios established by the Royal Dutch Meteorological Institute (KNMI) have been used: no climate change, G scenario ('low end'), and W + scenario ('high end') (van den Hurk et al. 2007). These scenarios cover a range of possible future climates in the Netherlands. The low-end scenario reflects moderate climate change with small increases in mean summer precipitation, while the high-end scenario reflects higher climate change with large decreases in mean summer precipitation. These scenarios have been combined with ten synthetic realizations of 100 years of annual precipitation and evaporation events from the KNMI Rainfall Generator (Beersma 2002) to generate a total ensemble of 30 possible futures (ten futures for the current climate, and ten for each of the other two climate change scenarios).

Step 2: Assess system vulnerabilities and opportunities, and establish adaptation tipping points

Each of the identified futures is then analysed against the desired target indicator using a fast Integrated Assessment Meta-Model (IAMM) (Haasnoot et al. 2012) to determine if and when current policies reach an ATP. Given that the average annual non-navigable duration (29 days) greatly exceeds the specified 2% target of 7 days over four consecutive years, it is little surprise that the IAMM reveals the system to already be performing unacceptably, such that the ATP is reached straight away in 2003 under all three climate change scenarios, and new actions need to be taken immediately.

Step 3: Identify alternative actions, and assess their adaptation tipping point conditions and timing

The next step is to establish a set of actions to address this vulnerability. The actions considered have been based on existing plans and potential strategies for navigation (low-flow management) in the Netherlands (Table 4.1). Their performance is then assessed using the IAMM, in order to determine the timing range of the ATP for each action. Given that the assessment is being carried out against 30 plausible futures, the uncertainties in relation to this timing can be conveyed using box–whisker plots. Figure 4.6 and the summarized data provided in Table 4.2 illustrate that all the actions are relatively effective except for the medium-sized ships, which will only be effective in the short term (to 2004). They also demonstrate that there is little variation in the timing of the ATPs, except for the small-scale dredging option, which is much less effective in the high-end scenario.

Step 4: Design and evaluate adaptation pathways

Table 4.1 Description of individual actions and their total costs for the next 100 years

Actions	Cost (M€)
1. Use of smaller ships (300 tonnes) to ensure navigation at low discharges	40
2. Use of medium ships (3000 tonnes) to ensure navigation at lower discharges	40
3. Small-scale dredging of the river bed to permit larger ships to navigate the river at lower discharges	0.015–0.02 (annually)
4. Large-scale dredging of the river bed to permit larger ships to navigate the river at low discharges	0.18–0.22 (annually)

Costs are based on the situation in the Netherlands. In the current situation, large ships of 6000 tonnes are used

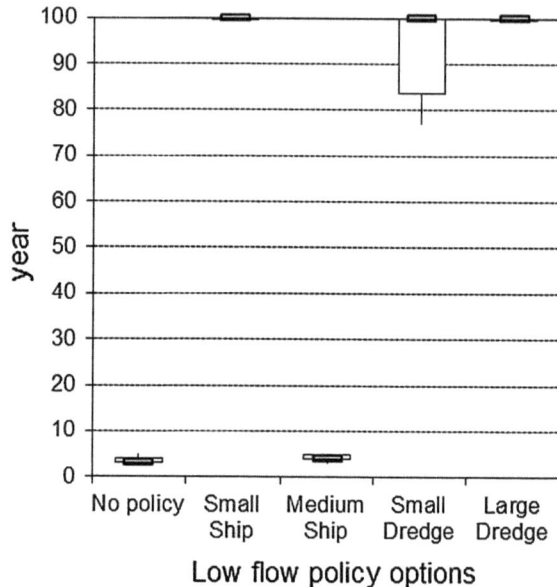

Fig. 4.6 Box–whisker plots of the ATP of the strategies based on the results of all 30 plausible climate futures. Median values for each climate change scenario are also presented (smaller rectangle within each range)

Median ATPs established in the previous step can be used to develop adaptation pathways. After a tipping point is reached, all the other relevant actions are considered. Figure 4.7 presents the possible adaptation pathways for low-flow management. Similar to a metro map, the circles indicate a transfer station to another policy; only here, it is not possible to go back, since the lines present a route through time. The blocks indicate a terminal station at which an ATP is reached. Starting from the current situation, targets begin to be not achieved after 4 years. Following the grey lines of the current policy, one can see that there are four possible options after this tipping point. With the small ships and large-scale dredging actions, targets are achieved for the next 100 years in all plausible futures. When medium ships are chosen, a second tipping point is soon reached, and a shift to one of the other three actions is needed

4 Dynamic Adaptive Policy Pathways (DAPP)

Table 4.2 (a) Median scenario-dependent timing of ATPs under the 10 climate realizations for each scenario and (b) minimum, median, and maximum timing for adaptation tipping points under all 30 climate futures

Action	(a) Scenario-dependent timing of ATP*		
	No climate change	Low end	High end
A. Small ships	>2100	>2100	>2100
B. Medium ships	2004	2004	2004
C. Small-scale dredging	>2100	>2100	2082
D. Large-scale dredging	>2100	>2100	>2100
Action	(b) Summary timing of ATP (all scenarios)		
	Minimum	Median	Maximum
A. Small ships	>2100	>2100	>2100
B. Medium ships	2004	2004	2004
C. Small-scale dredging	2077	>2100	>2100
D. Large-scale dredging	>2100	>2100	>2100

*Median values used

to achieve the target by following the yellow lines. With small-scale dredging, a shift to other actions is needed much later in the century. The dashed yellow and green lines represent a combination of both small-scale dredging and medium ships, which when implemented together also achieves the target until the end of the century in the high-end scenario. Each of the possible pathway routes can then be evaluated on its performance and on other criteria.

Table 4.3 presents a scorecard for this example for the high-end scenario. According to the scorecard, Pathways 6–8 would be the preferred pathways, given their relatively low costs, deferred investments, relative benefits, and limited negative co-benefits. One could also select Pathways 1 and 9; however, these both reduce the flexibility of the plan and lead to locked-in situations. Moreover, Pathway 1 has large upfront costs, and Pathway 9 implements a more intensive version of the dredging policy before it is needed. Pathways 2, 3, and 5 are all too expensive, and there is little value in implementing the medium ship policy in the short term.

Step 5: Design the adaptive strategy

The preferred adaptive plan consists of undertaking small-scale dredging in the short term and leaving options open to shift to either large-scale dredging, small ships, or a combination of small-scale dredging and medium ships in future. Enabling actions that could be enacted to improve the robustness of this plan could include the establishment of a monitoring programme to measure the frequency of low flows; enacting the necessary environmental protections to mitigate any negative effects

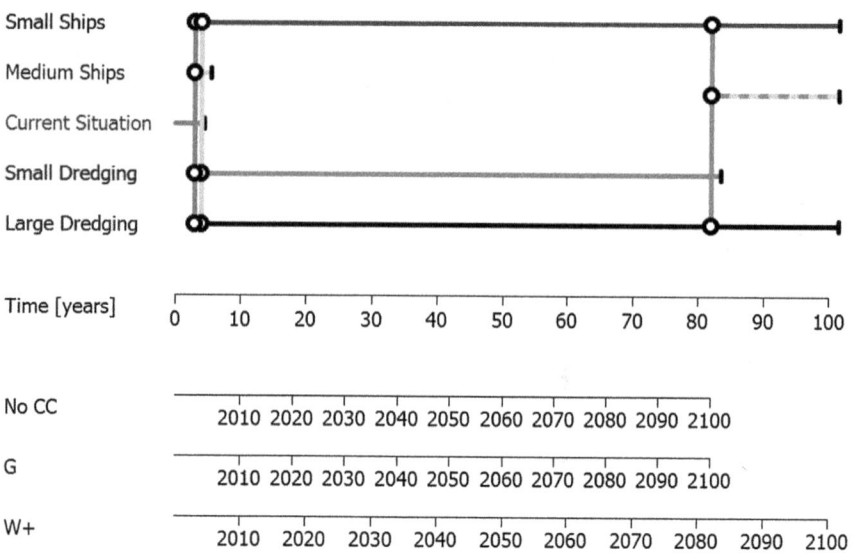

Fig. 4.7 Adaptation pathways map for low-flow management based on the median values for ATP of actions for all plausible climate futures

Table 4.3 Scorecard to evaluate the performance of each of the possible pathways for the high-end scenario

Possible Pathways (W+ scenario)	Costs	Benefits	Co-benefits
1. ○	+++	+	0
2. ○→○	+++++	0	0
3. ○→○→○○	+++++	0	0
4. ○→○→○	+++	0	0
5. ○→○	++++	0	0
6. ○→○	+++	0	-
7. ○→○○	+++	0	-
8. ○→○	+	+	---
9. ○	++	+	---

of dredging; identifying appropriate locations for dredging soil disposal; and (later) regulating shipping companies to replace aging vessels with smaller craft; should monitoring suggest the next tipping point is approaching (and if either Pathway 6 or 7 becomes the preferred longer-term plan).

Step 6: Implement the plan, and Step 7: Monitor the plan

With the completion of the plan, it can then be implemented. Small-scale dredging activities, along with any of the identified immediate enabling actions, are commenced, and the monitoring programme is set up to measure the trajectory of low flows in the region. This information is then collected, and the plan is periodically reviewed and revised as new information comes to light. Once the monitoring indicates the climate trajectory that is occurring, the enabling actions for the preferred long-term options can then start being implemented, and a shift to the preferred policy enacted when a trigger point is reached.

4.4 Under What Conditions Is This Approach Useful?

The strengths of the DAPP method are that it facilitates decision making by offering intuitively understandable visualizations of policy options, and that it stimulates planners preparing the decisions to acknowledge uncertainties and to include adaptation over time explicitly in their plans—to explicitly think about the actions that may need to be taken now to keep long-term options open, and the decisions that can be postponed until the situation becomes clearer. In doing so, the inevitable changes to plans become part of a larger, recognized process, and are not forced to be made repeatedly on an ad hoc basis. Similarly, the path dependencies among different actions are rendered explicit, which helps to limit the emergence of stranded assets and inflexible 'lock-ins'—those situations where prior decisions severely limit the number of actions left open for future adaptations. Planners, through monitoring and corrective actions, can try to keep the system headed towards the original goals.

By unravelling an initial plan (e.g. one derived from Dynamic Adaptive Planning (Chap. 3) or Robust Decision Making (Chap. 2)) into adaptation or development pathways, DAPP can also serve as an enabler for action. The identification of adaptation and opportunity tipping points, and the development of pathways, stimulates planners, policymakers, and stakeholders to discuss and consider the possible limits of adaptation to uncertain changing conditions, as well as the potential longer-term impacts for future generations. They are encouraged to think more broadly than the traditional types of actions, and in particular to consider the possibilities for larger, transformative actions to cope with large changes to the system that go beyond the confines of present practices. The decision space can thus be widened and future difficulties anticipated. This helps to challenge societies to prepare for the future—potentially large—changes.

Adaptive approaches such as DAPP are particularly useful when the degree of uncertainty over the planning horizon is relevant for decision making, when the implementation time is relatively short compared to the rate of change (i.e. there is time to implement the action and for the system to adapt); when there is flexibility in the solutions (Maier et al. 2016); when alternative decision/actions or staged decision making is possible; when the functional lifetime of the decision is shorter than the planning horizon (i.e. tipping points are reached); when there are relevant path dependencies; when the implications of the decisions cover a long lifetime; and when trends can be signalled. They are especially useful in complex coupled physical–technical–human systems, where there are large uncertainties and multiple management decisions to be taken over long time periods and where the path dependency of these decisions is high. In such systems, long-term investment decisions are particularly exposed to the uncertain drivers of change. Without considering the long term and path dependency, maladaptive actions can result, leaving assets stranded long before their design lives are reached. Adaptation pathways can help to limit these risks by helping to identify the most appropriate sequence of measures with flexibly to adapt to the changing conditions.

Finally, over time and with other anthropogenic and environmental changes, problems and opportunities may become more visible or urgent than what is envisaged today, and alternative adaptation options to solve these issues may become available.

4.5 Recent Advances

The concept of adaptation pathways is relatively new. As such, the approach is constantly being elaborated further by scholars and practitioners alike. Some recent advances include the work of Kwakkel et al. (2015), who developed a model-driven multi-objective robust optimization approach to identify and evaluate the most promising adaptation pathways directly using an Integrated Assessment Meta-Model (IAMM). This approach establishes a computational technique for searching in a structured way for a diverse set of robust adaptation pathways. It substantially prunes the set of candidate pathways, yielding only those that can best cope with the multiplicity of futures that span the various uncertainties.

Additional efforts have expanded the concept of tipping points. ATPs remain central to the DAPP method, but in some instances it may be useful to also consider the *opportunity tipping points* of actions (e.g. Bouwer et al. under review). These are points at which a particular action becomes feasible or attractive, for example because of lower costs of actions or technical developments. Also, certain actions can have required conditions on actions (e.g. successful pilot), on long lead times (e.g. the planning and construction of a reservoir). As a result, they cannot be considered as viable actions for the current conditions, but can be considered for implementation at a later point. For example, the implementation of a nature-based solution may require certain sediment and water depth conditions, or some actions may require a successful pilot before it is possible to implement it on a large scale.

Accounting for transfer costs in the economic evaluation of pathways has been another recent development. Standard economic evaluation studies are skewed towards short-term returns on investment by using a short time horizon. Thus, these evaluations have severe limitations in assessing the economic impact of present-day decisions for long-lived infrastructure and its performance over time under significant uncertainty (OECD 2013). Traditional economic approaches focus on single decisions rather than sequences of decisions and are not well equipped to deal with deep uncertainty (i.e. uncertainty that cannot be quantified with probability distributions). An exception is Real Options Analysis (ROA); in its present applications, it does not take into account path dependency of decisions (Smet 2017), although a variation of ROA was used to evaluate pathways in the New Zealand case study (see Chap. 9, Sect. 9.4). The approach of Engineering Options Analysis (EOA) considers a path-dependent decision tree for different environmental changes (Smet 2017 and Chap. 6). Haasnoot et al. (under review) show that when transfer costs are taken into account the economic costs associated with certain pathways can increase significantly such that these become less attractive. They also show that these impacts are particularly sensitive to both the timing of the ATP and the discount rate.

Research has also been carried out into the design of appropriate adaptation signals to include within DAPP monitoring plans (Haasnoot et al. 2018; Stephens et al. 2018). Haasnoot et al. (2015, 2018) present a framework to ensure that pathways do not tend towards being maladaptive or leave action until too late. Similarly, they advocate for efforts being put into searching for strong signals (often different from the overarching performance condition) that indicate that an ATP is approaching. These are signals that exhibit little noise in relation to the ensemble of possible futures and that provide an early indication of the scenario trajectory being followed. Finally, a typology of adaptation pathways to sea level rise has been developed for six typical coastal archetypes (urban/rural open coasts, urban/rural deltas, and urban/rural estuaries), with a focus on how best to manage the challenges of rising sea levels (Brown et al. 2016). These recognize that the impacts of sea level rise could be potentially severe, and that adaptation will be essential and may require extensive transformative actions.

4.6 Links with Other DMDU Approaches

Since the original specification of DAPP, various researchers have contributed to its further development. A first line of work focuses on the model-based design of adaptation pathways. Kwakkel et al. (2015) present a many-objective robust optimization (MORO) approach and demonstrate this using the Waas case. In a follow-up study, they compare this optimization approach with robust decision-making (RDM) and conclude that the iterative stress testing of RDM is a complementary model-based approach for designing pathways (Kwakkel et al. 2016). Along a similar line, Zeff et al. (2016) combine MORDM and adaptation pathways, illustrating this using a water resource application in the USA. Along a different dimension, there is ongoing research aimed at combining Decision Scaling (DS) and DAPP. These are combined in the Collaborative Risk-Informed Decision Analysis (CRIDA) approach (http://agwaguide.org/CRIDA). CRIDA is a risk-informed approach similar in its planning steps to DAPP (establish the decision context, perform a bottom-up vulnerability analysis, formulate actions, evaluate alternatives, and institutionalize decisions). Compared to DAPP, CRIDA offers scalable practical guidance depending on the outcome of the stress test: the plausibility of adverse impacts (termed future risks) and analytical uncertainty. Under higher risks and analytical uncertainty, deviations from existing planning practice are proposed, such as the development of staged flexible strategies.

A second line of research is focused on comparing DAPP with other approaches. As noted above, Kwakkel et al. (2016) compare DAPP and RDM, highlighting their complementarity. Gersonius et al. (2015) and Buurman and Babovic (2016) both report on using DAPP with ROA.

4.7 Future Challenges

As detailed above, DAPP is a DMDU approach that is constantly evolving and being elaborated. Some of these advances have naturally emerged as it has been applied in different ways and in different jurisdictions and contexts. Several challenges in relation to the method remain, however, which can form the basis for future scholarly work. First, challenges remain in demonstrating to practitioners and policymakers the value of the approach. It demands a shift in mindset and practice away from more traditional planning procedures, so tools are needed that can encourage its wider application, for example using simulation games to embed a change in thinking. Chapter 9 offers a potential way forward in this regard, in relation to the adoption of DAPP in local government in New Zealand. Somewhat related to this point, much of the work related to DAPP has been carried out for large, well-resourced, national-scale planning studies (e.g. the Dutch Delta Programme and the Bangladesh Delta Plan). Practice has shown that the approach lends itself well for tailoring to context-specific conditions. Further application of the approach to more resource-constrained local planning challenges is needed, in particular with respect to the implementation and monitoring of the resulting plans.

This chapter has also described several ways to identify ATPs and develop pathways. At present, many applications of the method have focused on computer-based approaches that demand large numbers of simulation runs. In many instances, this will not be feasible, in which case more qualitative, stakeholder-driven approaches are another area ripe for future development. Challenges also remain in relation to both the economic evaluation of pathways and the identification of appropriate monitoring signals. The work on the economic evaluation of pathways (outlined above) has revealed it is not a straightforward process and is one that will require further refinement, in particular due to the sensitivity of the evaluation to variation in the discount rate and the timing of tipping points. Likewise, studies into appropriate monitoring signals are very much at a nascent stage, particularly when one considers the myriad of external conditions for which signals may need to be found across the biophysical and socio-economic spheres.

Another challenge is to support the design of adaptation pathways in complex multi-stakeholder contexts effectively. Different stakeholders are responsible for different actions and have different preferences. What is desirable or robust for one stakeholder might not be robust for another, giving rise to robustness conflicts (Herman et al. 2014; Bosomworth et al. 2017). At present, little attention has been given to these issues in the design of pathways. Addressing this challenge involves both a governance component, focused on the alignment of incentives among stakeholders and the establishment of trust, and a policy analytic component, aimed at identifying robustness conflicts. The emergence of robustness conflicts might be conditional on specific actions, driven by changing conditions, or be defined by individual preferences and culture. The overarching challenge is to bring into focus the societal robustness of pathways in addition to robustness with respect to the future.

References

Beersma, J. J. (2002). *Rainfall generator for the Rhine basin: Description of 1000-year simulations*. De Bilt, The Netherlands: KNMI

Bosomworth, K., Leith, P., Harwood, A., & Wallis, P. J. (2017). What's the problem in adaptation pathways planning? The potential of a diagnostic problem-structuring approach. *Environmental Science & Policy, 76,* 23–28. https://doi.org/10.1016/j.envsci.2017.06.007.

Bouwer, L. M., Burzel, A., Winsemius, H. C., Ward, P. J., Jeuken, A., & M. Haasnoot. (under review). When to adapt: informing flood risk management by analysing adaptation opportunities in Europe.

Brown, S., Haasnoot, M., Bucx, T., van de Brugge, R., Wadey, M., & Nicholls, R. J., et al. (2016). *Report on generic adaptation roadmap with reduced impact uncertainty (per selected case studies and with an across scales component)*. RISES-AM EU research project, Deliverable 4.2.

Buurman, J., & Babovic, V. (2016). Adaptation pathways and real options analysis—An approach to deep uncertainty in climate change adaptation policies. *Policy and Society, 35*(2), 137–150. https://doi.org/10.1016/j.polsoc.2016.05.002.

Campos, I. S., Alves, F., Dinas, J., Truninger, M., Vizinho A., & Penha-Lopes, G. (2016) Climate adaptation, transitions, and socially innovative action-research approaches. *Ecology and Society, 21*(13).

Gersonius, B., Ashley, R., Jeuken, A., Pathinara, A., & Zevenbergen, C. (2015). Accounting for uncertainty and flexibility in flood risk management: Comparing Real-In-Options optimisation and Adaptation Tipping Points. *Journal of Flood Risk Management, 8,* 135–144.

Haasnoot, M. (2013). *Anticipating change—Sustainable water policy pathways for an uncertain future*. Ph.D. dissertation, Twente University, Enschede, Netherlands. https://doi.org/10.3990/1.9789036535595.

Haasnoot, M., Kwakkel, J. H., Walker, W. E., & ter Maat, J. (2013). Dynamic adaptive policy pathways: A method for crafting robust decisions for a deeply uncertain world. *Global Environmental Change, 23*(2), 485–498.

Haasnoot, M., Middelkoop, H., Offermans, A., van Beek, E., & van Deursen, W. P. A. (2012). Exploring pathways for sustainable water management in river deltas in a changing environment. *Climatic Change, 115*(3–4), 795–819.

Haasnoot, M., Schellekens, J., Beersma, J. J., Middelkoop, H., & Kwadijk, J. C. J. (2015). Transient scenarios for robust climate change adaptation illustrated for water management in the Netherlands. *Environmental Research Letters*. https://doi.org/10.1088/1748-9326/10/10/105008.

Haasnoot, M., van Aalst, M., Rozenberg, J., Dominique, K., Matthews, J., & Bouwer, L. M., et al. (under review). Investments under non-stationarity: Economic evaluation of adaptation pathways.

Haasnoot, M., van Klooster, S., & van't Alphen, J. (2018). Designing a monitoring system to detect signals to adapt to uncertain climate change. *Global Environmental Change, 52,* 273–285. https://doi.org/10.1016/j.gloenvcha.2018.08.003.

Haasnoot, M., van Deursen, W. P. A., Guillaume, J. H. A., Kwakkel, J. H., van Beek, E., & Middelkoop, H. (2014). Fit for purpose? Building and evaluating a fast, integrated model for exploring water policy pathways. *Environmental Modelling & Software, 60,* 99–120.

Hermans, L. M., Haasnoot, M., Kwakkel, J. H., & ter Maat, J. (2017). Designing monitoring arrangements for collaborative learning about adaptation pathways. *Environmental Science & Policy, 69,* 29–38.

Herman, J. D., Zeff, H. B., Reed, P. M., & Characklis, G. (2014). Beyond optimality: Multistakeholder robustness tradeoffs for regional water portfolio planning under deep uncertainty. *Water Resources Research, 50,* 7692–7713.

Katsman, C., Hazeleger, W., Drijfhout, S., van Oldenborgh, G., & Burgers, G. (2008). Climate scenarios of sea level rise for the northeast Atlantic ocean: A study including the effects of ocean dynamics and gravity changes induced by ice melt. *Climatic Change, 91,* 351–374. https://doi.org/10.1007/s10584-008-9442-9.

Kuijken, W. (2010). *The Delta Programme in the Netherlands: The Delta Works of the Future.* Retrieved January 18, 2013, from https://www.deltacommissaris.nl/documenten/toespraken/2010/09/29/the-delta-programme-in-the-netherlands-the-delta-works-of-the-future.

Kwadijk, J. C. J., Haasnoot, M., Mulder, J. P. M., Hoogvliet, M., Jeuken, A., van der Krogt, R., et al. (2010). Using adaptation tipping points to prepare for climate change and sea level rise: A case study in the Netherlands. *Wiley Interdisciplinary Review Climate Change, 1*(5), 729–740. https://doi.org/10.1002/wcc.64.

Kwakkel, J. H., Haasnoot, M., & Walker, W. E. (2015). Developing dynamic adaptive policy pathways: A computer-assisted approach for developing adaptive strategies for a deeply uncertain world. *Climatic Change, 132,* 373–386. https://doi.org/10.1007/s10584-014-1210-4.

Kwakkel, J. H., Haasnoot, M., & Walker, W. E. (2016). Comparing robust decision-making and dynamic adaptive policy pathways for model-based decision support under deep uncertainty. *Environmental Modelling and Software, 86,* 168–183. https://doi.org/10.1016/j.envsoft.2016.09.017.

Kwakkel, J. H., Walker, W. E., & Marchau, V. A. W. J. (2010). Adaptive airport strategic planning. *European Journal of Transport and Infrastructure Research, 10*(3), 249–273.

Lawrence, J., & Haasnoot, M. (2017). What it took to catalyse uptake of dynamic adaptive pathways planning to address climate change uncertainty. *Environmental Science & Policy, 68,* 47–57. https://doi.org/10.1016/j.envsci.2016.12.003.

Maier, H. R., Guillaume, J. H. A., van Delden, H., Riddell, G. A., Haasnoot, M., Kwakkel, J. H. (2016). An uncertain future, deep uncertainty, scenarios, robustness and adaptation: How do they fit together? *Environmental Modelling & Software, 81.* https://doi.org/10.1016/j.envsoft.2016.03.014.

OECD. (2013). *Water and climate change adaptation: Policies to navigate uncharted waters.* Paris: OECD Publishing.

Offermans, A., Haasnoot, M., & Valkering, P. (2011). A method to explore social response for sustainable water management strategies under changing conditions. *Sustainable Development, 19,* 312–324. https://doi.org/10.1002/sd.439

Ranger, N., Reeder, T., & Lowe, J. (2013). Addressing 'deep' uncertainty over long-term climate in major infrastructure projects: four innovations of the Thames estuary 2100 Project. *EURO Journal on Decision Processes, 1,* 233–262. https://doi.org/10.1007/s40070-013-0014-5.

Reeder, T., & Ranger, N. (2010). How do you adapt in an uncertain world? Lessons from the Thames Estuary 2100 project. *World Resources Report*, Washington DC. Available online at http://www.worldresourcesreport.org, pp. 16.

Smet, K. (2017). *Engineering options: A proactive planning approach for aging water resource infrastructure under uncertainty.* Cambridge: Harvard University.

Stephens, S., Bell, R., & Lawrence, J. (2018). Developing signals to trigger adaptation to sea-level rise. *Environmental Research Letters, 13*(10). https://doi.org/10.1088/1748-9326/aadf96.

Van den Hurk, B., Klein Tank, A., Lenderink, G., Van Ulden, A., Van Oldenborgh, G., Katsman, C., et al. (2007). New climate change scenarios for the Netherlands. *Water Science and Technology, 56*, 27–33.

van der Brugge, R., & Roosjen, R. (2016). An institutional and sociocultural perspective on the adaptation pathways approach. *Journal of Water and Climate Change, 6*(4), 743–758. https://doi.org/10.2166/wcc.2015.001.

Walker, W. E. (1988). Generating and screening alternatives (Chap. 6). In H. J. Miser & E. S. Quade (Eds.), *Handbook of systems analysis: Craft issues and procedural choices.* North Holland, NY.

Walker, W. E., Rahman, S. A., & Cave, J. (2001). Adaptive policies, policy analysis, and policy-making. *European Journal of Operational Research, 128*(2), 282–289.

Zeff, H. B., Herman, J. D., Reed, P. M., & Characklis, G. (2016). Cooperative drought adaptation: Integrating infrastructure development, conservation, and water transfers into adaptive policy pathways. *Water Resources Research, 52*, 7327–7346. https://doi.org/10.1002/2016WR018771.

Dr. Marjolijn Haasnoot is a Senior Researcher and Advisor at Deltares and Associate Professor in adaptive delta management at Utrecht University, the Netherlands. Her research focus is on water management, integrated assessment modeling, and decisionmaking under deep uncertainty. She is working and consulting internationally on projects assessing impacts of climate change and socio-economic developments, and alternative management options to develop robust and adaptive plans. Marjolijn developed the Dynamic Adaptive Policy Pathways planning approach. She has a Ph.D. in engineering and a Masters in environmental science.

Andrew Warren, M.Sc. (Deltares) is an Advisor and Researcher at Deltares. His work is focused on integrated water resources management and climate change adaptation. He works and consults internationally on projects to develop robust and adaptive plans through the assessment of climate change impacts and socio-economic developments. He also concentrates on the development of interactive methods and communication tools to engage stakeholders in technical decision-making processes. He holds a Masters in Civil Engineering from Delft University of Technology.

Dr. Jan H. Kwakkel is Associate Professor at the Faculty of Technology, Policy and Management (TPM) of Delft University of Technology. He is also the Vice President of the Society for Decision Making under Deep Uncertainty, and a member of the editorial boards of *Environmental Modeling and Software*, *Futures*, and *Foresight*. His research focuses on model-based support for decision making under uncertainty. He is involved in research projects on smart energy systems, high-tech supply chains, transport, and climate adaptation. Jan is the developer of the Exploratory Modeling Workbench, an open-source software implementing a wide variety of state-of-the-art techniques for decisionmaking under uncertainty.

Open Access This chapter is licensed under the terms of the Creative Commons Attribution 4.0 International License (http://creativecommons.org/licenses/by/4.0/), which permits use, sharing, adaptation, distribution and reproduction in any medium or format, as long as you give appropriate credit to the original author(s) and the source, provide a link to the Creative Commons licence and indicate if changes were made.

The images or other third party material in this chapter are included in the chapter's Creative Commons licence, unless indicated otherwise in a credit line to the material. If material is not included in the chapter's Creative Commons licence and your intended use is not permitted by statutory regulation or exceeds the permitted use, you will need to obtain permission directly from the copyright holder.

Chapter 5
Info-Gap Decision Theory (IG)

Yakov Ben-Haim

Abstract

- Info-Gap (IG) Decision Theory is a method for prioritizing alternatives and making choices and decisions under deep uncertainty.
- An "info-gap" is the disparity between what *is known* and what *needs to be known* for a responsible decision.
- Info-gap analysis does not presume knowledge of a worst-case or of reliable probability distributions.
- Info-gap models of uncertainty represent uncertainty in parameters and in the shapes of functional relationships.
- IG Decision Theory offers two decision concepts: *robustness* and *opportuneness*.
- The *robustness* of an alternative is the greatest horizon of uncertainty up to which that alternative satisfies critical outcome requirements.
- The *robustness strategy* satisfices the outcome and maximizes the immunity to error or surprise. This differs from outcome optimization.
- The *robustness function* demonstrates the trade-off between immunity to error and quality of outcome. It shows that knowledge-based predicted outcomes have no robustness to uncertainty in that knowledge.
- The *opportuneness* of a decision alternative is the lowest horizon of uncertainty at which that decision enables better-than-anticipated outcomes.
- The *opportuneness strategy* seeks windfalls at minimal uncertainty.
- We discuss "innovation dilemmas" in which the decisionmaker must choose between two alternatives, where one is putatively better but more uncertain than the other.
- Two examples of info-gap analysis are presented, one quantitative that uses mathematics and one qualitative that uses only verbal analysis.

Y. Ben-Haim (✉)
Technion—Israel Institute of Technology, Haifa, Israel
e-mail: yakov@technion.ac.il

5.1 Info-Gap Theory: A First Look

Info-Gap (IG) is a non-probabilistic decision theory for prioritizing alternatives and making choices and decisions under deep uncertainty (Ben-Haim 2006, 2010). They might be operational alternatives (design a system, choose a budget, decide to launch or not, etc.) or more abstract alternatives (choose a model structure, make a forecast, formulate a policy, etc.). Decisions are based on data, scientific theories, empirical relations, knowledge, and contextual understanding, all of which I'll refer to as one's *models*, and these models often recognize and quantify uncertainty.

IG theory has been applied to decision problems in many fields, including various areas of engineering (Kanno and Takewaki 2006; Chinnappen-Rimer and Hancke 2011; Harp and Vesselinov 2013), biological conservation (Burgman 2005), economics (Knoke 2008; Ben-Haim 2010), medicine (Ben-Haim et al. 2012), homeland security (Moffitt et al. 2005), public policy (Hall et al. 2012), and more (see www.info-gap.com). IG robust satisficing has been discussed non-mathematically elsewhere (Schwartz et al. 2011; Ben-Haim 2012a, b, 2018; Smithson and Ben-Haim 2015).

Uncertainty is often modeled with probability distributions. If the probability distributions are correct and comprehensive, then one can exploit the models exhaustively to reliably achieve stochastically optimal outcomes, and one doesn't need IG theory. However, if one's models will be much better next year when new knowledge has become available (but you must decide now), or if processes are changing in poorly known ways, or if important factors will be determined beyond your knowledge or control, then one faces deep uncertainty and IG theory might help. This section presents an intuitive discussion of two basic ideas of IG theory: satisficing and robustness. A more systematic discussion of IG robustness appears in Sect. 5.2. Simple examples are presented in Sects. 5.3 and 5.4, and more detailed examples appear in Chap. 10.

Knight (1921) distinguished between what he called "risk" (for which probability distributions are known) and "true uncertainty" (for which probability distributions are not known). Knightian ("true") uncertainty reflects ignorance of many things, including underlying processes, functional relationships, strategies or intentions of relevant actors, future events, inventions, discoveries, surprises, and so on. Info-gap models of uncertainty provide a non-probabilistic quantification of Knightian uncertainty. An info-gap is the disparity between what you *do know* (or think to be true) and what you *need to know* for making a reliable or responsible decision (though what is needed may be uncertain). An info-gap is not ignorance per se, but rather those aspects of one's Knightian uncertainty that bear on a pending decision and the quality of its outcome.

An info-gap model of uncertainty is particularly suitable for representing uncertainty in the shape of a function. For instance, one might have an estimate of the stress–strain curve for forces acting on a metal, or of the supply and demand curves for a new product, or of a probability density function (pdf), but the shape of the function (e.g., the shape of the elastic–plastic transition curve in the first case, or the shapes of the supply and demand curves, or the tails of the pdf) may be highly uncertain. Info-gap models are also widely used to represent uncertainty in parameters or vectors, or sometimes uncertainty in sets of such entities.

Decisionmakers often try to optimize the outcome of their decisions. That is usually approached by using one's best models to predict the outcomes of the various alternatives, and then choosing the one whose predicted outcome is best. The aspiration for excellence is commendable, but outcome optimization may be costly, or one may not really need the best possible outcome. Schwartz (2004) discusses the irrelevance of optimal outcomes in many situations.

Outcome optimization—using one's models to choose the decision whose predicted outcome is best—works fine when the models are pretty good, because exhaustively exploiting good models will usually lead to good outcomes.

However, when one faces major info-gaps one's models contain major errors or lacunae, and exhaustively exploiting the models can be unrealistic, unreliable, and can lead to undesired outcomes (Ben-Haim 2012a). Under deep uncertainty, it is better to ask: What outcomes are critical and must be achieved? This is the idea of *satisficing* introduced by Simon (1956): achieving a satisfactory or acceptable, but not necessarily optimal, outcome—"good enough" according to an explicitly stated set of criteria.

Planners, designers, and decisionmakers in all fields have used the language of optimization (the lightest, the strongest, the fastest) for ages. In practice, however, satisficing is very widespread, although not always recognized as such. Engineers satisfy design specifications (light enough, strong enough, fast enough). Stock brokers, business people, and investors of all sorts do not really need to maximize profits; they only need to beat the competition, or improve on last year, or meet the customer's demands. Beating the competition means satisficing a goal. The same can be said of the public official who must reduce crime below a legislative target, or the foreign aid planner who must raise public health to international standards.

Once the decisionmaker identifies the critical goals or outcomes that must be achieved, the next step is to decide about something or choose an action that will achieve those goals despite current ignorance or unknown future surprises. A decision has high *robustness* if it satisfices the performance requirements over a wide range of unanticipated contingencies. Conversely, a decision has low robustness if even small errors in our knowledge can prevent achievement of the critical goals. The *robust-satisficing* decisionmaker prioritizes the alternatives in terms of their robustness against uncertainty for achieving the critical goals. The decision methodology of IG robust satisficing is often motivated by the pernicious potential of the unknown. However, uncertainty can be propitious, and IG theory offers a method for prioritizing one's alternatives with respect to the potential for favorable surprises. The idea of "windfalling" supplements the concept of satisficing. The opportune windfalling decisionmaker prioritizes the alternatives in terms of their potential for exploiting favorable contingencies. This is illustrated in the example in Sect. 5.4.

Min-max or worst-case analysis is a widely used alternative to outcome optimization when facing deep uncertainty and bears some similarity to IG robustness. Neither min-max nor IG presumes knowledge of probabilities. The basic approach behind these two methods is to find decisions that are robust to a range of different contingencies. Wald (1947) presented the modern formulation of min-max, and it has been applied in many areas (e.g., Hansen and Sargent 2008). The decisionmaker considers a bounded family of possible models, without assigning probabilities to their occurrence. One then identifies the model in that set which, if true, would result in a worse outcome than any other model in the family. A decision is made that minimizes this maximally bad outcome (hence "min-max"). Min-max is attractive because it attempts to insure against the worst anticipated outcome. However, min-max has been criticized for two main reasons. First, it may be unnecessarily costly to assume the worst case. Second, the worst usually happens rarely and therefore is poorly understood. It is unreliable (and perhaps even irresponsible) to focus the decision analysis on a poorly known event (Sims 2001).

Min-max and IG methods both deal with Knightian uncertainty, but in different ways. The min-max approach is to choose the decision for which the contingency with the worst possible outcome is as benign as possible: Identify and ameliorate the worst case. The IG robust-satisficing approach requires the planner to think in terms of the worst consequence that can be tolerated, and to choose the decision whose outcome is no worse than this, over the widest possible range of contingencies. Min-max and IG both require a prior judgment by the planner: Identify a worst model or contingency (min-max) or specify a worst tolerable outcome (IG). These prior judgments are different, and the corresponding policy selections may, or may not, agree. Ben-Haim et al. (2009) compare min-max and IG further.

5.2 IG Robustness: Methodological Outline

This section is a systematic description of the IG robustness methodology for decisionmaking under deep uncertainty. The following two sections present examples.

IG robustness is the attribute of satisfying critical requirements even when the situation is, or evolves, differently from what is expected. A decision is robust to uncertainty if it remains acceptable even if the understanding that was originally available turns out to be substantially wrong. The robustness is assessed by answering the question: How wrong can our current model be so that the outcome of this decision will still be acceptable? In other words, how immune to our current ignorance is this decision? A decision is highly robust if it remains acceptable throughout a wide range of deviation of reality from the original understanding. More robustness to uncertainty is better than less, so decisionmakers prefer the more robust decision over one with less robustness to uncertainty. We will now formulate these ideas more rigorously. Sections 5.3 and 5.4 present two simple examples.

5.2.1 Three Components of IG Robust Satisficing

The IG analysis of robustness to uncertainty rests on three components: the model, the performance requirements, and the uncertainty model. These components constitute the framing of the analysis. Furthermore, this framing may be done in a quick exploratory mode, or may be more fundamental and advanced. Likewise, decisions based on the analysis may be short-term actions, long-term options, or adaptive strategies that combine the short and long term.

The *model*, as defined earlier, is our understanding of the system or situation that must be influenced, its temporal dynamics, the evidence, the environment, and any other available relevant knowledge.

The *performance requirements* are specified in response to the question: What do we need to achieve in order for the outcome of the decision to be acceptable? The economist may require inflation within specified bounds; the engineer may require an operational lifetime exceeding a given value; the military commander may require a substantial decrease in insurgent violence, etc. The analyst together with the decisionmaker makes judgments of the required or acceptable levels of performance. More demanding performance entails greater vulnerability to uncertainty and hence lower levels of robustness, as we will see.

Uncertainty model. We have best estimates of our model: the knowledge, understanding, and evidence relating to the situation. However, these estimates may be wrong or incomplete. We may also be uncertain about the performance requirements (how much inflation or insurgent violence is acceptable?). There are many specific forms of info-gap models of uncertainty, encoding different information about the uncertainty. However, they all express the intuition that we do not know how wrong our best estimates are. The info-gap model also includes insights and contextual understanding about this uncertainty. An info-gap model of uncertainty expresses what we do know, as well as the unbounded horizon of uncertainty surrounding our knowledge. It expresses the idea that we cannot confidently identify a realistic worst case.

5.2.2 IG Robustness

The analyst must formulate and prioritize candidate decisions and will do so based on their robustness against uncertainty for achieving acceptable outcomes: More robustness is preferred over less robustness. We evaluate the robustness to uncertainty of any candidate decision by combining these three components: the model, the performance requirements, and the uncertainty model. We do this by addressing two questions, the first regarding the putative performance; the second addressing the robustness against uncertainty.

Putative performance. The putative performance of a decision is the outcome that is predicted by the best understanding and the evidence in hand (the model). Given a candidate decision, we ask if its putative performance satisfies our performance requirements. That is, we ask whether or not this outcome is acceptable (according to the performance requirements) as assessed by the best understanding that we have (our model). If the answer is negative, then we reject this decision alternative. At this stage, we are ignoring uncertainty.

Robustness. Given a candidate decision that has gotten a positive answer to the question regarding putative performance, we now ask: How much could the model change without violating the performance requirements? This explicitly addresses the uncertainty. We are asking: What is the greatest horizon of uncertainty (in the uncertainty model) up to which the performance requirements are guaranteed by this decision? How much could reality deviate from our understanding and evidence (our model) so that the decision we are contemplating would still satisfy the requirements? Could the decision tolerate *any* error in the model, up to some large degree, without violating the performance requirements (implying large robustness)? Or is there *some* small error in the model that would jeopardize the requirements (implying low robustness)?

5.2.3 Prioritization of Competing Decisions

More robustness is better than less robustness. This means that, given two alternative decisions that both putatively satisfy the performance requirements, we prefer the decision that satisfies the performance requirements throughout a larger range of uncertainty. This prioritization is called *robust satisficing*, because it selects the decision that is more robust against uncertainty while also satisfying the performance requirements.

Note that the robust-satisficing decision optimizes the robustness against uncertainty rather than optimizing the substantive quality of the decision's outcome. We do not optimize the outcome. The outcome must be satisfactory, though the analyst can choose the satisficing level to be more or less demanding. We optimize the robustness against uncertainty, and we satisfice the outcome.

5.2.4 How to Evaluate Robustness: Qualitative or Quantitative?

Many decisions under uncertainty are amenable to quantitative analysis (i.e., using mathematics). Many other situations depend on conceptual models and verbal formulations that cannot be captured with equations. IG theory has been applied in both qualitative and quantitative analyses, as we illustrate in Sects. 5.3 and 5.4, respectively. These examples are independent and one can read either or both.

5.3 IG Robustness: A Qualitative Example

We begin with a simple math-free qualitative example. In this section, we examine five conceptual proxies for the concept of robustness (Ben-Haim and Demertzis 2016) and then discuss a simple example based on qualitative reasoning.

5.3.1 Five Conceptual Proxies for Robustness

Like all words, "robustness" has many connotations, and its meanings overlap with the meanings of other words.[1] We discuss five concepts that overlap significantly with the concept of robustness against uncertainty, and that are useful in the qualitative assessment of decisions under uncertainty. Each of these five concepts emphasizes a different aspect of robustness, although they also overlap. The *five proxies for robustness* are resilience, redundancy, flexibility, adaptiveness, and comprehensiveness. A decision, policy, action, or system is highly robust against uncertainty if it is strong in some or all of these attributes; it has low robustness if it is weak in all of them. We will subsequently use the term "system" quite broadly, to refer to a physical or organizational system, a policy for formulating or implementing decisions, or a procedure for political foresight or clinical diagnosis, etc.

Resilience of a system is the attribute of rapid recovery of critical functions. Adverse surprise is likely when facing deep uncertainty. A system is robust against uncertainty if it can rapidly recover from adverse surprise and achieve critical outcomes.

Redundancy of a system is the attribute of providing multiple alternative solutions. Robustness to surprise can be achieved by having alternative responses available.

Flexibility (sometimes called agility) of a system is the ability for rapid modification of tools and methods. Flexibility or agility, as opposed to stodginess, is often useful in recovering from surprise. A physical or organizational system, a policy, or a decision procedure is robust to surprise if it can be modified in real time.

Adaptiveness of a system is the ability to adjust goals and methods in the mid- to long-term. A system is robust if it can be adjusted as information and understanding change. Managing Knightian uncertainty is rarely a once-through procedure. We often must re-evaluate and revise assessments and decisions. The emphasis is on the longer time range, as distinct from on-the-spot flexibility.

Comprehensiveness of a system is its interdisciplinary system-wide coherence. A system is robust if it integrates relevant considerations from technology, organizational structure and capabilities, cultural attitudes and beliefs, historical context, economic mechanisms and forces, or other factors. A robust system addresses the multi-faceted nature of the problem.

---

[1] The representation of knowledge with words is fraught with info-gaps (Ben-Haim 2006, Sect. 13.2).

5.3.2 Simple Qualitative Example: Nuclear Weapon Safety

Nuclear weapons play a role in the national security strategy of several countries today. Like all munitions, the designer must assure effectiveness (devastating explosion during wartime use) together with safety (no explosion during storage, transport, or abnormal accident conditions). The "always/never" dilemma is "the need for a nuclear weapon to be safe and the need for it to be reliable …. A safety mechanism that made a bomb less likely to explode during an accident could also, during wartime, render it more likely to be a dud. … Ideally, a nuclear weapon would always detonate when it was supposed to—and never detonate when it wasn't supposed to" (Schlosser 2013, pp. 173–174). There are many quantitative methods for assessing effectiveness, safety, and the balance between them, but there remains a great need for human judgment based on experience. We briefly illustrate the relevance of the five qualitative proxies for assessing and achieving robustness to uncertainty.

Nuclear weapon safety is assured, in part, by the requirement for numerous independent actions to arm and detonate the weapon. Safety pins must be removed, secret codes must be entered, multiple activation keys controlled by different individuals must be inserted and turned, etc. This *redundancy* of safety features is a powerful concept for assuring the safety of weapon systems. On the other hand, the wartime detonation of the weapon is prevented if any of these numerous redundant safety features gets stuck and fails to activate the device. Redundancy for safety is a primary source of the "always/never" dilemma.

Resilience of the weapon system is the ability to recover critical functions—detonation during wartime in the present example—when failure occurs. For example, resilience could entail the ability to override safety features that fail in the locked state in certain well-defined circumstances. This override capability may be based on a voting system of redundant safety features, or on human intervention, or on other functions. The robustness to uncertainty is augmented by redundant safety features together with a resilient ability to countervail those safety features in well-defined situations where safety features have failed in the locked mode.

Sometimes, the critical function of a system is not a physical act, like detonation, but rather the act of deciding. A command and control hierarchy, like those controlling nuclear weapon use, needs to respond effectively to adverse surprise. The decision to initiate the use of nuclear weapons in democratic countries is usually vested exclusively in the highest civilian executive authority. A concern here is that a surprise "decapitation" strike against that civilian authority could leave the country without a nuclear-response capability. The decisionmaking hierarchy needs *flexibility* against such a surprise: the ability to exercise the critical function of deciding to use (or not to use) nuclear weapons after a decapitating first strike by an adversary. Flexibility could be attained by a clearly defined line of succession after incapacitation of the chief executive, together with both physical separation between the successors and reliable communication among them. It is no simple matter to achieve this finely balanced combination of succession, separation, and communication. The concept of flexibility assists in assessing alternative implementations in terms of the resulting robustness against uncertainty in hierarchical decisionmaking.

Hierarchical decisionmaking needs to be *adaptive* in response to changing circumstances in the mid- to long-term. For example, the US line of Presidential succession is specified in the US Constitution, and this specification has been altered by amendment and clarified by legislation repeatedly over time to reflect new capabilities and challenges.

The *comprehensiveness* of a decision is the interdisciplinary scope of the mechanisms and interactions that it accounts for and the implications it identifies. The uncertainties regarding nuclear weapons are huge, because many mechanisms, interactions, and implications are unknown or poorly understood. This means that the potential for adverse surprise is quite large. Comprehensiveness of the decision analysis is essential in establishing robustness against uncertainty. Thinking "outside the box" is a quintessential component in achieving comprehensiveness, and human qualitative judgment is of foremost importance here.

We now use these ideas to schematically prioritize two alternative hypothetical strategies for supervising nuclear weapons in a liberal democracy such as the USA, based on the proxies for robustness.

The first strategy is based on current state-of-the-art (SotA) technologies, and authority is vested in the President as commander in chief. Diverse mechanisms assure redundancy of safety features as well as resilience and flexibility of the systems to assure only operational detonation, and adaptability in response to changes over longer times. The system of controls is comprehensive, but primarily based on human observation, communication, and decision.

The second strategy is new and innovative (NaI) and extensively exploits automated sensor- and computer-based access, authentication, communication, control, and decision. The strategy employs "big data" and artificial intelligence in assessing threats and evaluating risks. Humans are still in the loop, but their involvement is supported to a far greater extent by new and innovative technologies.

Our best understanding of these strategies—SotA and NaI—predicts that the second strategy would provide better safety and operability. However, deep uncertainties surround both strategies, and more so for the innovative second strategy because of its newness. The *innovation dilemma* is that the putatively preferable innovative alternative is more uncertain, and hence potentially worse, than the standard alternative. Two properties of robustness assist in resolving this dilemma: zeroing and trade-off.

The *zeroing* property of an IG robustness assessment states that the predicted performances have no robustness to uncertainty. This is because even small errors or lacunae in the knowledge or understanding (upon which predictions are based) could result in outcomes that are worse than predicted. Hence, prioritizing the strategies based on the predictions is unreliable and irresponsible. We must ask what degrees of safety and operability are essential for acceptable performance. That is, we satisfice the performance, rather than trying to optimize it.

We then note that more demanding performance requirements can fail in more ways and thus are more vulnerable to uncertainty. This implies a *trade-off* between performance and robustness to uncertainty: Greater robustness is obtained only by accepting more modest performance. In high-consequence systems, such as nuclear weapons, the performance requirements are very demanding. Nonetheless, the trade-off is irrevocable and it is wishful thinking to ignore it. The robustness of each strategy is assessed by its strength in the conceptual proxies. The *robust-satisficing* preference is for the strategy that satisfies the performance requirements at greater robustness.

Suppose that the predicted performance of the SotA strategy only barely satisfies the performance requirements. The proxies for robustness of the SotA will have low strength, because small errors can jeopardize the adequacy of the performance. This may in fact have motivated the search for the NaI strategy whose predicted performance exceeds the requirements. In this case, the robust preference will be for NaI, although consideration must be given to the strength of its proxies for robustness. If the proxies for robustness of NaI are also weak, then neither alternative may be acceptable.

Alternatively, suppose that the SotA satisfies the performance requirements by a wide margin. Its proxies for robustness will be strong and probably stronger than for NaI. In this case, the robust preference is for SotA, illustrating the potential for a *reversal of preference* between the strategies: NaI is putatively preferred (based on predicted outcomes), but SotA is more robust, and hence SotA is preferred (based on robustness) over NaI. This emphasizes the difference between robustly satisficing the performance requirements (which leads to either SotA or NaI, depending on the requirements) as distinct from prioritizing based on predicted outcomes (which leads to the putatively better alternative, NaI).

5.4 IG Robustness and Opportuneness: A Quantitative Example

We have discussed qualitative concepts for assessing deep uncertainty and for prioritizing the options facing a decisionmaker: trade-off between robustness and performance requirements, zero robustness of predicted outcomes, innovation dilemmas, and preference reversals. These concepts are embodied in mathematical theorems of IG theory, and they have quantitative realizations, as we illustrate in this section with a simple mechanical example. The one-dimensional linear model of a gap-closing electrostatic actuator with uncertainty in a single parameter stands in for other systems, often significantly more complex and uncertain in their models and predictions.

The nonlinear force–displacement relation for the gap-closing electrostatic actuator (a type of electric switch) in Fig. 5.1 is fairly well represented by:

$$F = kx - \frac{\varepsilon A V^2}{2(g-x)^2} \qquad (5.1)$$

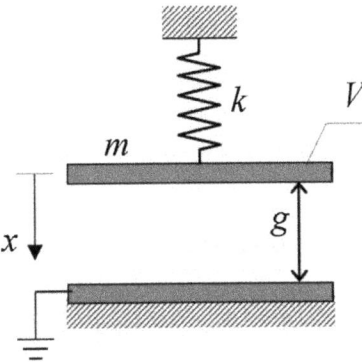

Fig. 5.1 Gap-closing electrostatic actuator. The figure is reproduced here with the permission of Prof. David Elata, head, Mechanical Engineering Micro Systems (MEMS) lab, Technion—Israel Institute of Technology

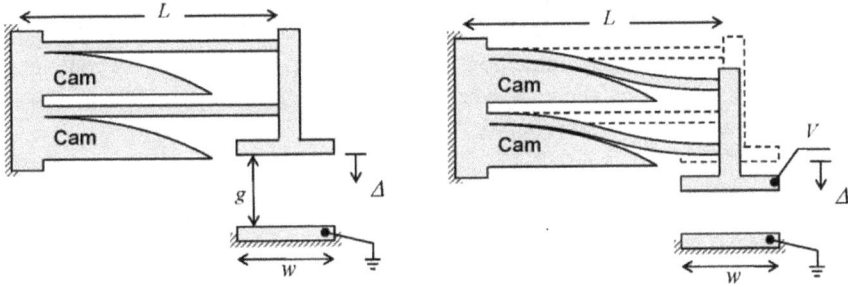

Fig. 5.2 Mechanically linearized gap-closing electrostatic actuator. The figure is reproduced here with the permission of Prof. David Elata

where F is the applied force, x is the displacement, ε is the dielectric constant, A is the area of the plates, V is the electric potential on the device, k is the spring stiffness, and g is the initial gap size.

Clever mechanical design can circumvent the complex nonlinearity of Eq. (5.1). Figure 5.2 shows a mechanically linearized modification of the device for which the force–displacement relation is, putatively, linear:

$$F = Kx \qquad (5.2)$$

where K is a constant stiffness coefficient whose value is uncertain because it depends on the precise shapes of the cams that may vary in manufacture. We will explore the robustness to uncertainty in the stiffness coefficient of the linearized device. We will also explore robustness to uncertainty in a probabilistic model. Finally, we will consider opportuneness. We assume F and x to be positive.

In our first approach to this problem, we suppose that our knowledge of the stiffness coefficient, K, is quite limited. We know an estimated value, $\widetilde{K}$, and we have an estimate of the error, s, but the most we can confidently assert is that the true stiffness, K, deviates from the estimate by $\pm s$ or more, although K must be positive. We do not know a worst-case or maximum error, and we have no probabilistic information about K.

There are many types of info-gap models of uncertainty (Ben-Haim 2006). A fractional-error info-gap model is suitable to this state of knowledge:

$$U(h) = \left\{ K \colon K > 0, \; \left| \frac{K - \widetilde{K}}{s} \right| \leq h \right\}, \quad h \geq 0 \qquad (5.3)$$

The info-gap model of uncertainty in Eq. (5.3) is an unbounded family of sets of possible values of the uncertain entity, which is the stiffness coefficient K in the present case. For any non-negative value of h, the set $U(h)$ is an interval of K values. Like all info-gap models, this one has two properties: *nesting* and *contraction*. "Nesting" means that the set $U(h)$ becomes more inclusive (containing more and more elements) as h increases. "Contraction" means that $U(h)$ is a singleton set containing only the known putative value $\widetilde{K}$ when $h = 0$. These properties endow h with its meaning as a "horizon of uncertainty."

5.4.1 IG Robustness

IG robustness is based on three components: a system model, an info-gap uncertainty model, and one or more performance requirements. In this present case, Eq. (5.2) is the system model and Eq. (5.3) is the uncertainty model. Our performance requirement is that the displacement, x, be no less than the critical value x_c.

The IG robustness is the greatest horizon of uncertainty h up to which the system model obeys the performance requirement:

$$\hat{h}(x_c) = \max \left\{ h \colon \left(\min_{K \in U(h)} x \right) \geq x_c \right\} \qquad (5.4)$$

Reading this equation from left to right, we see that the robustness $\hat{h}$ is the maximum horizon of uncertainty h up to which all realizations of the uncertain stiffness K in the uncertainty set $U(h)$ result in displacement x no less than the critical value x_c.

Robustness is a useful decision support tool because more robustness against uncertainty is better than less. Given two options that are approximately equivalent in other respects but one is more robust than the other, the robust-satisficing decisionmaker will prefer the more robust option. In short, "bigger is better" when prioritizing decision options in terms of robustness.

Fig. 5.3 Robustness curve of Eq. (5.5). $F/s = 3$, $\widetilde{K}/s = 1$

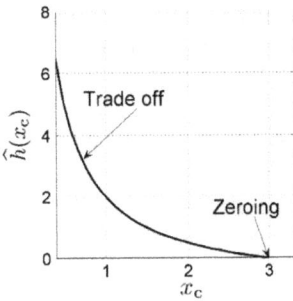

Derivation of the robustness function is particularly simple in this case. From the system model, we know that $x = F/K$. Let $m(h)$ denote the inner minimum in Eq. (5.4), and note that this minimum occurs, at horizon of uncertainty h, when $K = \widetilde{K} + sh$. The robustness is the greatest value of h up to which $m(h)$ is no less than x_c:

$$m(h) = \frac{F}{\widetilde{K} + sh} \geq x_c \Rightarrow \hat{h}(x_c) = \frac{1}{s}\left(\frac{F}{x_c} - \widetilde{K}\right) \tag{5.5}$$

or zero if this is negative which occurs if the performance requirement x_c is too large to be obtained with the putative system.

5.4.2 Discussion of the Robustness Results

The robustness function in Eq. (5.5) demonstrates two fundamental properties that hold for all IG robustness functions: *trade-off* and *zeroing*, illustrated in Fig. 5.3.

The performance requirement is that the displacement x be no less than the critical value x_c. This requirement becomes more demanding as x_c increases. We see from Eq. (5.5) and Fig. 5.3 that the robustness decreases as the requirement becomes more demanding. That is, robustness trades off against performance: The robustness can be increased only by relaxing the performance requirement. The negative slope in Fig. 5.3 represents the *trade-off between robustness and performance*: Strict performance requirements, demanding very good outcome, are less robust against uncertainty than lax requirements. This trade-off quantifies the intuition of any healthy pessimist: More demanding requirements are more vulnerable to surprise and uncertainty than lower requirements.

The second property illustrated in Fig. 5.3 is *zeroing*. Our best estimate of the stiffness is $\widetilde{K}$, so the predicted displacement is $x = F/\widetilde{K}$. Equation (5.5) shows that the robustness becomes zero precisely at the value of the critical displacement x_c that is predicted by the putative model: $x_c = F/\widetilde{K}$, which equals 3 for the parameter values in Fig. 5.3. Stated differently, the zeroing property asserts that best-model predictions have no robustness against error in the model. Like trade-off, this is true for all info-gap robustness functions.

Models reflect our best understanding of the system and its environment. Nonetheless, the zeroing property means that model predictions are not a good basis for design or planning decisions, because those predictions have no robustness against errors in the models. Recall that we are discussing situations with large info-gaps as represented in Eq. (5.3): The putative value of the stiffness $\widetilde{K}$ is known, but the size of its deviation from the true stiffness K is unknown. If your models are correct (no info-gaps), then you do not need robustness against uncertainty. However, robustness is important when facing deep uncertainty.

The zeroing property asserts that the predicted outcome is not a reliable characterization of the system. The trade-off property quantifies how much the performance requirement must be reduced in order to gain robustness against uncertainty. The slope of the robustness curve reflects the cost of robustness: What decrement in performance "buys" a specified increment in robustness. Outcome quality can be exchanged for robustness, and the slope quantifies the cost of this exchange. In Fig. 5.3, we see that the cost of robustness is very large at large values of x_c and decreases as the performance requirement is relaxed.

The robustness function is a useful response to the pernicious side of uncertainty. In contrast, the opportuneness function is useful in exploiting the potential for propitious surprise. We now discuss the info-gap opportuneness function.

5.4.3 IG Opportuneness

IG opportuneness is based on three components: a system model, an info-gap model of uncertainty, and a performance aspiration. The performance aspiration expresses a desire for a better-than-anticipated outcome resulting from propitious surprise. This differs from the performance requirement for robustness, which expresses an essential or critical outcome without which the result would be considered a failure.

We illustrate the opportuneness function with the same example, for positive F and x. The robust-satisficing decisionmaker requires that the displacement be no less than x_c. The opportune windfalling decisionmaker recognizes that a larger displacement would be better, especially if it exceeds the anticipated displacement, $F/\widetilde{K}$. For the opportune windfaller, the displacement would be wonderful if it is as large as x_w, which exceeds the anticipated displacement. The windfaller's aspiration is not a performance requirement, but it would be great if it occurred.

Achieving a windfall requires a favorable surprise, so the windfaller asks: What is the lowest horizon of uncertainty at which windfall is possible (though not necessarily guaranteed)? The answer is the opportuneness function, defined as:

$$\hat{\beta}(x_w) = \min\left\{h : \left(\max_{K \in U(h)} x\right) \geq x_w\right\} \quad (5.6)$$

5 Info-Gap Decision Theory (IG)

Reading this equation from left to right, we see that the opportuneness $\hat{\beta}$ is the minimum horizon of uncertainty h up to which at least one realization of the uncertain stiffness K in the uncertainty set $U(h)$ results in displacement x at least as large as the wonderful windfall value x_w. The opportuneness function $\hat{\beta}(x_w)$ is the complement of the robustness function $\hat{h}(x_c)$ in Eq. (5.4). The min and max operators in these two equations are reversed. This is the mathematical manifestation of the inverted meaning of these two functions. Robustness is the *greatest* uncertainty that *guarantees* the *required* outcome, while opportuneness is the *lowest* uncertainty that *enables* the *aspired* outcome.

Opportuneness is useful for decision support, because a more opportune option is better able to exploit propitious uncertainty than a less opportune option. An option whose $\hat{\beta}$ value is small is opportune, because windfall can occur even at low horizon of uncertainty. The opportune windfaller prioritizes options according to the smallness of their opportuneness function values: An option with small $\hat{\beta}$ is preferred over an option with large $\hat{\beta}$. That is, "smaller is better" for opportuneness, unlike robustness for which "bigger is better." We again note the logical inversion between robustness and opportuneness.

Whether a decisionmaker prioritizes the options using robustness or opportuneness is a methodological decision that may depend on the degree of risk aversion of the decisionmaker. Furthermore, these methodologies may or may not prioritize the options in the same order.

The opportuneness function is derived in a manner analogous to the derivation of Eq. (5.5), yielding:

$$\hat{\beta}(x_w) = \frac{1}{s}\left(\tilde{K} - \frac{F}{x_w}\right) \qquad (5.7)$$

or zero if this is negative, which occurs when x_w is so small, modest, and unambitious that it is possible even with the putative design and does not depend on the potential for propitious surprise.

5.4.4 Discussion of Opportuneness Results

The robustness and opportuneness functions, Eqs. (5.5) and (5.7), are plotted in Fig. 5.4. The opportuneness function displays zeroing and trade-off properties, whose meanings are the reverse of those for robustness. The opportuneness function equals zero at the putative outcome $x = F/\tilde{K}$ like the robustness function. However, for the opportuneness function this means that favorable windfall surprise is not needed for enabling the predicted outcome. The positive slope of the opportuneness function means that greater windfall (larger x_w) is possible only at larger horizon of uncertainty.

The robustness and opportuneness functions may respond differently to proposed changes in the design solution, as we now illustrate. From Eq. (5.5), we note that

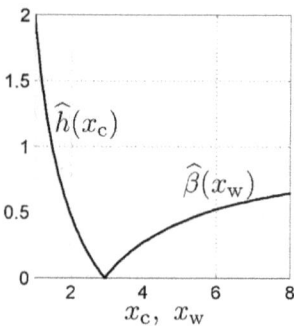

Fig. 5.4 Robustness and opportuneness curves of Eqs. (5.5) and (5.7)

$\widehat{h}$ decreases as the putative stiffness $\widetilde{K}$ increases. From Eq. (5.7), we see that $\widehat{\beta}$ increases as $\widetilde{K}$ increases:

$$\frac{\partial \widehat{h}}{\partial \widetilde{K}} < 0, \quad \frac{\partial \widehat{\beta}}{\partial \widetilde{K}} > 0 \qquad (5.8)$$

Recall that "bigger is better" for robustness while "smaller is better" for opportuneness. We see that any increase in $\widetilde{K}$ will make both robustness and opportuneness worse, and any decrease in $\widetilde{K}$ will improve them both. In summary, robustness and opportuneness are *sympathetic* with respect to change in stiffness.

Now consider the estimated error, s in the info-gap model of Eq. (5.3). A smaller value of s implies greater confidence in the estimate $\widetilde{K}$ while a larger s implies a greater propensity for error in the estimate. From Eq. (5.5), we see that robustness improves ($\widehat{h}$ increases) as s decreases: Better estimate of $\widetilde{K}$ implies greater robustness against uncertainty in s. In contrast, from Eq. (5.7) we see that opportuneness gets worse ($\widehat{\beta}$ increases) as s decreases: a lower opportunity for windfall as the uncertainty of the estimate declines. In short:

$$\frac{\partial \widehat{h}}{\partial s} < 0, \quad \frac{\partial \widehat{\beta}}{\partial s} < 0 \qquad (5.9)$$

A change in the estimated error acts differently on robustness and opportuneness: By reducing the error of the estimated stiffness, one increases the robustness but diminishes the opportuneness; increasing the error acts in the reverse. In short, robustness and opportuneness are *antagonistic* with respect to the error in the estimated stiffness.

5.4.5 An Innovation Dilemma

An innovation dilemma occurs when the decisionmaker must choose between two options, where one is putatively better but more uncertain than the other.

Fig. 5.5 Robustness curves for innovative and state-of-the-art options

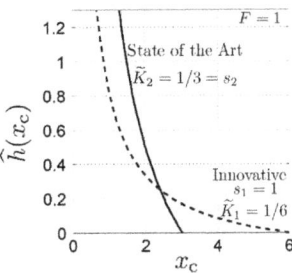

Technological innovations provide the paradigm for this dilemma. An innovation is supposedly better than the current state of the art, but the innovation is new so there is less experience with it and in practice it may turn out worse than the current state of the art. We will illustrate an innovation dilemma with the previous example, demonstrating its resolution using the robustness functions of the two options.

Consider two alternative designs (Option 1 and Option 2) of the linear elastic system (Eq. 5.2), one of which has lower estimated stiffness than the other:

$$\widetilde{K}_1 < \widetilde{K}_2 \tag{5.10}$$

Both designs will operate under the same positive force F, so the predicted displacement, $x = F/\widetilde{K}$, is greater with Option 1. Thus, Option 1 is preferred based on the estimated stiffnesses and the requirement for large displacement.

However, the putatively better Option 1 is based on innovations for which the actual stiffness in operation is more uncertain than for Option 2, which is the state of the art. Referring to the uncertainty estimate, s, in the info-gap model of Eq. (5.3), we express this as:

$$s_1 > s_2 \tag{5.11}$$

The dilemma is that Option 1 is putatively better (Eq. 5.10) but more uncertain (Eq. 5.11). This dilemma is manifested in the robustness functions for the two options, which also leads to a resolution, as we now explain. To illustrate the analysis, we evaluate the robustness function for each option (Eq. 5.5) with the following parameter values: $F = 1$, $\widetilde{K}_1 = 1/6$, $s_1 = 1$, $\widetilde{K}_2 = 1/3 = s_2$. The robustness curves are shown in Fig. 5.5.

The innovative Option 1 in Fig. 5.5 (dashed curve) is putatively better than the state-of-the-art Option 2 (solid curve), because the predicted displacement of Option 1 is $F/\widetilde{K}_1 = 6$, while the predicted displacement for Option 2 is only 3. However, the greater uncertainty of Option 1 causes a stronger trade-off between robustness and performance than for Option 2. The cost of robustness is greater for Option 1 for x_c values exceeding about 2, causing the robustness curves to cross one another at about $x_c = 2.4$.

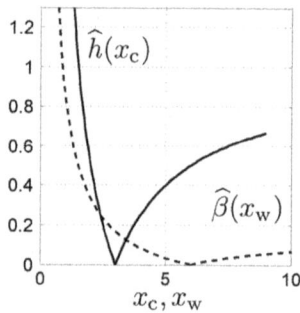

Fig. 5.6 Robustness and opportuneness curves for innovative (dashed) and state-of-the-art (solid) options

The innovation dilemma is manifested graphically by the intersection of the robustness curves in Fig. 5.5, and this intersection is the basis for the resolution. Option 1 is more robust than Option 2 for highly demanding requirements ($x_c > 2.4$), and hence, Option 1 is preferred for this range of performance requirements. Likewise, Option 2 is more robust for more modest requirements ($x_c < 2.4$), and hence, Option 2 is preferred for this lower range of performance requirements.

The robust-satisficing designer will be indifferent between the two options for performance requirements at or close to the intersection value of $x_c = 2.4$. Considerations other than robustness can then lead to a decision. Figure 5.6 shows the robustness curves from Fig. 5.5 together with the opportuneness curves (Eq. 5.7) for the same parameter values. We note that the innovative (dashed) Option is more opportune (smaller $\widehat{\beta}$) than the state-of-the-art Option (solid) for all values of x_w exceeding the putative innovative value. Designers tend to be risk averse and to prefer robust satisficing over opportune windfalling. Nonetheless, opportuneness can "break the tie" when robustness does not differentiate between the options at the specified performance requirement.

5.4.6 Functional Uncertainty

We have discussed the IG robustness function and its properties of trade-off, zeroing, and cost of robustness. We have illustrated how these concepts support the decision process, especially when facing an innovation dilemma. We have described the IG opportuneness function and its complementarity to the robustness function. These ideas have all been illustrated in the context of a one-dimensional linear system with uncertainty in a single parameter. In most applications with deep uncertainty, the info-gaps include multiple parameters as well as uncertainty in the shapes of functional relationships. We now extend the previous example to illustrate the modeling and management of functional or structural uncertainty in addition to the parametric uncertainty explored so far. This will also illustrate how uncertain probabilistic models can be incorporated into an IG robust-satisficing analysis.

5 Info-Gap Decision Theory (IG)

Let the stiffness coefficient K in Eq. (5.2) be a random variable whose estimated probability density function (pdf) $\tilde{p}(K)$ is normal with mean μ and variance σ^2. We are confident that this estimate is accurate for K within an interval around μ of known size $\pm \delta_s$. However, outside of this interval of K values, the fractional error of the pdf is unknown. In other words, we are highly uncertain about the shape of the pdf outside of the specified interval. This uncertainty derives from lack of data with extreme K values and absence of fundamental understanding that would dictate the shape of the pdf. We face "functional uncertainty" that can be represented by info-gap models of many sorts, depending on the type of information that is available. Given the knowledge available in this case, we use the following fractional-error info-gap model:

$$U(h) = \left\{ p(K) : \int_{-\infty}^{\infty} p(K) dK = 1, \ p(K) \geq 0 \text{ for all } K, \right.$$
$$p(K) = \tilde{p}(K) \text{ for } |K - \mu| \leq \delta_s,$$
$$\left. \left| \frac{p(K) - \tilde{p}(K)}{\tilde{p}(K)} \right| \leq h \text{ for } |K - \mu| > \delta_s \right\}, \quad h \geq 0 \quad (5.12)$$

The first row of this info-gap model states that the set $U(h)$ contains functions $p(K)$ that are normalized and non-negative (namely, mathematically legitimate pdfs). The second line states that these functions equal the estimated pdf in the specified interval around the mean, μ. The third line states that, outside of this interval, the functions in $U(h)$ deviate fractionally from the estimated pdf by no more than h. In order to avoid some technical complications, we assume that the pdfs in $U(h)$ are non-atomic: containing no delta functions. In short, this info-gap model is the unbounded family of nested sets, $U(h)$, of pdfs that are known within the interval $\mu \pm \delta_s$ but whose shapes are highly uncertain beyond it. This is one example of an info-gap model for uncertainty in the shape of a function.

The system fails if $x < x_c$ where $x = F/K$ and F is a known positive constant. x is now a random variable (because K is random) so the performance requirement is that the probability of failure not exceed a critical value P_c. We will explore the robustness function. We consider the special case that $F/x_c > \mu \pm \delta_s$, meaning that the failure threshold for K lies outside the interval in which the pdf of K is known. The probability of failure is:

$$P_f(p) = \text{Prob}(x < x_c) = \text{Prob}(K > F/x_c) = \int_{F/x_c}^{\infty} p(K) dK \quad (5.13)$$

For the estimated pdf, $\tilde{p}(K)$, one finds the following expression for the estimated probability of failure:

Fig. 5.7 Robustness curve for Eq. (5.16)

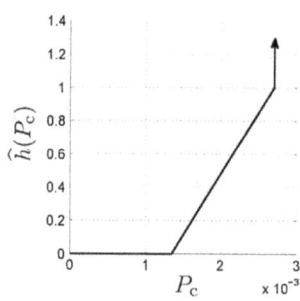

$$P_f(\tilde{p}) = 1 - \Phi\left(\frac{(F/x_c) - \mu}{\sigma}\right) \qquad (5.14)$$

where $\Phi(\cdot)$ is the cumulative pdf of the standard normal variate.

The robustness function, $\hat{h}(P_c)$, is the greatest horizon of uncertainty h up to which all pdf's $p(K)$ in the uncertainty set $U(h)$ do not have failure probability $P_f(p)$ in excess of the critical value P_c:

$$\hat{h}(P_c) = \max\left\{h : \left(\max_{p \in U(h)} P_f(p)\right) \leq P_c\right\} \qquad (5.15)$$

After some algebra, one finds the following expression for the robustness:

$$\hat{h}(P_c) = \begin{cases} 0 & \text{if } 0 \leq P_c < P_f(\tilde{p}) \\ \frac{P_c}{P_f(\tilde{p})} - 1 & \text{if } P_f(\tilde{p}) \leq P_c \leq 2P_f(\tilde{p}) \\ \infty & \text{otherwise} \end{cases} \qquad (5.16)$$

The robustness function in Eq. (5.16) is illustrated in Fig. 5.7 for $[(F/x_c) - \mu]/\sigma = 3$, meaning that the failure threshold is 3 standard deviations above the mean. Hence, the estimated probability of failure is $P_f(\tilde{p}) = 1.35 \times 10^{-3}$. The trade-off property is evident in this figure: Lower (better) required probability of failure P_c entails lower (worse) robustness $\hat{h}(P_c)$. Note the discontinuous jump of robustness to infinity at $P_c = 2P_f(\tilde{p})$. This is because the actual probability of failure $P_f(p)$ cannot exceed more than twice the estimated value $P_f(\tilde{p})$ resulting from constraints on the pdfs in the info-gap model of Eq. (5.12). The zeroing property is expressed by robustness becoming zero when the performance requirement P_c equals the estimated value $P_f(\tilde{p})$.

5.5 Conclusion and Future Challenges

We live in an innovative world. Our scientific optimism embodies the belief that knowledge and understanding will continue to grow, perhaps at an ever-increasing rate. An inescapable implication of scientific optimism is that we are currently quite ignorant, and that we will be repeatedly and profoundly surprised in the near and not-so-near future. Because of this uncertainty, the planner, designer, or decisionmaker faces a profound and unavoidable info-gap: the disparity between what is currently known (or thought to be true) and what needs to be known for making a responsible decision (but is still hidden in the future). IG theory provides two complementary methodologies for managing this uncertainty. Robust satisficing helps in protecting against pernicious surprise and in achieving critical outcomes. Opportune windfalling helps in exploiting favorable surprise and in facilitating windfall outcomes.

We have illustrated the info-gap analysis in two examples, one quantitative and using mathematics, and one qualitative and using only verbal analysis. We discussed the trade-off between robustness and outcome requirements, showing that enhanced robustness against uncertainty is obtained only by relaxing the outcome requirements. In quantitative analysis, this allows an explicit assessment of the cost (in terms of reduced robustness) of making the requirements more demanding.

We also showed that predicted outcomes—based on the best available models and understanding—have no robustness against uncertainty in that knowledge. This has a profound implication for the decisionmaker and for the conventional conception of optimization. One's alternatives cannot be responsibly prioritized with their predicted outcomes, because those predictions have no robustness against uncertainty and surprise. Attempting to optimize the outcome, based on zero-robustness predictions, is not recommended when facing deep uncertainty. Instead, one should prioritize the alternatives according to their robustness for achieving critical outcomes, supplemented perhaps by analysis of opportuneness for windfall. This is particularly pertinent when facing an innovation dilemma: the choice between a new and putatively better alternative that is more uncertain due to its newness, and a more standard state-of-the-art alternative. The info-gap analysis of robustness enables the decisionmaker to assess the implications of uncertainty and to prioritize the alternatives to robustly achieve critical goals. We showed that this can lead to a reversal of the preference from the putatively optimal choice.

Many challenges remain. The quantitative analysis of robustness and opportuneness of large and complicated systems often faces algorithmic or numerical difficulties resulting from high dimensionality of the computations. Another challenge arises in response to new types of information and new forms of uncertainty. Many different mathematical forms of info-gap models of uncertainty exist (here we have examined only a few). However, analysts sometimes need to construct new types of info-gap models of uncertainty.

Another challenge is in bridging the gap between mathematics and meaning: between quantitative and qualitative analysis. Mathematics is a powerful tool that has facilitated the exploration of everything under the sun. However, the incorporation of mathematics is difficult when knowledge is predominantly verbal, and when the meanings of subtle concepts are crucial to the decisionmaking process. A mathematical equation expresses a structural relationship between abstract entities. Meaning can be attributed to an equation, but meaning is not inherent in the equation. Witness the fact that the same equation can describe diverse and unrelated phenomena. Bridging the divide between mathematics and meaning is challenging in both directions. Quantitative analysts are often challenged to appreciate the limitations of their tools, while qualitative analysts often find it difficult to appreciate the contribution that mathematics can make.

The analysis and management of deep uncertainty faces many challenges, but our scientific optimism, tempered by recognition of our persistent ignorance, will carry us through as we acquire new understanding and face new surprises. As John Wheeler wrote (1992): "We live on an island of knowledge surrounded by a sea of ignorance. As our island of knowledge grows, so does the shore of our ignorance."

References

Ben-Haim, Y. (2006). *Info-gap decision theory: Decisions under severe uncertainty* (2nd ed.). London: Academic Press.

Ben-Haim, Y. (2010). *Info-gap economics: An operational introduction*. London: Palgrave-Macmillan.

Ben-Haim, Y. (2012a). Doing our best: Optimization and the management of risk. *Risk Analysis, 32*(8), 1326–1332.

Ben-Haim, Y. (2012b). Why risk analysis is difficult, and some thoughts on how to proceed. *Risk Analysis, 32*(10), 1638–1646.

Ben-Haim, Y. (2018). *Dilemmas of wonderland: Decisions in the age of innovation*. Oxford: Oxford University Press.

Ben-Haim, Y., Dacso, C. C., Carrasco, J., & Rajan, N. (2009). Heterogeneous uncertainties in cholesterol management. *International Journal of Approximate Reasoning, 50,* 1046–1065.

Ben-Haim, Y., & Demertzis, M. (2016). Decision making in times of Knightian uncertainty: An info-gap perspective, *Economics*, The Open-Access, Open Assessment e-journal, Special Issue Radical Uncertainty and Its Implications for Economics. No. 2016-23: 1–29. http://www.economics-ejournal.org/economics/journalarticles/2016-23.

Ben-Haim, Y., Zetola, N. M., & Dacso, C. (2012). Info-gap management of public health policy for TB with HIV-prevalence. *BMC Public Health, 12,* 1091. https://doi.org/10.1186/1471-2458-12-1091.

Burgman, M. (2005). *Risks and decisions for conservation and environmental management*. Cambridge: Cambridge University Press.

Chinnappen-Rimer, S., & Hancke, G. P. (2011). Actor coordination using info-gap decision theory in wireless sensor and actor networks. *International Journal of Sensor Networks, 10*(4), 177–191.

Hall, J. W., Lempert, R. J., Keller, K., Hackbarth, A., Mijere, C., & McInerney, D. J. (2012). Robust climate policies under uncertainty: A comparison of robust decision making and info-gap methods. *Risk Analysis, 32*(10), 1657–1672.

Hansen, L. P., & Sargent, T. J. (2008). *Robustness*. Princeton: Princeton University Press.

Harp, D. R., & Vesselinov, V. V. (2013). Contaminant remediation decision analysis using information gap theory. *Stochastic Environmental Research and Risk Assessment, 27*(1), 159–168.

Kanno, Y., & Takewaki, I. (2006). Robustness analysis of trusses with separable load and structural uncertainties. *International Journal of Solids and Structures, 43*(9), 2646–2669.

Knight, F. H. (1921). *Risk, uncertainty and profit*. Houghton Mifflin Co. (Re-issued by University of Chicago Press, 1971).

Knoke, T. (2008). Mixed forests and finance—Methodological approaches. *Ecological Economics, 65*(3), 590–601.

Moffitt, L. J., Stranlund, J. K., & Field, B. C. (2005). Inspections to avert terrorism: Robustness under severe uncertainty. *Journal of Homeland Security and Emergency Management, 2*(3). http://www.bepress.com/jhsem/vol2/iss3/3.

Schlosser, E. (2013). *Command and control: Nuclear weapons, the Damascus accident, and the illusion of safety*. New York: Penguin Books.

Schwartz, B. (2004). *Paradox of choice: Why more is less*. New York: Harper Perennial.

Schwartz, B., Ben-Haim, Y., & Dacso, C. (2011). What makes a good decision? Robust satisficing as a normative standard of rational behaviour. *The Journal for the Theory of Social Behaviour, 41*(2), 209–227.

Simon, H. (1956). Rational choice and the structure of the environment. *Psych. Rev., 63*(2), 129–138.

Sims, C. A. (2001). Pitfalls of a minimax approach to model uncertainty. *American Economic Review, 91*(2), 51–54.

Smithson, M., & Ben-Haim, Y. (2015). Reasoned decision making without math? Adaptability and robustness in response to surprise. *Risk Analysis, 35*(10), 1911–1918.

Wald, A. (1947). *Sequential analysis*. J. Wiley & Sons (re-issued by Dover Publications, 1973).

Wheeler, J. A. (1992). Quoted in *Scientific American*, December, 1992, p. 20.

Prof. Yakov Ben-Haim (Technion—Israel Institute of Technology) is a professor of mechanical engineering and holds the Yitzhak Moda'i Chair in Technology and Economics. He initiated and developed Info-Gap (IG) Decision Theory for modeling and managing deep uncertainty. IG theory has impacted the fundamental understanding of uncertainty in human affairs and is applied by scholars and practitioners around the world in engineering, biological conservation, economics, project management, climate change, natural hazard response, national security, medicine, and other areas (see info-gap.com). He has been a visiting scholar in many countries and has lectured at universities, technological and medical research institutions, public utilities, and central banks. He has published more than 100 articles and 6 books.

Open Access This chapter is licensed under the terms of the Creative Commons Attribution 4.0 International License (http://creativecommons.org/licenses/by/4.0/), which permits use, sharing, adaptation, distribution and reproduction in any medium or format, as long as you give appropriate credit to the original author(s) and the source, provide a link to the Creative Commons licence and indicate if changes were made.

The images or other third party material in this chapter are included in the chapter's Creative Commons licence, unless indicated otherwise in a credit line to the material. If material is not included in the chapter's Creative Commons licence and your intended use is not permitted by statutory regulation or exceeds the permitted use, you will need to obtain permission directly from the copyright holder.

Chapter 6
Engineering Options Analysis (EOA)

Richard de Neufville and Kim Smet

Abstract This chapter presents and explains Engineering Options Analysis (EOA) in two ways.

- First, we present it for what it is: an approach to quantitative analysis for planning, design, and management of engineering systems over time, in the context of uncertainty. We offer a brief introduction to the approach, and then go through each of the broad methodological steps.
- Second, we underline important differences between EOA and Real Options Analysis (ROA), and contrast EOA briefly with other methods presented in this book. While the chapter synthesizing the DMDU approaches covers this topic in a general way (Chap. 15), it is useful for clarity to cover it briefly here. Since EOA and ROA sound so similar, readers might easily assume they are really very close if not identical—but they are not! In a nutshell, ROA assumes that we can estimate future uncertainties sufficiently accurately, is generally limited to analyzing a single option at a time, and aims to develop single monetary values for options. EOA deals with deeper uncertainties, handles multiple options simultaneously, and allows for all sorts of measures of benefits and values.

6.1 Introduction

Engineering Options Analysis (EOA) refers to the process of assessing the value of including flexibility in the design and management of technical systems. It consists of procedures for calculating the value of options (e.g., the benefits due to flexibility in the timing, size, and location of changes in the engineering system) in terms of the distribution of additional benefits due to the options. EOA presents these benefits to decisionmakers in terms of average expectations, extreme possibilities, and initial

---

R. de Neufville (✉)
Massachusetts Institute of Technology, Cambridge, MA, USA
e-mail: ardent@mit.edu

K. Smet
University of Ottawa, Ottawa, Canada

capital expenses (Capex). Decisionmakers can choose whether (or to what extent) to monetize these values.

Chess provides a useful analogy to EOA. How a game of chess will develop is deeply uncertain. So, how do chess masters deal with this situation? In general, they explore the possible combinations, and choose opening and subsequent moves to give themselves the position, as the game develops in time and space, to respond effectively as opportunities and threats arise. They do not have a fixed strategy; they develop a flexible strategy that maximizes their chances of success. Moreover, the immediate product of the analysis is the recommended first or immediate decision. So it is with EOA. EOA explores the context widely, compares the range of possibilities, and proposes strategies, and thus opening decisions, that are most likely to be successful in the long run.

EOA is genealogically related to Real Options Analysis (ROA). It is fair to say that it is an offshoot of the corpus of ROA as Luenberger and Trigeorgis codified then existing practice in their texts (Luenberger, 1996; Trigeorgis, 1996). In the intervening years, however, EOA has grown significantly apart from ROA. Many researchers and practitioners have evolved the practice of EOA through their explorations of its application to a wide range of situations, such as the deployment of satellites (de Weck et al. 2004), the development of oil fields (Lin et al. 2013), hospital design (Lee et al., 2008), the implementation of factories (Cardin et al., 2015), the design of military ships (Page, 2011), and the renewal of facilities (Esders et al., 2015). At present, the books by de Neufville and Scholtes (2011) and Geltner and de Neufville (2018) are the primary textbook presentations of the EOA approach.

The goal of this chapter is to present the core elements of EOA so as to situate it properly among the range of approaches we can use to consider and shape adaptive policies to deal with deep uncertainties. The idea is to leave you, the reader, with an overall appreciation for what it is, what kinds of problems it can address effectively, and what it can achieve in terms of useful insights for the management of engineering systems. The companion chapter in this book (Chap. 11) illustrates the application and use of EOA in detail.

EOA is based on computer simulation of the interaction of the possible uncertainties, and of the thoughtful managerial responses to events as they unfold. The logic is that this is an efficient way forward, given the computational immensity of many problems, and the lack of acceptable simplifying assumptions (such as path independence) that might enable a feasible way to compute a best solution explicitly. The idea is to examine the consequences of a sequence of scenarios that combine both possible circumstantial outcomes (how fast might sea level rise, for example) and the reasonable managerial responses (when to install new pumping stations). For each combination of uncertain events and subsequent management decision, EOA calculates the associated consequences. By examining the range of possibilities, we obtain the range of outcomes. Furthermore, by sampling the range of uncertain possibilities according to their estimated distributions (in cases of deep uncertainty we do not have to fixate on a single one), we develop distributions of the possible outcomes. In short, the analysis imposes distributions on the system, and obtains distributions of outcomes from the system—as modulated by the complexity of what happens within the system.

Finally, EOA efficiently organizes the immense data it inherently provides. In general, the analysis generates thousands of data points associated with a range of alternative strategies. EOA makes sense of this information through the way it structures the analysis itself, and the way it contrasts the results.

6.2 Methodology of Engineering Options Analysis

To explain the use and value of EOA, this section takes three perspectives. First, it sets the scene: it sketches the kinds of issues that EOA can address effectively for the benefit of both decisionmakers and the larger public of stakeholders. Second, it describes the main steps of the analysis itself, emphasizing their roles, rationale, and usefulness in the procedure. Finally, it indicates the kinds of results we can expect from EOA: clear, compact representations of the results that support insights into good strategies for dealing with deep uncertainty.

6.2.1 Setting the Scene

A most important feature of EOA is that the method caters directly to the informational needs of stakeholders. It is "fit for purpose" for the decisionmakers, managers, and others concerned with, and by, an issue. These participants need more than a ranking or overall assessment of a strategy. They need to:

- Understand the range of possible outcomes of any strategy for dealing with uncertainty. What is the worst that can happen? What is the best?
- Appreciate what may occur in terms of the likelihood of the possible amounts, the possible timing and location of these events, and distribution of beneficiaries.
- Have information, both about extreme possibilities and about benefits versus costs, about the "value for money" of alternatives.
- Obtain insights into their problem: which strategy provides the best outcomes and why?

EOA explicitly supports all of these requirements. EOA recognizes that analysts of modern engineering systems face deep uncertainties. This goes with the territory of analyzing and designing the complex, often large-scale, interconnected technological developments of our era. The design, rollout, and management of these systems require us to project into futures that we can only imagine. Consider the design of new technologies such as smartphones, or of major infrastructure projects such as the tunnel under the English Channel. The outcomes of such ventures depend on all kinds of unknowns. The case of the double-decker Airbus provides a cogent example. This aircraft is a marvel of engineering, but as a system it has failed to meet its financial objectives, a victim of unanticipated changes in airline routes and travel patterns. The fact is that the design of engineering systems is full of uncertainties, even if the detailed engineering of artifacts can be precise. We need to deal with this reality of deep uncertainties, and EOA is one way to fulfill this need.

Importantly, EOA deals explicitly with the great number of alternatives for exercising flexibility. It easily and routinely considers the many ways that managers could alter the capacity of an engineered system, both in terms of its physical size and its functionality. This ability to choose the size, the timing, and the location of a change in the system implies an exponentially combinatorial set of possibilities. The number is daunting. The simple problem of considering 4 possible sizes, over 3 periods, in 5 locations implies 64 to the power of 5 or over 10 million logical combinations. Yet EOA can address these issues in a reasonable amount of time. It achieves this by sampling the space sufficiently to obtain a useful distribution of the possible outcomes.

Most significantly in the context of options analysis, EOA has a feature of immense practical importance: it enables users to consider the value of multiple options simultaneously. This is a realistic requirement for dealing with real-world issues: at any time, managers can decide what kind of adaptation they should adopt. As discussed more fully in the section contrasting engineering options and real options below (Sect. 6.4), such analyses are not possible in routine financially based options analysis.

It is important at this stage to be clear about the definition of an "option" in the context of options analyses, be they engineering, real, or financial. In this context it has a well-established, precise meaning—more specialized than the general sense in which it is often used in everyday language. To maintain consistency with the vast literature on options analysis, our discussion needs to keep to this tight definition.

6.2.2 Definition of an Option

An option (in the context of options analysis) is:

- The right, but not the obligation
- To take some action (such as to expand a plant, because you have the space)
- At some cost (such as the purchase of machines)
- Over some time frame (may be limited or not).

This definition is much more specific than the popular meaning, which interprets the word "option" very generously to include "policies, strategies, plans, alternatives, courses of action" as done elsewhere in some of the publications on decisionmaking under deep uncertainty. Please keep this in mind as we discuss the details of EOA.

6.2.3 Main Steps of Analysis

In brief, EOA involves the following steps:

1. Formulate the (engineering system) problem or opportunity.
2. Specify the objectives and outcomes of interest.
3. Develop a computationally efficient analytic model of the system.
4. Generate a set of options.
5. Specify the relevant uncertainties, and generate a sufficient number of future scenarios (typically in the thousands, for instance).
6. Calculate system performance across the range of scenarios, using computer simulation to couple the variation in external forces and the management responses in terms of exercising their options along the way (i.e., engage in "Exploratory Modeling").
7. Reduce the resulting data for stakeholder consideration, using suitable target curves and multi-objective tables to illustrate trade-offs.
8. Support the choice of a preferred starting decision and plan for its adoption, using the insights the analysis provides.
9. Implement a monitoring system that tracks variables that may trigger future adjustment of the current decision.

These steps fit the generic description of the steps of DMDU as presented in Chap. 1:

- Frame the analysis—corresponding mostly to EOA steps 1–3;
- Perform exploratory uncertainty analyses—as covered by EOA steps 4–7; and
- Choose short-term actions and long-term contingent actions—EOA steps 8 and 9.

However, the EOA approach differs subtly, but importantly, from the view that the framing of the analysis should specify a priori possible strategies. From the perspective of EOA, it is premature to define strategies before the exploration of the range of combinations. Indeed, often the practice to identify alternative choices early lead to overly simplistic visions of the possibilities. For example, many regions have set out the "airport development issue" as whether "to build a new airport or not", yes or no. Such a frame fails to consider more complex, yet potentially more desirable solutions, such as "acquire land for potential new facilities, but defer action until the future becomes more clear" (de Neufville, 2007; de Neufville and Odoni, 2013).

6.2.4 Details of Each Step

Step 1: Formulate the (engineering system) problem or opportunity. This step names the problem at the level suitable for the principal stakeholders. For example, it might be: "How do we best expand the water supply for Singapore?" A related but different statement might be "How do we ensure that available supply in Singapore meets realized demand?"—this implies a different constellation of solutions including demand management, and different impacts.

Step 2: Specify the objectives and outcomes. This process names and quantifies these factors. For the water supply case, we might focus not only on the quantity of water (cubic meters, say) but also its rate (cubic meters per hour), its quality (potable or other), and its reliability in face of drought. The outcomes would reflect other important issues, such as capital and the operating costs, the environmental impacts, and so on.

Step 3: Develop a computationally efficient analytic model of the system. Some exploratory modeling may be necessary in this regard. It is not necessarily obvious how best to characterize and explore the system. EOA requires the development or availability of an analytic model that is both technically acceptable and computationally efficient.

- The model must of course represent the technical, environmental, and economic reality correctly. But very detailed or precise models are not required when we are dealing with deep uncertainty. When the input values are obviously imprecise, it is pointless to strive for great detail in the technical models. Moreover, such detail may impede the analysis.
- Most importantly, EOA needs computationally efficient models that can run a great many scenarios—routinely thousands—within the period available for analysis. Sometimes such models are readily available and analysts can run them quickly on the fast computers we now use. Other times, analysts will have to create surrogate, simpler models that mimic the essential features of the detailed models. Analysts will then use these surrogates as "screening models" (i.e., as test beds) to "screen out" unattractive strategies and to identify the most attractive strategies. In this case, the analysts need to check the results from the screening models by testing the resulting strategies on the full engineering models.

Step 4: Generate options. For example, for the issue of providing Singapore with adequate water, one option might be to add a desalination plant, another might be to extend a reservoir. Being options, system managers could exercise them when, where, and if needed. The analyst should choose options that are most effective in dealing with the uncertainties that have the greatest effect on the system. [A cautionary note in this regard: the parameters that are most uncertain are often not those that have the greatest effect on the system. Some parameters have significant leverage, so that their small variations can have disproportionate consequences].

Step 5: Specify the relevant uncertainties and generate scenarios. In EOA, the characterization of uncertainties should focus on identifying broad trends and likely ranges. In other words, what factors shape the uncertainty? Is it a "random walk," as is appropriate for commodity prices in a fluid market? Is it cyclical behavior, as is common for real estate markets? Is it likely to be monotonic, as one would expect for sea level rise? Is the distribution normal, or skewed in some direction? We can also estimate likely ranges based on a combination of past experience and an understanding of limiting factors, such as the competitiveness of alternatives, or physical constraints. This approach differs from that of financial options and ROA, where the object is to define the price or value of an option (using the technical definition of this term as a "right but not an obligation"). For those analyses it is correspondingly important to work with a well-calibrated model of uncertainty, usually derived from extensive empirical observations. In EOA, however, we estimate distributions and ranges of possible outcomes, and do not require precise descriptions of uncertainties.

Having characterized the uncertainties, we then use computer programs to generate scenarios automatically by combining random selections from the sample space. EOA thus commonly examines "all" scenarios, insofar as its thousands of scenarios sample the whole space.

Step 6: Calculate system performance across the range of scenarios. We use standard simulation:

- To derive the performance of the system, using the analytic model from Step 3,
- Considering the alternative uncertain scenarios from Step 5,
- In conjunction with the alternative strategies managers might use to exercise the options defined in Step 4.

Good practice in EOA explicitly addresses the needs of decisionmakers and stakeholders. Typically, the idea is to use the analysis to compare alternative engineering fixed plans and flexible design strategies. For example, analysts might compare the relative advantages of modular and massive approaches to development, or of centralized and decentralized strategies. The decisionmaking under deep uncertainty (DMDU) community often refers to this assessment of different possible courses of action across diverse future conditions as Exploratory Modeling (Bankes et al., 2013).

Step 7: Reduce the resulting data for stakeholder consideration. The object is to inform decisionmakers and stakeholders, to empower them to make good decisions (Step 8). The goal is to summarize the data into forms that end users can process. EOA typically provides:

- Graphical summary numerical representations of the computed distributions, either as probability or cumulative distributions. Either of these can usefully portray the extent to which some alternatives are more efficient, or less risky, or both (Geltner and de Neufville, 2018).
- Tables of data showing such indices as the average performance, the values at risk (or gain), the robustness (as indicated by the standard deviation), the present value of the costs, the net present values of the alternatives (if appropriate), and other factors associated with the alternative strategies.

In short, EOA is "fit for purpose": it directly informs decisionmakers and stakeholders quantitatively about their strategic choices.

Step 8: Support the choice of a preferred starting decision and plan for its adoption. EOA does not tell you what you should do. It provides information to support the decisionmaking process. A complete EOA provides much more than numerical results: it provides insights and guidance. It indicates which development strategy offers the best advantages, and provides the rationale behind this choice. For example, consider the analysis of a common engineering solution to meet future demands, namely a single large facility to take advantage of economies of scale. The EOA would recognize the great uncertainties in the projection of future demands, and would examine the value of the option of modular capacity expansion. It might then provide insights, such as:

- Uncertainty in future demand leads to substantial risks of loss for a large plant;
- Modular plants reduce this risk, especially if timed to match the actual growth of demand;
- Modular plants can also deal profitably with greater demand than expected, should it occur.

Such insights do not typically depend on precise descriptions of the deep uncertainties. Indeed, analysts can vary the input probability distributions and identify the ranges over which one approach is preferable to another. The projections of the uncertainties do of course influence the strength of the effect, but often not the essential nature of the insights. The general conclusion then is that a modular strategy, flexible as to the location and timing of modules, can provide greater expected gains and lower risks than the default single large facility, all at substantial saving of initial capital expenditures (Capex). This is a win-win strategy at less cost—a great result!

Step 9: Implement a monitoring plan that tracks variables that may trigger future adjustment. An EOA is not a one-shot deal. It is a process of monitoring developments and readjusting.

EOA is very much like the game of chess in this respect. At any point in the game, chess masters explore the possible combinations, and choose their next and subsequent moves to give themselves the position, as the game develops in time and space, to respond effectively as opportunities and threats arise. They do not have a single plan; they develop a flexible approach that maximizes their chances of success. Moreover, the immediate product of the analysis is the recommended first or immediate decision.

So it is with EOA. EOA explores the context widely, compares the range of possibilities, and proposes strategies, and thus opening decisions, that are most likely to be successful. Subsequently, as time goes by, as the future clarifies, we should take note, extend the EOA, and act as appropriate.

6.3 A Simple Example: A Parking Garage

This section provides a thumbnail illustration of the EOA process. A journal article provides details on this case (de Neufville et al., 2006). The case also features as the running example in the textbook on EOA (de Neufville and Scholtes, 2011). More extended cases illustrate EOA in the companion EOA chapter in this book (Chap. 11).

Step 1: Formulate the Opportunity. The general opportunity is to design infrastructures to provide better, less risky financial returns over time. Standard engineering practice is to design such facilities based on fixed forecasts of future demands. The opportunity lies in improving performance by recognizing and dealing with the evident uncertainty and variability in demand. The specific opportunity in our example concerns a multi-story parking garage, modeled on an actual design in England.

Step 2: Specify the Objectives. The prime objective in this case was to "Maximize Expected Net Present Value (ENPV)," that is, the net present value (NPV) as a probabilistic average over the possible scenarios of demand for parking spaces. The secondary objectives were to improve the probabilistic distribution of the possible outcomes; that is, to reduce the maximum downside, to increase the maximum upside, while reducing the initial capital expenditure.

Step 3: Develop the Model. The basic model was an Excel® spreadsheet of revenues and expenses over time, as a function of the demand. It set up a standard discounted cash flow analysis. We enhanced the model with IF/THEN statements to enable the exercise of options when demand exceeded thresholds[1] justifying expansion of the garage.

Step 4: Specify the Options. The options were to add extra floors to the existing structure of the parking garage. To enable this possibility, the basic structure had to have the strength to carry additional loads. Simply put, fat columns and bigger footings made it possible to add one or more floors to the garage at any time. The garage system then had flexibility over capacity and time.

Step 5: Specify the Relevant Uncertainties and Generate Scenarios. The analysis considered that the actual demand in any year was a normal distribution around the forecast, both at the initial level and over time. The Monte Carlo simulation generated thousands of scenarios (or cases) in seconds. As in the real world, some scenarios started and remained low, others started low and went high, others simply varied up and down, etc.

Step 6: Calculate the Performance. The spreadsheet calculated the performance of the systems, accounting for both the variation in demand and, most significantly, for how intelligent management would exercise the options (that is, when they would add floors in response to high demands).

Step 7: Reduce the Data for Stakeholder Consideration. The analysis reduced the data on the thousands of scenarios to present "target curves" of both the frequency and cumulative distributions of the results, and to provide comparisons or trade-offs among objectives in table form, as in Table 6.1.

Step 8: Choose a Preferred Strategy. Given the results, it was obvious that the preferred solution was a flexible design that started with a smaller structure than would seem optimal in the conventional designs, but that included extra strength in columns and footings to enable the options to expand when conditions were right according to the stakeholders. The smaller starting structure reduced the possible losses. The option to expand increased the possible gains in case of high demand.

---

[1] *Note* 'Threshold values' in EOA serve a similar function to that of the 'triggers' or 'adaptation tipping points' used in other DMDU approaches.

Table 6.1. A summary of some EOA results for the parking garage example

Objective	Performance (millions of dollars)		Comparison
	Fixed design	Flexible design	
	Base solution	EOA solution	
Expected NPV	3.5	10.5	EOA better
Minimum NPV	−18.0	−13.1	EOA better
Maximum NPV	8.3	29.8	EOA better
Capex	21.7	18.1	EOA better

Overall, the flexible solution increased ENPV (a win), and lowered Capex by building smaller (another win). The EOA produced a clear win-win solution.

Step 9: Implement the Preferred Strategy and a Monitoring System for Future Action. In practice, management would monitor the demand for parking spaces, and expand when extra capacity appeared desirable.

6.4 Contrasting Engineering Options Analysis with Real Options Analysis

This section highlights the differences between ROA and EOA. These arise from the fact that the two approaches fulfill different roles in quite different situations. Let us first consider the different professional contexts of these methods, and then pass on to the more significant detailed differences.

6.4.1 Different Professional Contexts

Most saliently in the context of this book, ROA and EOA deal with different kinds of uncertainty. ROA assumes that analysts can estimate uncertainties with reasonable accuracy from existing or available data. That is, ROA is suitable in a situation in which one knows, or acts as if one knows, the probability distributions (sometimes called the domain of Level 2 uncertainty—see Table 1.1 in Chap. 1). By contrast, EOA deals with futures that are largely unknowable, that concern conditions unparalleled in previous experience. EOA deals with deep uncertainly (Levels 3 or 4 uncertainty).

ROA and EOA also differ dramatically in terms of their end objectives. ROA, as a sub-field of financial options analysis, seeks to determine the proper monetary value for a special kind of financial or similar opportunity known as an option or contingent claim. The objective of ROA is to find the correct price to pay for the purchase of an option. EOA by contrast seeks to determine the best design—the best strategy for implementation in a system. For EOA, the calculations are a means to determine which plans are best and, more generally, by how much. Whereas for ROA the calculation of the most appropriate expected price is the main object of the analysis, for EOA the calculation of this average is incidental—the objective of EOA is to help decisionmakers and stakeholders understand alternative plans, and to choose the one that best meets their needs.

Best practice in engineering systems design recognizes that system planners typically need to balance average expectations with downside risks. Indeed, good engineering understands that a focus on average value can lead to bad policy recommendations and to inappropriate advice. A simple example of a contract for automobile insurance makes the point. (Formally, this contract is a "put" option on the potential loss; if the value of the car crashes, we sell it for the insured amount). When we think about it, we know that buying such an ordinary insurance contract does not maximize expected value. Insurance companies pay out less than they take in—they thus cover their expenses and earn some profit. However, even though buying insurance does not maximize one's expected value, it does not therefore follow that you and I should not buy insurance. By buying insurance, we trade small periodic losses, and an overall expected loss, for some peace of mind and protection against disastrous consequences. We come to this conclusion because we consider the entire range of possible outcomes, as does EOA.

6.4.2 Some Specific Differences

Following on the above, we first remark that ROA is not "fit for purpose" for planners. ROA is focused on producing a market price for options. In general, this is not what decisionmakers or stakeholders concerned with deep uncertainty need. First, they need to develop an understanding of the range of possible outcomes of any strategy for dealing with future uncertainties. Second, their prime objectives may not be monetary; instead, they may be concerned with the risk of floods (climate change), morbidity and mortality rates (hospitals), reliability (water supply), or the likelihood of intercepting missiles (military).

A proper ROA also differs from EOA in that it requires detailed statistical data in order to proceed. Indeed, the analytic beauty and power of financial options analysis lies in the way it continuously adjusts the implicit discount rate it uses in its analysis to the risk content of the ongoing evolution of the system. This is remarkably clever and theoretically powerful. But it depends on correct descriptions of the uncertainties. In practice, ROA invokes substantial data on past events (such as the historic price of copper), processes this information statistically to define their distributions, then projects these patterns ahead to obtain the value of financial options on near term projects (such as the purchase of a copper mine). (See for example Tufano and Moel, 1997) EOA does not suffer from such constraints.

The greatest limitation of ROA is its general inability to deal with path-dependent situations, and thus with multiple options simultaneously. ROA is an excellent mechanism for calculating the answer to straightforward yes–no questions, such as: is it desirable to acquire this option now or not? Should we pay more than some stated amount? Is it time to exercise or cash in on an option? However, ROA does not enable planners to explore the value of several options to use at any time (for example, by exercising desalination or recycling options to increase reliability of water supply). The technical explanation is that ROA assumes "path independence" for the value of options: the assumption is that the value of an option (for example, to buy oil at a given future price) is independent of the way the uncertainty (that is, the price of oil) evolves. While path independency is reasonable when individuals deal with large, liquid markets, the assumption is not realistic for managers of important engineering projects. This is because intelligent managers respond to what happens by adapting the system in some way, so that the eventual value of an option does depend on the evolutionary path of uncertainties. The practical consequence of this fact is that major standard analytic methods of financial options analysis (specifically, dynamic programming and lattice analysis) are not suitable for dealing with realistic managerial issues. EOA does not suffer from these limitations of ROA; EOA deals smoothly with path dependencies. Here again, this is a further instance in which EOA is better suited to addressing the issues associated with dealing with deep uncertainty.

6.5 Contrasting Engineering Options Analysis with Other Approaches in This Book

6.5.1 *Engineering Options Analysis as a Planning Approach*

As a DMDU approach, EOA focuses on the investment, design, and management of technical systems. This is distinct from the majority of the other methods in this Part presented in this book, which can be used more broadly to develop adaptive strategies for both infrastructural and non-technical applications alike. For instance, a study of water security on the Colorado River (U.S. Bureau of Reclamation, 2012) used Robust Decision Making to explore both structural investments (such as construction of desalination plants) as well as non-structural courses of action (such as watershed management in the form of forest management). In contrast, an EOA of the same case study would focus in more detail on the technical aspects of the problem.

This focus on technical systems somewhat narrows the scope of applicability of EOA as compared to the other DMDU approaches in this book. However, when considering different courses of action, the focus of EOA is on exploring the performance of diverse alternatives in more detail, each with varying degrees of flexibility in the form of options. Its more detailed analysis of different possible interventions provides unique insights, distinct from and complementary to those obtained from other approaches.

For example, Dutch researchers have used the Dynamic Adaptive Policy Pathways (DAPP) approach to explore different possible policy responses to ensure continued flood protection and adequate fresh water supply in the Rhine Delta in the Netherlands (Haasnoot et al., 2013). They explored both structural alternatives, such as raising dikes and increasing pumping capacity, and non-structural measures, such as altered flow regimes or reducing the demand for water. Exploring these different types of measures using a DAPP approach provides insights about how long different measures stay effective under different future conditions, and demonstrates the impact of sequencing of different courses of action. In contrast, an EOA of the same problem could focus in more depth on the technical alternatives of dike raising and expanding pumping capacity, exploring how the performances of different structural designs compare. For instance, possible options could include reserving land adjacent to the dike or building it with a wider base, both of which support easy heightening of the dike later. When looking at the pumping capacity, options could include the addition of extra pump bays at the outset, enabling easy capacity expansion later on, or investing in pumps that remain functional over a wider range of water heads. The subsequent analysis would seek to compare how designs with and without an option compare, providing insights about the added value of incorporating flexibility in the form of options within structural design decisions. Ultimately, the EOA approach helps discover and indicate the options whose potential future benefits outweigh possible up-front added costs, if any. [Indeed, one of the benefits of options, and of flexible planning in general, is that they allow us to defer commitments until we need to do so, and thus in many cases to avoid their expense completely.] Similar to DAPP, EOA explicitly considers the timing of actions, and its analyses demonstrate those cases where we should take actions now versus waiting until later.

In summary, two crucial characteristics of EOA emerge; these characteristics are central in differentiating EOA from other approaches:

- its in-depth focus on technical systems, providing unique insights, distinct from and complementary to those obtained from other approaches.
- it actively expands the decision space of possible alternatives under examination. It explicitly seeks more flexible variants of previously conceived courses of action.

6.5.2 Engineering Options Analysis as a Computational Decision-Support Tool

Finally, EOA is a computational decision-support tool that relies on systematic analysis. In this sense it is clearly an "engineering" approach, insofar as it "runs the numbers" on the performance of an engineering system. (Among the methods presented in this book, Robust Decision Making (Chap. 2) and Decision Scaling (Chap. 12) can also be considered "engineering" approaches). It computes a distribution of possible outcomes for a desired set of measures of or assumptions about future conditions, and provides detailed results that document the advantages and disadvantages of alternative strategies or "game plans" for dealing with future uncertainties. (Exploratory Modeling can also be considered a computational decision-support tool, which is used in practically all of the DMDU approaches).

6.6 Conclusions

The practice of EOA is still in its infancy. While it already has a substantial and expanding list of successful applications across many different fields, we still have a lot to learn about how to do it best and most efficiently. It is worthwhile to review two open questions.

First, it seems clear to many of us that we should shape applications of EOA to the particularities of the substantive issues and needs of practitioners in specific fields. Different problems have different time scales and structures of costs, and different drivers of most effective design. For example, contrast the fields of water resources and real estate development. Water resources planning faces long-term, gradual, reasonably steady uncertainties (in the evolution of population, water usage, and climate), whereas real estate development must deal with cyclical trends punctuated by sudden market crashes. These two fields also differ in terms of their criteria of performance and their ability to deal with uncertainties. Whereas water resource planners greatly focus on reliability of supply, and have the general ability to ration use to some degree in case of need, real estate developers mostly focus on economic profit and principally concern themselves with the timing of market entry and exit. Thus, while EOA has already proven useful to both fields, analysts need to shape its detailed practice differently, according to the characteristics of each field.

In terms of technique, a most pressing issue for EOA is that of defining a coherent catalog of computationally efficient screening models (also known as Fast Simple Models). The objective is to identify suitably simple yet representative models of complex systems. Researchers have already tried many possibilities with success in various applications. Practitioners have, for example, variously used functional forms they derived statistically from the outputs of a detailed model (called "meta-models"), used mass balance formulations to complex flows, and simplified complex models by dropping higher order terms or using partial linearization. No single approach will work everywhere, and it is sure that there will be limits to what we can do. That said, we can expect substantial progress in this facet of EOA (and DMDU in general).

Our assessment is that the prospects for EOA are good. So far, applications have focused on "low-hanging fruit"—on opportunities to demonstrate analytically the potential value of adaptive strategies as compared to fixed strategies. Analysts are constantly uncovering new opportunities in new fields. Moreover, they are collectively documenting the need to change the paradigm of systems engineering and design from one of starting with fixed requirements, to one that recognizes that we need to acknowledge that conditions and opportunities change over time, and that good system design should be able to adapt to new circumstances, as they evolve or suddenly occur.

References

Bankes, S., Walker, W. E., and Kwakkel, J. H. (2013). Exploratory modeling and analysis, entry. In S. I. Gass & M. C. Fu (Eds.), *Encyclopedia of Operations Research and Management Science* (3rded., pp. 532–537). New York: Springer.

Cardin, M.-A., Bourani, M., & de Neufville, R. (2015). Improving the lifecycle performance of engineering projects with flexible strategies: example of on-shore LNG production design. *Systems Engineering, 18*(3), 253–268. https://doi.org/10.1002/sys.21301.

de Neufville, R. (2007). Low-cost airports for low-cost airlines: Flexible design to manage the risks. *Transportation Planning and Technology, 31*(1), 35–68 [Special Issue: 'Approaches to Developing the Airport of the Future', In W. Walker, V. Marchau, & O. Van De Riet (Eds.)].

de Neufville, R., & Odoni, A. (2013). *Airport systems planning, design, and management* (2nd ed.). New York: McGraw-Hill.

de Neufville, R., Scholtes, S., & Wang, T. (2006). Valuing options by spreadsheet: parking garage case example. *ASCE Journal of Infrastructure Systems, 12*(2), 107–111.

de Neufville, R., & Scholtes, S. (2011). *Flexibility in engineering design*. Cambridge, MA, USA: MIT Press.

de Weck, O., Chaize, M., & de Neufville, R. (2004). Enhancing the economics of communications satellites via orbital reconfigurations and staged development. *Journal of Aerospace Computing, Information, and Communication, 1*(3), 119–136.

Esders, M., Della Morte, N., & Adey, B. (2015). A methodology to ensure the consideration of flexibility and robustness in the selection of facility renewal projects. *International Journal of Architecture, Engineering and Construction, 4*(3), 126–139.

Geltner, D., & de Neufville, R. (2018). *Flexibility and real estate valuation under uncertainty: a practical guide for developers*. Cambridge, UK and New York, USA: Wiley-Blackwell.

Haasnoot, M., Kwakkel, J. H., Walker, W. E., & ter Maat, J. (2013). Dynamic adaptive policy pathways: a method for crafting robust decisions for a deeply uncertain world. *Global Environmental Change, 23*(2), 485–498.

Lee, Y.S., Scholtes, S., and de Neufville, R. (2008). Using flexibility to improve value-for-money in PFI projects. *Symposium on Redefining healthcare infrastructure: Integrating services, technologies and the built environment*, London: Tanaka Business School, Imperial College.

Lin, J., de Weck, O., de Neufville, R., & Yue, H. (2013). Enhancing the value of oilfield developments with flexible subsea tiebacks. *Journal of Petroleum Science and Engineering, 102*(2), 73–83. https://doi.org/10.1016/j.petrol.2013.01.003.

Luenberger, D. (1996). *Investment science*. New York: Oxford University Press.

Page, J. (2011). Flexibility in early stage design of us navy ships: an analysis of options. *Master of Science thesis, MIT System Design and Management and Naval Engineer Programs*.

Trigeorgis, L. (1996). *Real options: managerial flexibility and strategy in resource allocation*. Cambridge, MA: MIT Press.

Tufano, P. & Moel. A. (1997). Bidding for Antamina, Case 297054-PDF-ENG, Harvard Business School, Boston, MA.

U.S. Bureau of Reclamation. (2012). *Colorado river basin water supply and demand study: study report*, Boulder City, Nevada [online]. Retrieved September 10, 2017 from https://www.usbr.gov/lc/region/programs/crbstudy.html.

Prof. Richard de Neufville (Massachusetts Institute of Technology, Institute for Data, Systems, and Society) has worked on decisionmaking under uncertainty since the 1970s. He focuses now on the practical application of Engineering Options Analysis. He co-authored: 'Flexibility in Engineering Design' (MIT Press, 2011) and 'Flexibility and Real Estate Evaluation under Uncertainty: A practical Guide for Developers' (Wiley, 2018). He applies these approaches internationally to the design of infrastructure systems and 'Airport Systems Planning, Design, and Management' (McGraw-Hill, 2013).

Dr. Kim Smet (University of Ottawa, Smart Prosperity Institute) is a Postdoctoral Researcher at the Smart Prosperity Institute, an environmental policy think tank and research network based at the University of Ottawa. She completed her Ph.D. in Environmental Engineering at Harvard University in 2017, where she explored how Engineering Options Analysis can be incorporated into the redesign of aging water resources infrastructure under uncertainty. Through her research, she has worked with the US Army Corps of Engineers and the Rijkswaterstaat in the Netherlands.

Open Access This chapter is licensed under the terms of the Creative Commons Attribution 4.0 International License (http://creativecommons.org/licenses/by/4.0/), which permits use, sharing, adaptation, distribution and reproduction in any medium or format, as long as you give appropriate credit to the original author(s) and the source, provide a link to the Creative Commons licence and indicate if changes were made.

The images or other third party material in this chapter are included in the chapter's Creative Commons licence, unless indicated otherwise in a credit line to the material. If material is not included in the chapter's Creative Commons licence and your intended use is not permitted by statutory regulation or exceeds the permitted use, you will need to obtain permission directly from the copyright holder.

Part II
DMDU Applications

Chapter 7
Robust Decision Making (RDM): Application to Water Planning and Climate Policy

David G. Groves, Edmundo Molina-Perez, Evan Bloom and Jordan R. Fischbach

Abstract

- Long-term water planning is increasingly challenging, since hydrologic conditions appear to be changing from the recently observed past and water needs in many locales are highly relative to existing and easily developed supplies.
- Globally coordinated policymaking to address climate change also faces large uncertainties regarding underlying conditions and possible responses, making traditional prediction-based planning approaches to be inadequate for the purpose.
- Methods for Decision Making under Deep Uncertainty (DMDU) can be useful for addressing long-term policy challenges associated with multifaceted, nonlinear, natural and socio-economic systems.
- This chapter presents two case studies using Robust Decision Making (RDM).
- The first describes how RDM was used as part of a seven-state collaboration to identify water management strategies to reduce vulnerabilities in the Colorado River Basin.
- The second illustrates how RDM could be used to develop robust investment strategies for the Green Climate Fund (GCF)—an international global institution charged with making investments supporting a global transition toward more sustainable energy systems that will reduce GHG emissions.
- Both case studies describe the development of robust strategies that are adaptive, in that they identify both near-term decisions and guidance for how these responses should change or be augmented as the future unfolds.

D. G. Groves (✉) · E. Bloom · J. R. Fischbach
RAND Corporation, Santa Monica, CA, USA
e-mail: groves@rand.org

E. Molina-Perez
Tecnológico de Monterrey, Monterrey, Mexico

© The Authors(s) 2019
V. A. W. J. Marchau et al. (eds.), *Decision Making under Deep Uncertainty*,
https://doi.org/10.1007/978-3-030-05252-2_7

7.1 Long-Term Planning for Water Resources and Global Climate Technology Transfer

Developing water resources to meet domestic, agricultural, and industrial needs has been a prerequisite for sustainable civilization for millennia. For most of history, "water managers" have harnessed natural hydrologic systems, which are inherently highly variable, to meet generally increasing but predictable human needs. More recently, water managers have worked to mitigate the consequences of such development on ecological systems. Traditional water resources planning is based on probabilistic methods, in which future water development needs are estimated based on a single projection of future demand and estimates of available supplies that are based on the statistical properties of recorded historical conditions. Different management strategies are then evaluated and ranked, generally in terms of cost effectiveness, for providing service at a specified level of reliability. Other considerations, such as environmental attributes, are also accounted for as necessary (Loucks et al. 1981).

While tremendous achievements have been made in meeting worldwide water needs, water managers are now observing climate change effects through measurable and statistically significant changes in hydrology (Averyt et al. 2013). Moreover, promising new strategies for managing water resources are increasingly based on new technologies (e.g., desalination, demand management controls), the integration of natural or green features with the built system, and market-based solutions (Ngo et al. 2016). These novel approaches in general do not yet have the expansive data to support accurate estimates of their specific effects on water management systems or their costs. The modern water utility thus must strive to meet current needs while preparing for a wide range of plausible future conditions. For most, this means adapting the standard probabilistic engineering methods perfected in the twentieth century to (1) consider hydrologic conditions that are likely changing in unknown ways, and (2) meet societal needs that are no longer increasing at a gradual and predictable pace.

Water utilities have begun to augment traditional long-term planning approaches with DMDU methods that can account for deep uncertainty about future hydrologic conditions, plausible paths of economic development, and technological advances. These efforts, first begun as research collaborations among policy researchers and innovative planners (e.g., Groves et al. 2008, 2013), are now becoming more routine aspects of utility planning. As evidence of this uptake, the Water Utility Climate Alliance (WUCA), comprised of ten large US water utilities, now recommends DMDU methods as best practice to support long-term water resources planning (Raucher and Raucher 2015).

The global climate clearly exerts tremendous influence on the hydrologic systems under management. Over the past decade, the global community has developed a consensus that the climate is changing in response to anthropogenic activities, most significantly the combustion of carbon-based fuels and the release of carbon dioxide and other greenhouse gasses into the atmosphere (IPCC 2014a). In 2016, nations around the world adopted the Paris Agreement, with a new consensus that development must be managed in a way that dramatically reduces net greenhouse gas emissions in order to stave off large and potentially disastrous changes to the climate (United Nations 2016). As part of this agreement, the Green Climate Fund (GCF) was established to steer funding toward activities that would encourage the shift from fossil fuels to more sustainable energy technologies. Governments will also need to prepare for climate changes that are unavoidable and help adapt many aspects of civilization (IPCC 2014b).

Climate mitigation and adaptation will require billions of dollars of investments in new ways to provide for society's needs—be it energy for domestic use, industry, and transportation, or infrastructure to manage rising seas and large flood events. Technological solutions will need to play a large role in helping facilitate the needed transitions. As with water resources planning, governments will need to plan for deep uncertainty, and design and adopt strategies that effectively promote the development and global adoption of technologies for climate mitigation under a wide range of plausible futures.

This chapter presents two case studies that demonstrate how DMDU methods, in particular Robust Decision Making (RDM) (see Chap. 2), can help develop robust long-term strategies. The first case study describes how RDM was used for the 2012 Colorado River Basin Study—a landmark 50-year climate change adaptation study, helping guide the region toward implementing a robust, adaptive management strategy for the Basin (Bloom 2015; Bureau of Reclamation 2012; Groves et al. 2013). This case study highlights one of the most extensive applications of an RDM-facilitated deliberative process to a formal planning study commissioned by the US Bureau of Reclamation and seven US states. The second case study presents a more academic evaluation of how RDM can define the key vulnerabilities of global climate policies and regimes for technology transfer and then identify an adaptive, robust policy that evolves over time in response to global conditions (Molina-Perez 2016). It further illustrates how RDM can facilitate a quantitative analysis that employs simple models yet provides important insights into how to design and implement long-term policies.

7.2 Review of Robust Decision Making

RDM was the selected methodology for the studies described here, as the decision contexts are fraught with deep uncertainty, are highly complex, and have a rich set of alternative policies (c.f. Fig. 1.3 Chap. 1). For such contexts, an "agree on decisions approach" is particularly useful, as it provides a mechanism for linking the analysis of deep uncertainty through the identification of preliminary decisions' vulnerabilities to the definition of an *adaptive strategy*—characterized by near-term low-regret actions, signposts, and triggers, and decision pathways for deferred actions. For both case studies, RDM provides an analytic framework to identify vulnerabilities of current strategies in order to identify robust, adaptive strategies—those that will succeed across the broad range of plausible futures.

The two case studies also illustrate how RDM can be applied in different sectors. The water management study emphasizes that RDM addresses both deep uncertainty and supports an iterative evaluation of robust, adaptive strategies designed to minimize regret across the uncertainties. In this case, it does so by selecting among numerous already existing investment proposals. The second case study describes a policy challenge different from natural resources and infrastructure management—promoting global climate technology transfer and uptake—that uses RDM as the means to discover and define policies anew. Both studies define how the strategies could adapt over time to achieve greater robustness than a static strategy or policy could, using information about the vulnerabilities of the tested strategies. This represents an important convergence between RDM and the Dynamic Adaptive Policy Pathways (DAPP) methodology described in Chaps. 4 and 9.

7.2.1 Summary of Robust Decision Making

As described in Chap. 2 (Box 2.1), RDM includes four key elements:

- Consider a multiplicity of plausible futures;
- Seek robust, rather than optimal strategies;
- Employ adaptive strategies to achieve robustness; and
- Use computers to facilitate human deliberation over exploration, alternatives, and trade-offs.

Figure 7.1 summarizes the steps to implement the RDM process, and Fig. 7.2 relates these steps to the DMDU framework introduced in Chap. 1. In brief, it begins with (1) a *Decision Framing* step, in which the key external developments (X), policy levers or strategies (L), relationships (R), and measures of performance (M),[1] are initially defined through a participatory stakeholder process, as will be illustrated in the case studies. Analysts then develop an experimental design, which defines a large set of futures against which to evaluate one or a few initial strategies. Next, the models are used to (2) Evaluate strategies in many futures and develop a large database of simulated outcomes for each strategy across the futures. These results are then inputs to the (3) Vulnerability analysis, in which Scenario Discovery (SD) algorithms and techniques help identify the uncertain conditions that cause the strategies to perform unacceptably. Often in collaboration with stakeholders and decisionmakers, these conditions are refined to describe "Decision-Relevant Scenarios". The next step of RDM (4) explores the inevitable trade-offs among candidate robust, adaptive strategies. This step is designed to be participatory, in which the analysis and interactive visualization provide decisionmakers and stakeholders with information about the range of plausible consequences of different strategies, so that a suitably *low regret*, robust strategy can be identified. The vulnerability and tradeoff information is used

---

[1]Chapter 1 (see Fig. 1.2) uses the acronym XPROW for external forces, policies, relationships, outcomes, and weights.

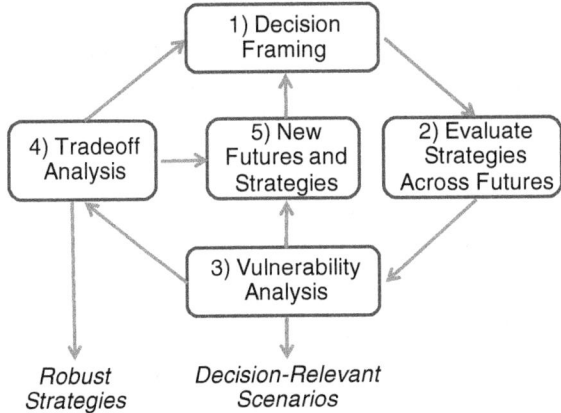

Fig. 7.1 RDM process (Adapted from Lempert et al. (2013b)

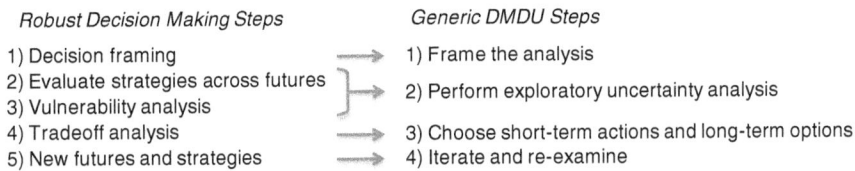

Fig. 7.2 Mapping of RDM to DMDU steps presented in Chap. 1

to design more robust and adaptive strategies, and to include additional plausible futures for stress testing in (5) the New futures and strategies step. This can include *hedging* actions—those that help improve a strategy's performance under the vulnerable conditions—and *shaping* actions—those designed to reduce the likelihood of facing the vulnerable conditions. *Signposts* and *triggers* are also defined that are used to reframe static strategies into adaptive strategies that are implemented in response to future conditions. RDM studies will generally implement multiple iterations of the five steps in order to identify a set of candidates' robust, adaptive strategies.

While both case studies highlight the key elements above, they also illustrate how RDM supports the design of adaptive strategies. Fischbach et al. (2015) provide a useful summary of how RDM and other DMDU approaches can support the development of adaptive plans from those that are simply *forward looking*, to those that include *plan adjustment*.

For example, a static water management strategy might specify a sequence of investments to make over the next several decades. In some futures, specific later investments might be unnecessary. In other futures, the investments may be implemented too slowly or may be insufficient. An adaptive water management strategy, by contrast, might be forward looking or include policy adjustments, consisting of simple decision rules in which investments are made only when the water supply and demand balance is modeled to be below a threshold (see Groves (2006) for an early example). The RDM process can then determine if this adaptive approach improves the performance and reduces the cost of the strategy across the wide range of futures. RDM can also structure the exploration of a very large set of alternatively specified adaptive strategies—those defining different signposts (conditions that are monitored) or triggers (the specific value(s) for the signpost(s) or conditions that trigger investments). Lastly, RDM can explore how a more "formal review and continued learning" approach (Bloom 2015; Swanson et al. 2010) to adaptive strategies might play out by modeling how decisionmakers might update their beliefs about future conditions and adjust decisions accordingly.

In the first case study, the RDM uncertainty exploration (Steps 2 and 3) is used to identify vulnerabilities, and then test different triggers and thresholds defining an adaptive plan. It further traces out different decision pathways that planners could take to revise their understanding of future vulnerabilities and act accordingly. This is done by tying the simulated implementation of different aspects of a strategy to the modeled observation of vulnerabilities. In the second case study, the RDM process identifies low-regret climate technology policy elements, and then defines specific pathways in which the policies are augmented in response to unfolding conditions.

7.3 Case Study 1: Using RDM to Support Long-Term Water Resources Planning for the Colorado River Basin

The Colorado River is the largest source of water in the southwestern USA, providing water and power to nearly 40 million people and water for the irrigation of more than 4.5 million acres of farmland across seven states and the lands of 22 Native American tribes, and is vital to Mexico providing both agricultural and municipal needs (Bureau of Reclamation 2012). The management system is comprised of 12 major dams, including Glen Canyon Dam (Lake Powell along the Arizona/Utah boarder) and Hoover Dam (Lake Mead, at the border of Nevada and Arizona). Major infrastructure also transports water from the Colorado River and its tributaries to Colorado's Eastern Slope, Central Arizona, and Southern California. The 1922 Colorado River Compact apportions 15 million acre-feet (maf) of water equally among four Upper Basin States (Colorado, New Mexico, Utah, and Wyoming) and three Lower Basin States (Arizona, California, and Nevada) (Bureau of Reclamation 1922). Adherence to the Compact is evaluated by the flow at Lee Ferry—just down-river of the Paria River confluence. A later treaty in 1944 allots 1.5 maf of water per year to Mexico, with up to 200 thousand acre-feet of additional waters during surplus conditions. The US Bureau of Reclamation works closely with the seven Basin States and Mexico to manage this resource and collectively resolve disputes.

The reliability of the Colorado River Basin system is increasingly threatened by rising demand and deeply uncertain future supplies. Recent climate research suggests that future hydrologic conditions in the Colorado River Basin could be significantly different from those of the recent historical period. Furthermore, climate warming already underway and expected to continue will exacerbate periods of drought across the Southwest (Christensen and Lettenmaier 2007; Nash and Gleick 1993; Seager et al. 2007; Melillo et al. 2014). Estimates of future streamflow based on paleoclimate records and general circulation models (GCMs) suggest a wide range of future hydrologic conditions—many which would be significantly drier than recent conditions (Bureau of Reclamation 2012).

In 2010, the seven Basin States and the US Bureau of Reclamation (Reclamation) initiated the Colorado River Basin Study (Basin Study) to evaluate the ability of the Colorado River to meet water delivery and other objectives across a range of futures (Bureau of Reclamation 2012). Concurrent with the initiation of the Basin Study, Reclamation engaged RAND to evaluate how RDM could be used to support long-term water resources planning in the Colorado River Basin. After completion of the evaluation study, Reclamation decided to use RDM to structure the Basin Study's vulnerability and adaptation analyses (Groves et al. 2013). This model of evaluating DMDU methods through a pilot study and then adopting it to support ongoing planning is a low-threshold approach for organizations to begin to use DMDU approaches in their planning activities (Lempert et al. 2013a, b).

The Basin Study used RDM to identify the water management vulnerabilities with respect to a range of different objectives, including water supply reliability, hydropower production, ecosystem health, and recreation. Next, it evaluated the performance of a set of portfolios of different management actions that would be triggered as needed in response to evolving conditions. Finally, it defined the key trade-offs between water delivery reliability and the cost of water management actions included in different portfolios, and it identified low-regret options for near-term implementation.

Follow-on work by Bloom (2015) evaluated how RDM could help inform the design of an adaptive management strategy, both in terms of identifying robust adaptation through signposts and triggers and modeling how beliefs and new information can influence the decision to act. Specifically, Bloom used Bayes' Theorem to extend the use of vulnerabilities to analyze how decisionmaker views of future conditions could evolve. The findings were then incorporated into the specification of a robust, adaptive management strategy.

7.3.1 Decision Framing for Colorado River Basin Analyses

A simplified XLRM chart (Table 7.1) summarizes the RDM analysis of the Basin Study and follow-on work by Bloom (2015). Both studies evaluated a wide range of future hydrologic conditions, some based on historical and paleoclimate records, others on projections from GCMs. These hydrologic conditions were combined with six demand scenarios and two operation scenarios to define 23,508 unique futures. Both studies used Reclamation's long-term planning model—Colorado River Simulation System (CRSS)—to evaluate the system under different futures with respect to a large set of measures of performance. CRSS simulates operations at a monthly

Table 7.1 Simplified XLRM framing matrix from Colorado River RDM analyses

Exogenous factors (X)	Policy levers (L)
Future water demand (6) Streamflow under different climate regimes (thousands) • Resampled historical record • Resampled paleoclimate record • Downscaled global climate model projections System operations (2)	Water management portfolios (4) • Up to 40 different water management actions • Signposts and triggers used to specify investments
Relationships (R)	Measures of performance (M)
Colorado River Simulation System (CRSS)	10-year average streamflow at Lees Ferry (Upper Basin reliability) Lake Mead pool elevation (Lower Basin reliability)

time-step from 2012 to 2060, modeling the network of rivers, demand nodes, and 12 reservoirs with unique operational rules.

Both the Basin Study and Bloom focused the vulnerability analysis on two key objectives: (1) ensuring that the 10-year running average of water flow from the Upper to Lower Basin meets or exceeds 7.5 maf per year (Upper Basin reliability objective), and (2) maintaining Lake Mead's pool elevation above 1,000 feet (Lower Basin reliability objective). Both studies then used CRSS to model how the vulnerabilities could be reduced through the implementation of alternative water management portfolios, each comprised of a range of possible water management options, including municipal and industrial conservation, agricultural conservation, wastewater reuse, and desalination.

7.3.2 Vulnerabilities of Current Colorado River Basin Management

Both the Basin Study and Bloom's follow-on work used CRSS to simulate Basin outcomes across thousands of futures. They then used SD methods (see Chap. 2) to define key vulnerabilities. Under the current management of the system, both the Upper Basin and Lower Basin may fail to meet water delivery reliability objectives under many futures within the evaluated plausible range. Specifically, if the long-term average streamflow at Lee Ferry falls below 15 maf and an eight-year drought with average flows below 13 maf occurs, the Lower Basin is vulnerable (Fig. 7.3). The corresponding climate conditions leading to these streamflow conditions include historical periods with less than average precipitation, or those projected climate conditions (which are all at least 2° F warmer) in which precipitation is less than 105% of the historical average (see Groves et al. 2013). The Basin Study called this vulnerability *Low Historical Supply*, noting that the conditions for this vulnerability had

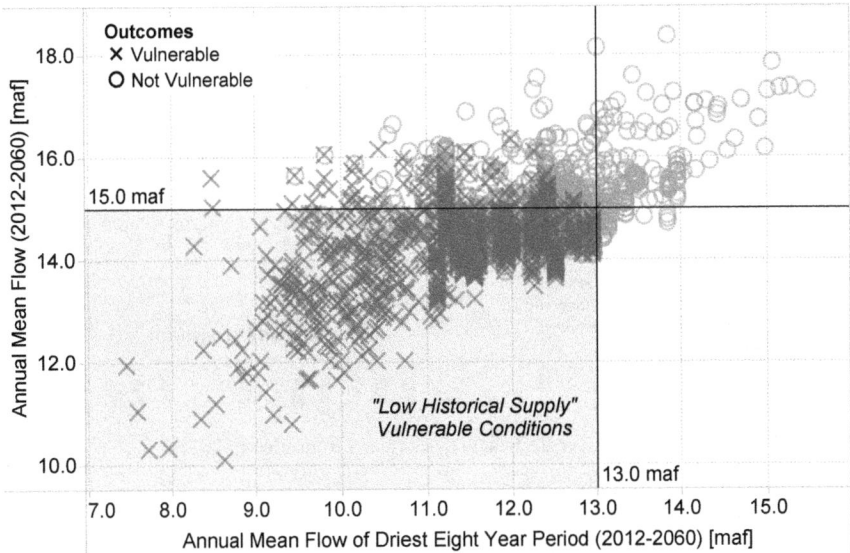

Fig. 7.3 One Colorado River lower basin vulnerability—low historical supply

been observed in the historical record. The Basin Study also identified a vulnerability for the Upper Basin defined by streamflow characteristics present only in future projections characterized by *Declining Supply*. Bloom further parsed the Basin's vulnerabilities to distinguish between *Declining Supply* and *Severely Declining Supply* conditions.

7.3.3 Design and Simulation of Adaptive Strategies

After identifying vulnerabilities, both studies developed and evaluated portfolios of individual management options that would increase the supply available to the Basin states (e.g., ocean and groundwater desalination, wastewater reuse, and watershed management), or reduce demand (e.g., agricultural and urban conservation). For the Basin Study, stakeholders used a Portfolio Development Tool (Groves et al. 2013) to define four portfolios, each including a different set of investment alternatives for implementation. Portfolio B (Reliability Focus) and Portfolio C (Environmental Performance Focus) represented two different approaches to managing future vulnerabilities. Portfolio A (Inclusive) was defined to include all alternatives in either Portfolios B or C, and Portfolio D (Common Options) was defined to include only those options in both Portfolios B and C. Options within a portfolio were prioritized based on an estimate of cost-effectiveness, defined as the average annual yield divided by total project cost.

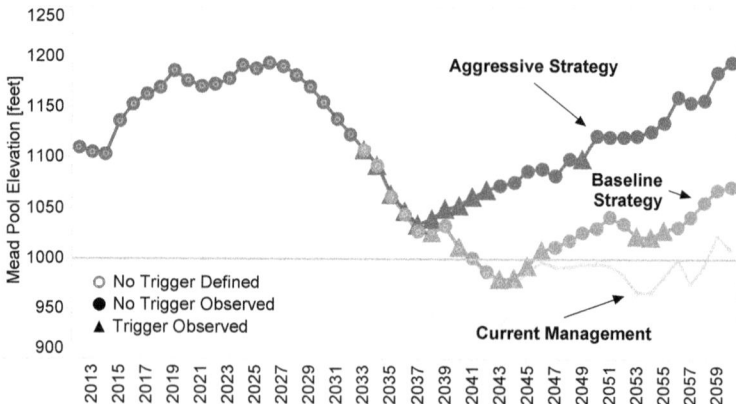

Fig. 7.4 Example Colorado River Basin adaptive strategy from Bloom (2015)

CRSS modeled these portfolios as adaptive strategies by simulating what investment decisions a basin manager, or agent, would take under different simulated Basin conditions. The agent monitors three key signpost variables on an annual basis: Lake Mead's pool elevation, Lake Powell's pool elevation, and the previous five years of observed streamflow. When these conditions drop below predefined thresholds, the agent implements the next available water management action from the portfolio's prioritized list. To illustrate how this works, Fig. 7.4 shows the effect of the same management strategy with two different trigger values in a single future. Triangles represent years in which observations cross the trigger threshold. In this future, the Aggressive Strategy begins implementing water management actions sooner than the Baseline Strategy, and Lake Mead stays above 1000 feet. For the Basin Study strategies, a single set of signposts and triggers were used. These strategies were then evaluated across the 23,508 futures. For each simulated future, CRSS defines the unique set of investments that would be implemented per the strategy's investments, signposts, and triggers. Under different signposts and triggers, different investments are selected for implementation.

The performance of each strategy is unique in terms of its reliability metric and cost. Figure 7.5 summarizes reliability across the *Declining Supply* futures, and shows the range of 50-year costs for each of the four adaptive strategies. In the figure, Portfolio A (Inclusive) consists of the widest set of options and thus has the greatest potential to prevent water delivery vulnerability. It risks, however, incurring the highest costs. It also contains actions that are controversial and may be unacceptable for some stakeholders. Portfolio D (Common Options) includes the actions that most stakeholders agree on—a subset of A. This limits costs, but also the potential to reduce vulnerabilities. Under these vulnerable conditions, Portfolios A and B lead to the fewest number of years with critically low Lake Mead levels. The costs, however, range from $5.5–$7.5 billion per year. Portfolio C, which includes more environmentally focused alternatives, performs less well in terms of Lake Mead

7 Robust Decision Making (RDM): Application …

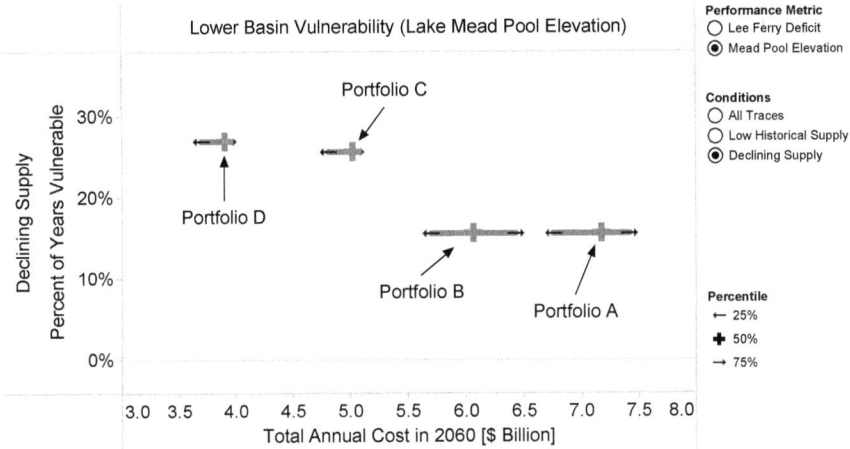

Fig. 7.5 Vulnerability versus cost trade-off under Declining Supply conditions for the four Basin Study portfolios

levels, but leads to a tighter range of costs—between $4.75 and $5.25 billion per year. Bloom expanded on this analysis by exploring across a range of different investment signposts and triggers, specified to be more or less aggressive in order to represent stakeholders and planners with different preferences for cost and reliability. A less aggressive strategy, for example, would trigger additional investments only under more severe conditions than a more aggressive strategy.

Figure 7.6 shows the trade-offs across two bookend static strategies ("Current Management" and "Implement All Actions") and five adaptive strategies based on Portfolio A (Inclusive), but with a range of different triggers from Bloom (2015). Note that the signposts and triggers for the Baseline strategy are those used for the Basin Study portfolios. The more aggressive the triggers, the fewer futures in which objectives are missed (left side of figure). However, the range of cost to achieve those outcomes increases.

7.3.4 Evaluating Regret of Strategies Across Futures

For this case study, we define a robust strategy as one that minimizes regret across a broad range of plausible future conditions. Regret, in this context, is defined as the additional amount of total annual supply (volume of water, in maf) that would be needed to maintain the Lake Mead level at 1000 feet across the simulation. The more supply that would be needed to maintain the Lake Mead level, the more regret.

The Basin Study adaptive strategies do not completely eliminate regret. Some strategies still underinvest in the driest futures and overinvest in the wettest futures. This result stems from both the use of relatively simplistic triggers and the inherent

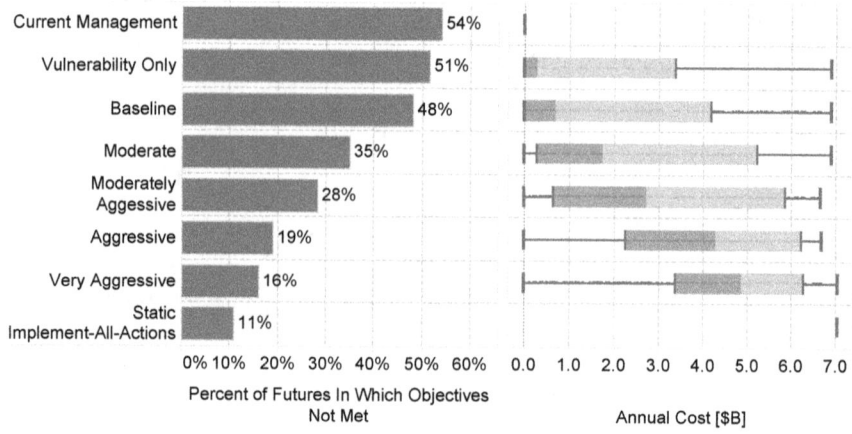

Fig. 7.6 Vulnerability/cost trade-offs across portfolios with different triggers from Bloom (2015)

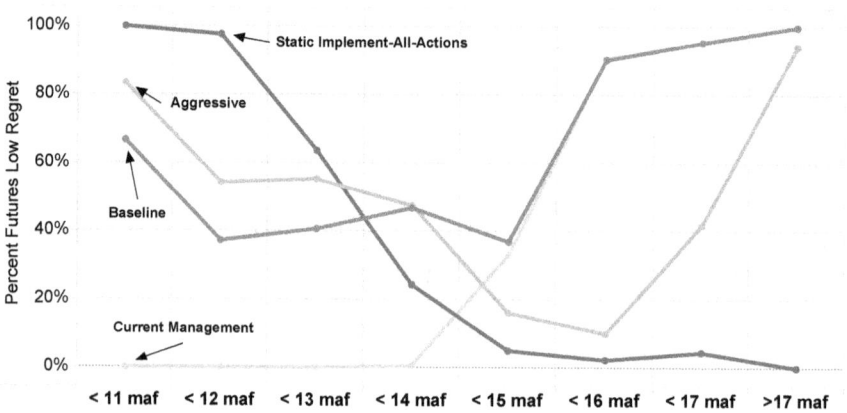

Fig. 7.7 Level of regret as a function of Colorado River long-term streamflow for four strategies

inability to predict future conditions based on observations. To illustrate this, Fig. 7.7 shows the percent of futures with low regret as a function of long-term streamflow for two static strategies ("Current Management" and "Implement All Actions") and two of the adaptive strategies shown in Fig. 7.6 ("Baseline" and "Aggressive"). The two static, non-adaptive strategies have the highest regret in either the high streamflow conditions ("Implement All Actions") or the low streamflow conditions ("Current Management"). By contrast, the two adaptive strategies are more balanced, although the "Baseline" adaptive strategy has moderate regret in the driest conditions, and the "Aggressive" strategy is low regret under wetter conditions, both adaptive strategies perform similarly.

7.3.5 Updating Beliefs About the Future to Guide Adaptation

There is no single robust choice among the automatic adjustment adaptive strategies evaluated by Bloom (2015). Each strategy suggests some tradeoff in robustness for one range of conditions relative to another. Furthermore, strategies with different triggers reflect different levels of risk tolerance with respect to reliability and costs. Therefore, a planner's view of the best choice will be shaped by her preference for the varying risks and perceptions about future conditions. An adaptive strategy based on formal review and continued learning would evolve not only in response to some predefined triggers but also as decisionmakers' beliefs about future conditions evolve over time. Bloom models this through Bayesian updating (see Lindley (1972) for a review of Bayes' Theorem).

Figure 7.8 shows how a planner with specific expectations about the likelihood of future conditions (priors) might react to new information about future conditions. The top panel (line) shows one plausible time series of decadal average streamflow measured at Lees Ferry. The bars show the running average flow from 2012 through the end of the specified decade. In this future, flows are around the long-term historical average through 2020 and then decline over the subsequent decades. The long-term average declines accordingly to about 12 maf/year—conditions that Bloom classifies as part of the *Severely Declining Supply* vulnerability.

The bottom panel shows how a decisionmaker might update her assessment of the type of future unfolding. In this example, her priors in 2012 were a 50% chance on the *Declining Supply* conditions, a 25% chance for the *Severely Declining Supply* conditions, and a 25% chance for all other futures. As low flows are observed over time, her beliefs adjust accordingly. While they do not adjust by 2040 all the way to the true state—*Severely Declining Supply*—her assessment of not being in the *Declining Supply* or *Severely Declining Supply* condition goes to zero percent. Another planner with higher priors for the Severely Declining Supply would have a higher posterior assessment of the *Severely Declining Supply* condition by 2040. Planners' priors can be based on the best scientific evidence available at the time and their own understanding of the policy challenge. Because climate change is deeply uncertain, planners may be unsure about their priors, and even a single planner may wish to understand the implications of a range of prior beliefs.

7.3.6 Robust Adaptive Strategies, and Implementation Pathways

Bloom then combined the key findings from the vulnerability analysis and the Bayesian analysis to describe a robust adaptive strategy in the form of a management strategy implementation pathway. Specifically, Bloom identified threshold values of beliefs and observations that would imply sufficient confidence that a particular vulnerability would be likely and that a corresponding management response should be

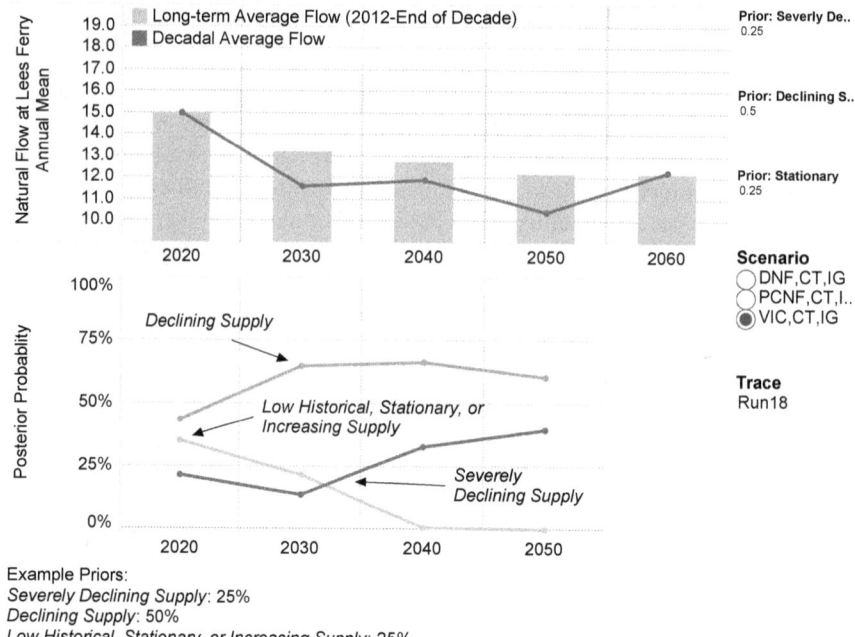

Fig. 7.8 Example of revising decisionmaker beliefs over time for 3 different prior expectations in response to declining streamflows

taken. He then used these thresholds to a suggest rule of thumb guide for interpreting new information and implementing of a robust adaptive strategy.

Figure 7.9 shows one example application of this information to devise a Basin-wide robust adaptive strategy that guides the investment of additional water supply yield (or demand reductions). In the figure, the horizontal axis shows each decade, and the vertical axis shows different identified vulnerabilities (ordered from less severe to more severe). Each box presents the 90th percentile of yield implemented for a low-regret strategy across all futures within the vulnerability. This information serves as recommendations for the amount of additional water supply or demand reduction that would need to be developed by each decade to meet the Basin's goals.

Figure 7.9 can be read as a decision matrix, where the dashed lines present one of many feasible implementation pathways through time. The example pathways show how Basin managers would develop new supply if the basin were on a trajectory consistent with the *Below Historical Streamflow With Severe Drought* vulnerability—similar to current Colorado River Basin conditions. By 2030, the Basin managers would have implemented options to increase net supply by 2.1 maf/year. The figure then highlights a 2030 decision point. At this point, conditions may suggest that the Basin is not trending as dry as *Below Historical Streamflow with Severe Drought*. Conditions may also suggest that system status is trending even worse. The hori-

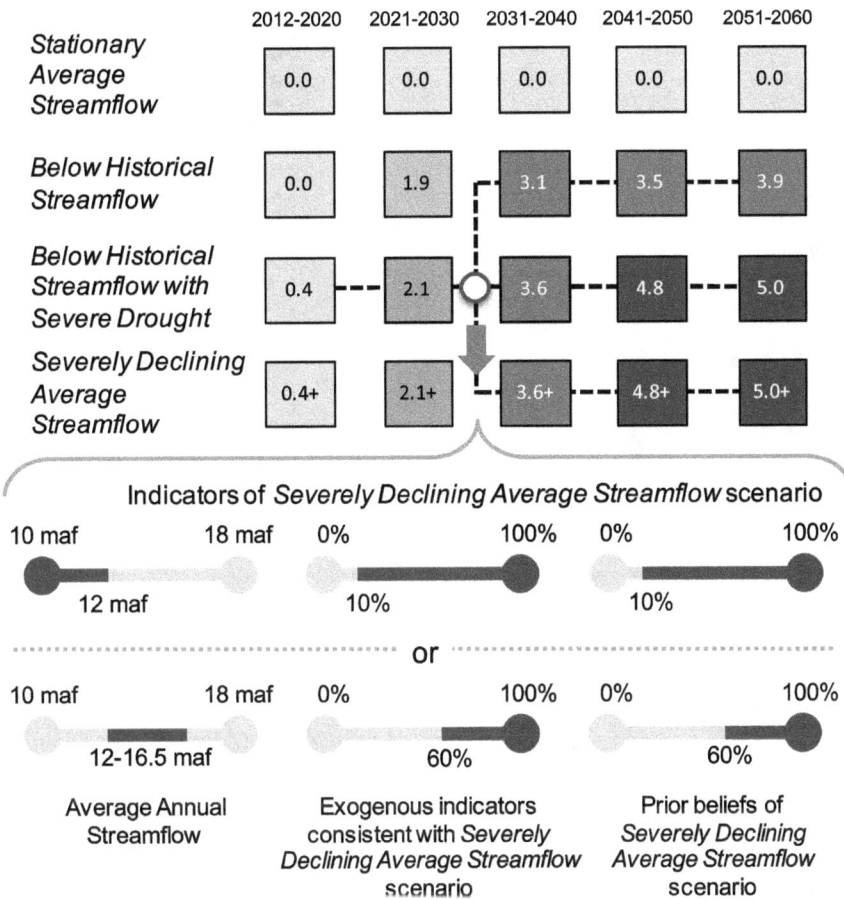

Fig. 7.9 Illustrative robust adaptive strategy for Colorado River from Bloom (2015)

zontal barbell charts in the bottom of the figure show a set of conditions and beliefs derived from the Bayesian analysis that are consistent with the *Severely Declining Supply* scenario. If these conditions are met and beliefs held, then the Basin would need to move down a row in the figure and increase net supply to more than 3.6 maf in the 2031–2040 decade. While such a strategy would likely not be implemented automatically, the analysis presented here demonstrates the conditions to be monitored and the level of new supply that would require formal review, and the range of actions that would need to be considered at such a point.

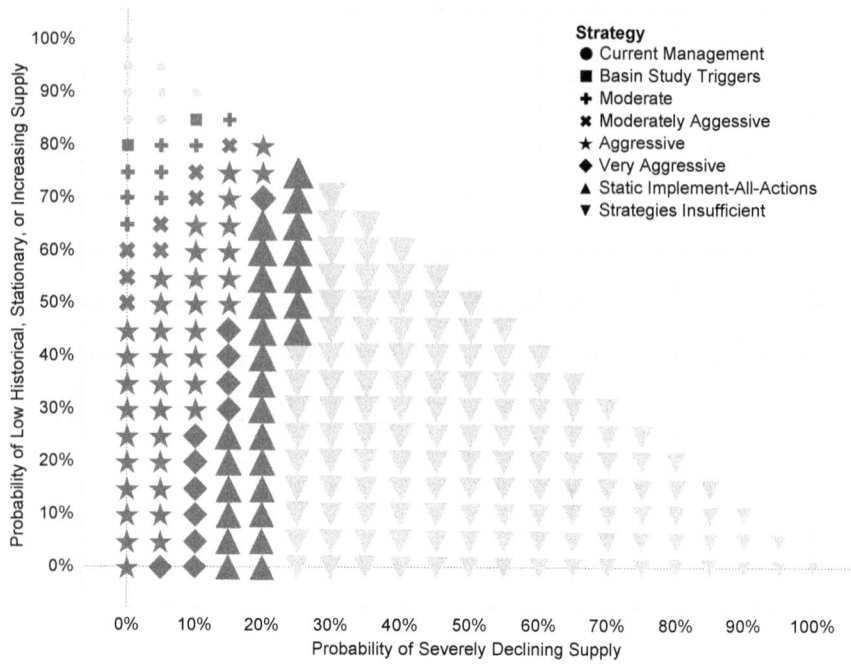

Fig. 7.10 Lowest-regret Colorado River Basin strategy for different probabilities of two vulnerabilities

7.3.7 Need for Transformative Solutions

The Basin Study analysis evaluated four adaptive strategies and explored the tradeoffs among them. Based on those findings, the Basin has moved forward on developing the least-regret options included in those portfolios (Bureau of Reclamation 2015). The follow-on work by Bloom (2015) concurrently demonstrated how the Basin could take the analysis further and develop a robust adaptive strategy to guide investments over the coming decade. One important finding from both works is that there are many plausible futures in which the water management strategies considered in the Basin Study would not be sufficient to ensure acceptable outcomes. Figure 7.10 shows the optimal strategy for different probabilities of facing a *Severely Declining Supply* scenario (horizontal axis), *Stationary or Increasing Supply* (vertical axis), or all other conditions (implied third dimension). In general, if the probability of facing the *Severely Declining Supply* scenario is 30% likely, then the options evaluated in the Basin Study will likely be insufficient and more transformative options, such as pricing, markets, and new technologies, will be needed. Additional iterations of the RDM analysis could help the Basin evaluate such options.

7.4 Case Study 2: Using RDM to Develop Climate Mitigation Technology Diffusion Policies

The historic 2016 Paris Climate Accord had two main policy objectives (United Nations 2016): (1) maintain global temperature increases below two degrees Celsius, and (2) stabilize greenhouse gas (GHG) emissions by the end of this century. The international community also realizes that meeting these aspirations will require significant and coordinated investment to develop and disseminate technologies to decarbonize the world's economies. Climate policies that achieve the Paris goals will likely include mechanisms to incentivize the use of renewables and energy efficiency through carbon taxes or trading schemes, as well as investments in the development and transfer to developing countries of new low-carbon technologies. These investments could take the form of price subsidies for technology or direct subsidies for research and development (R&D).

The Green Climate Fund (GCF) was designated to coordinate the global investments needed to help countries transition from fossil energy technologies (FETs) toward sustainable energy technologies (SETs). In this context, SETs include all technologies used for primary energy production that do not result in greenhouse gas emissions. Examples include: photovoltaic solar panels, solar thermal energy systems, wind turbines, marine current turbines, tidal power technologies, nuclear energy technologies, geothermal energy technologies, hydropower technologies, and low-GHG-intensity biomass.[2]

Technology-based climate mitigation policy is complex because of the disparity in social and technological conditions of advanced and emerging nations, the sensitivity of technological pathways to policy intervention, and the wide set of policies available for intervention. This policy context is also fraught with deep uncertainty. On the one hand, the speed and extent of future climate change remains highly uncertain. On the other hand, it is impossible to anticipate the pace and scope of future technological development. Paradoxically, these highly uncertain properties are critical to determine the strength, duration, and structure of international climate change mitigation policies and investments.

The presence of both climate and technological deep uncertainty makes it inappropriate to use standard probabilistic planning methods in this context, and instead calls for the use of DMDU methods for long-term policy design. This case study summarizes a study by Molina-Perez (2016), the first comprehensive demonstration of how RDM methods could be used to inform global climate technology policymaking. The case study shows how RDM can be used to identify robust adaptive policies (or strategies) for promoting international decarbonization despite the challenges posed by the inherent complexity and uncertainty associated with climate change and technological change. Specifically, the study helps illuminate under what conditions the GCF's investments and climate policy can successfully enable the international diffusion of SETs and meet the two objectives of the Paris Accord.

---

[2]Note that the SET group of technologies does not include carbon removal and sequestration technologies from power plant emissions, such as carbon capture and storage (CCS).

Table 7.2 Simplified XLRM framing matrix for climate technology policy analyses

Exogenous factors (X)	Policy levers (L)
Climate uncertainty (12 scenarios) Technological factors (300): • Elasticity of substitution • R&D returns SETs • R&D returns FETs • Innovation propensity SETs • Innovation propensity FETs • Transferability SETs • Transferability FETs	(P0) Current policies (P1) Independent carbon tax [Both] (P2) Independent carbon tax + Independent Tech/R&D [Both] (P3) Harmonized carbon tax + Co-Tech [GCF] + R&D [AR] (P4) Harmonized carbon tax + Co-Tech [GCF] + I. R&D [Both] (P5) Harmonized carbon tax + Co-R&D [GCF] + Tech [AR] (P6) Harmonized carbon tax + Co-R&D [GCF] + I. Tech [Both] (P7) Harmonized carbon tax + Co-Tech-R&D [GCF] System
Relationships (R)	Measures of performance (M)
Exploratory Dynamic Integrated Assessment Model (EDIAM)	• End-of-century temperature rise • Stabilization of GHG emissions • Economic costs of policy intervention

Note SETs = Sustainable Energy Technologies; FETs = Fossil Energy Technologies; GCF = Green Climate Fund

7.4.1 Decision Framing for Climate Technology Policy Analysis

To summarize the analysis, Table 7.2 shows a simplified XLRM chart from the Decision Framing step. The uncertain, exogenous factors include twelve scenarios for the evolution of the climate system based on endogenously determined global GHG emission levels ("Climate Uncertainty"), and seven factors governing the development and transfer of technologies ("Technological Factors"). The study considered seven different policies in addition to a no-additional action baseline ("Policy Levers"). The policies include a mixture of carbon taxes and subsidies for technology, research, and development. The first two policies consider no cooperation between regions, whereas the third through seventh policies consider some level of cooperation on taxes and subsidies. An integrated assessment model is used to estimate global outcome indicators (or measures of performance), including end-of-century temperatures, GHG emissions, and economic costs of the policies.

7.4.2 Modeling International Technological Change

Molina-Perez (2016) used an integrated assessment model specifically developed to support Exploratory Modeling and the evaluation of dynamic complexity. This

Exploratory Dynamic Integrated Assessment Model (EDIAM) focuses on describing the processes of technological change across advanced and emerging nations, and how this process connects with economic growth and climate change.

The EDIAM framework depends upon three main mechanisms. The first mechanism determines the volume of R&D investments that are captured by SETs or FETs (based on principles presented in Acemoglu (2002) and Acemoglu et al. (2012)). Five forces determine the direction of R&D:

- The "direct productivity effect"—incentivizing research in the sector with the more advanced and productive technologies
- The "price effect"—incentivizing research in the energy sector with the higher energy prices
- The "market size effect"—pushing R&D toward the sector with the highest market size
- The "experience effect"—pushing innovative activity toward the sector that more rapidly reduces technological production costs
- The "innovation propensity effect"—incentivizing R&D in the sector that more rapidly yields new technologies

The remaining two mechanisms describe how technologies evolve in each of the two economic regions—advanced and emerging. In the advanced region, entrepreneurs use existing technologies to develop new technologies, an incremental pattern of change commonly defined in the literature as "building on the shoulders of giants" (Acemoglu 2002; Arthur 2009). In the emerging region, innovative activity is focused on closing the technological gap through imitation and adaptation of foreign technologies. Across both regions, the speed and scope of development is determined by three distinct technological properties across the FETs and SETs sectors: R&D returns, innovation propensity, and innovation transferability. Finally, EDIAM uses the combination of the three mechanisms to endogenously determine the optimal policy response for mitigating climate change (optimally given all of the assumptions), thus making the system highly path-dependent.

7.4.3 Evaluating Policies Across a Wide Range of Plausible Futures

Molina-Perez used EDIAM to explore how different policies could shape technology development and transfer, and how they would reduce GHG emissions and mitigate climate change across a range of plausible futures. The study evaluated seven alternative policies across a diverse ensemble of futures that capture the complex interactions across the fossil and sustainable energy technologies. The ensemble is derived using a mixed experimental design of a full factorial sampling of 12 climate model responses (scenarios) and a 300-element Latin Hypercube sample across the seven technological factors shown in Table 7.2.

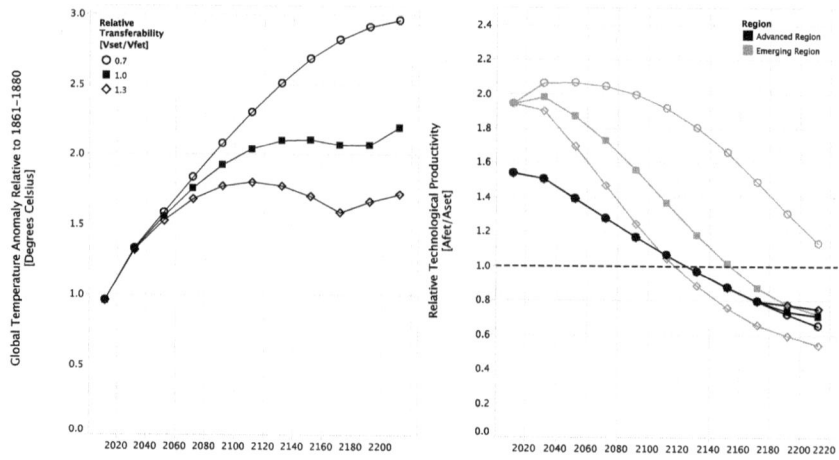

Fig. 7.11 Temperature and technological productivity outcomes for the optimal climate technology policy P2 for three futures

Figure 7.11 presents simulation results for the optimal policy for just three of these futures in which R&D returns are varied and all other factors are held constant. The left pane shows the global temperature anomaly over time for the best policy for three values of the relative R&D returns parameter. Only in two of the three cases is the optimal policy able to keep global temperature increases below 2 °C through 2100. The right pane shows the evolution of technological productivity for the advanced and emerging regions. As expected, higher relative R&D returns can lead to parity in technology productivity under the optimal policies and favorable exogenous conditions.

Molina-Perez also considered the role of climate uncertainty in determining the structure and effectiveness of the optimal policy response. For this study, the climate parameters of EDIAM were calibrated using the CMIP5 climate ensemble (IPCC 2014b; Taylor et al. 2012; Working Group on Coupled Modeling 2015). Each general circulation model used by the IPCC and included in the CMIP5 ensemble uses different assumptions and parameter values to describe the atmospheric changes resulting from growing anthropogenic GHG emissions and, as a result, the magnitude of the estimated changes varies greatly among different modeling groups. For example, Molina-Perez shows that for a more abrupt climate scenario, such as MIROC-ESM-CHEM, it is possible that the optimal policy would use a higher mix of carbon taxes, research subsidies, and technology subsidies than in the case of a less abrupt climate scenario (e.g., NorESM1-M).

7.4.4 Key Vulnerabilities of Climate Technology Policies

Following the RDM methodology, Molina-Perez used Scenario Discovery (SD) methods to isolate the key uncertainties that drive poor policy performance. Specifically, he expanded on prior SD methods (such as PRIM) by using high-dimensional stacking methods (Suzuki et al. 2006; Taylor et al. 2006; LeBlanc et al. 1990) to summarize the vulnerabilities.

In SD with dimensional stacking, the process starts by decomposing each uncertainty dimension into basic categorical levels. These transformed uncertainty dimensions are then combined into "scenario cells" that represent the basic elements of the uncertainty space. These scenario cells are then combined iteratively with other cells, yielding a final map of the uncertainty space. Following the SD approach, and in a similar way to Taylor et al. (2006), coverage and density statistics are estimated for each scenario cell such that only high-density and high-coverage cells are visualized. Finally, for each axis in the scenario map, the stacking order is determined using principal component analysis (i.e., the loadings of each principal component determine which dimensions can be stacked together). Uncertainty dimensions that can be associated with a principal component are stacked in contiguous montages.

Figure 7.12, for example, shows the vulnerabilities with respect to the two-degree temperature target identified using high-dimensional stacking for the second policy (P2), which includes independent carbon taxes, and subsidies for technology and R&D. The first horizontal montage describes three climate sensitivity bins (low, medium, and high); the second-order montage describes two relative innovation propensity bins (high and low); and the third-order montage describes two relative transferability bins (high and low). The first vertical montage describes three substitution levels (high, medium, and low); and the second-order montage describes two R&D returns bins (high and low). The cells are shaded according to the relative number of futures in which the policy meets the temperature target. The figure shows that policy P2 is more effective than the others in meeting the target in the low- and medium-sensitivity climate futures. Evaluations of the vulnerabilities for the other policies show that each policy regime works better under a different area of the uncertainty space. For instance, P7, which includes GCF funding for technology and research and development, meets the temperature objective in futures with low elasticity of substitution.

The evaluations of vulnerabilities for all seven policies suggest that the implementation of an international carbon tax (homogenous or not) can induce international decarbonization, but under many scenarios it fails to do this at a rate that meets the Paris objectives. As also shown in previous studies, the elasticity of substitution between SETs and FETs is a key factor determining the success (with respect to the Paris objectives) of the optimal policy response. However, this study also shows that technological and climate uncertainties are equally influential. In particular, the technological progress of SETs relative to FETs can affect the structure and effect of optimal environmental regulation to the same degree as climate uncertainty.

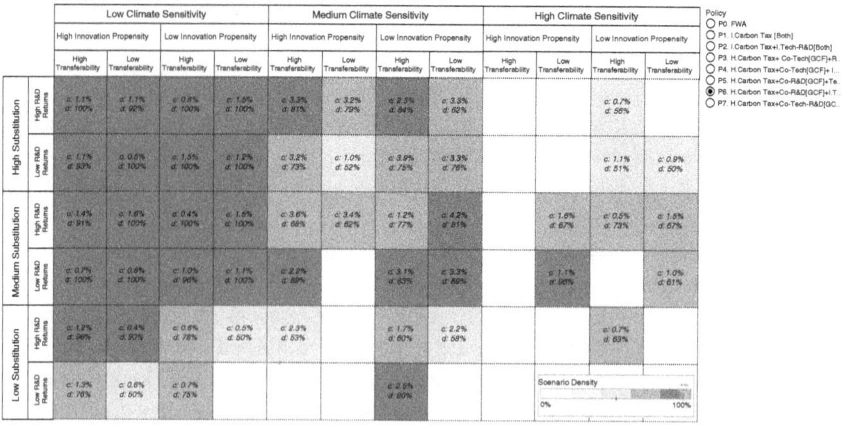

Fig. 7.12 Vulnerability map for policy P2 with respect to not exceeding the two-degree temperature rise target

7.4.5 Developing a Robust Adaptive Climate Technology Policy

Molina-Perez used a two-step process to identify a robust adaptive climate policy. In the first step, he identified the policies with the least regret that also do not exceed a 10% cost threshold across all the uncertainties.[3] Policy regret was defined as the difference in consumer welfare[4] between a specific policy and the best policy for a given future. Figure 7.13 depicts the lowest-regret policy that meets the cost threshold for the same uncertainty bins used to characterize the vulnerabilities. The label and the color legend indicate the corresponding least-regret policy for each scenario cell. The dark cells refer to the GCF-based policies (P3-P7), while the light cells denote the independent policy architectures (P1 and P2). Cells with no shading represent futures for which no policy has costs below the 10% cost threshold.

These results show that the elasticity of substitution of the technological sectors and climate sensitivity are the key drivers of this vulnerability type. However, the results also highlight the important role of some of the technological uncertainty dimensions. For instance, the figure shows that for the majority of scenario cells with low relative transferability, it is not possible to meet cost and climate targets, regardless of the other uncertainty dimensions. This exemplifies the importance that the pace of development of SETs in the emerging region has for reducing the costs

---

[3]Note that the 10% threshold is arbitrarily chosen, and could be the basis for additional exploratory analysis.

[4]The utility function of the representative consumers considers as inputs both the quality of the environment (inverse function of temperature rise) and consumption (proxy for economic performance).

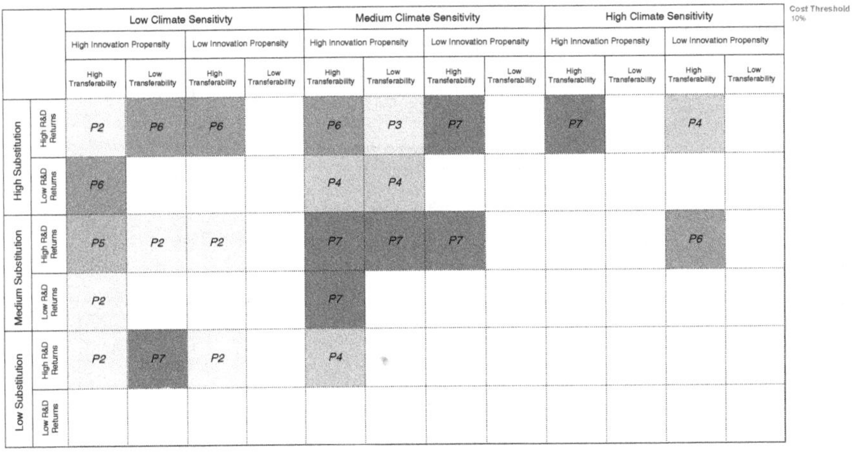

Fig. 7.13 Least-regret policies for different bins of uncertainty

of the best environmental regulation. Similarly, the R&D returns of SETs are also shown to be an important driver of vulnerability.

As with the Colorado River Basin case study, no single policy is shown to be robust across all futures. Therefore, in the second step, Molina-Perez defined transition rules to inform adaptive policies for achieving global decarbonization at a low cost. To do so, he traced commonalities among the least-regret policy identified in each of the scenario cells described in Fig. 7.13. This identifies the combination of uncertainties that signal a move from one policy option to another. A simple recursive algorithm is implemented to describe moves from P1 (independent carbon tax) to successively more comprehensive policies, including significant GCF investments. The algorithm proceeds by: (1) listing valid moves toward each individual scenario cell from external scenario cells (i.e., pathways), (2) listing the uncertainty conditions for each valid move (i.e., uncertainty bins), (3) searching for common uncertainty conditions across these moves, and (4) identifying transition nodes for the different pathways initially identified.

Figure 7.14 shows a set of pathways that represents how the global community might move from one policy to another in response to the changing conditions that were identified in the vulnerability analysis. In these pathways, independent carbon taxation in P1 is a first step, because this policy directs the economic agents' efforts toward the sustainable energy sector, and because this policy is also necessary to fund the complementary technology policy programs in both regions (a result from EDIAM's modeling framework). In subsequent phases, different pathways are triggered depending on the unfolding climate and technological conditions. For example, if climate sensitivity is extremely high ($\beta > 5$, abrupt climate change), then it would be necessary to implement environmental regulations that are above the 10% cost threshold. In comparison, if climate sensitivity is low ($\beta < 4$, slow climate change),

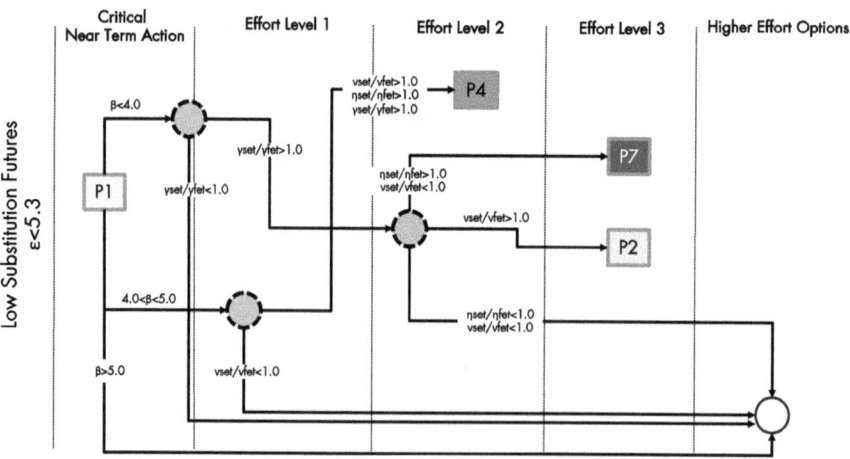

Fig. 7.14 Dynamic, adaptive policy for low elasticity of substitution futures

then the least-regret policy depends on SETs' R&D relative returns ($\gamma set/\gamma fet$). If SETs' R&D returns are below FETs' R&D returns ($set/\gamma fet < 1$, more progress in FETs), then only high cost policies would be available. On the contrary, if SETs outperform the R&D returns of FETs, two policy alternatives are available: If SETs' transferability is higher than FETs' transferability ($\upsilon set/\upsilon fet > 1$, more rapid update of SETs in emerging nations), then cooperation is not needed, and the individual comprehensive policy (P2) would meet the policy objectives at a low cost. In the opposite case ($\upsilon set/\upsilon fet < 1$, more rapid update of FETs in emerging nations), some cooperation through the GCF (P7) would be required to meet the climate objectives below the 10% cost threshold. Note that these pathways consider only the low elasticity of substitution futures. Other pathways are suitable for higher elasticity of substitution futures. Other pathways are suitable for higher elasticity of substitution futures.

These pathways illustrate clearly that optimal environmental regulation in the context of climate change mitigation is not a static concept but rather a dynamic one that must adapt to changing climate and technological conditions. This analysis also provides a means to identify and evaluate appropriate policies for this dynamic environment. The analysis shows that different cooperation regimes under the GCF are best suited for different combinations of climate and technological conditions, such that it is possible to combine these different architectures into a dynamic framework for technological cooperation.

7.5 Reflections

The complexity and analytic requirements of RDM have been seen by many as an impediment to more widespread adoption. While there have been many applications of RDM that use simple models or are modest in the extent of exploration, the first case study on Colorado River Basin planning certainly was not simple. In fact, this application is notable because it was both extremely analytically intensive, requiring two computer clusters many months to perform all the needed simulations, and it was integrated fully into a multi-month stakeholder process.

As computing becomes more powerful and flexible through the use of high-performance computing facilities and the cloud, however, near real-time support of planning processes using RDM-style analytics becomes more tenable. To explore this possibility, RAND and Lawrence Livermore National Laboratory put on a workshop in late 2014 with some of the same Colorado River Basin planners in which a high-performance computer was used to perform over lunch the equivalent simulations and analyses that were done for the Basin Study over the course of an entire summer (Groves et al. 2016). Other more recent studies have also begun to use cloud computing to support large ensemble analyses that RDM studies often require (Cervigni et al. 2015; Kalra et al. 2015; World Bank Group 2016). Connecting this near-simultaneous analytic power to interactive visualizations has the potential to transform planning processes.

These two case studies also highlight how the iterative, exploratory nature of RDM can support the development of adaptive strategies—an obvious necessity for identifying truly robust strategies in water management, international technology policymaking, and many other domains. A simple recipe might describe the approach presented here in the following way:

1. Define low regret options for an exhaustive evaluation of automatically adjusting strategies across a wide range of futures;
2. Identify the vulnerabilities for a strategy that implements just the low-regret options;
3. Define and evaluate different signposts and triggers (which could include decisionmaker beliefs about evolving conditions) for making the needed policy adjustments if vulnerabilities appear probable; and
4. Present visualizations of pathways through the decision space to stakeholders and decisionmakers.

This approach to designing adaptive strategies represents a step toward a convergence of two DMDU methods presented in this book—RDM (Chap. 2) and Dynamic Adaptive Policy Pathways (Chap. 4). Continued merging of DMDU techniques for exploration of uncertainties and policies, and techniques and visualizations for defining policy pathways, could yield additional benefit.

References

Acemoglu, D. (2002). Directed technical change. *The Review of Economic Studies, 69*(4), 781–809.

Acemoglu, D., Aghion, P., Bursztyn, L., & Hemous, D. (2012). The Environment and directed technical change. *American Economic Review, 102*(1), 131–166.

Arthur, W.B. (2009). The Nature of Technology: What It Is and How It Evolves: Penguin UK.

Averyt, K., Meldrum, J., Caldwell, P., Sun, G., McNulty, S., Huber-Lee, A., et al. (2013). Sectoral contributions to surface water stress in the coterminous United States. *Environmental Research Letters, 8,* 35046.

Bloom, E., (2015). Changing midstream: providing decision support for adaptive strategies using Robust Decision Making: applications in the Colorado River Basin. Pardee RAND Graduate School. Retrieved July 13, 2018 from http://www.rand.org/pubs/rgs_dissertations/RGSD348.html.

Bureau of Reclamation. (1922). Colorado River Compact. Retrieved July 11, 2018 from https://www.usbr.gov/lc/region/g1000/pdfiles/crcompct.pdf.

Bureau of Reclamation. (2012). Colorado River Basin water supply and demand study: study report United States Bureau of Reclamation (Ed.). Retrieved July 11, 2018 from http://www.usbr.gov/lc/region/programs/crbstudy/finalreport/studyrpt.html.

Bureau of Reclamation. (2015). Colorado River Basin stakeholders moving forward to address challenges identified in the Colorado River Basin study–phase 1 report. Retrieved July 11, 2018 from https://www.usbr.gov/lc/region/programs/crbstudy/MovingForward/Phase1Report.html.

Cervigni, R., Liden, R., Neumann. J. E., and Strzepek, K. M. (2015). Enhancing the climate resilience of Africa's infrastructure: the power and water sectors. World Bank, Washington, D.C. https://doi.org/10.1596/978-1-4648-0466-3.

Christensen, N. S., & Lettenmaier, D. P. (2007). A multimodel ensemble approach to assessment of climate change impacts on the hydrology and water resources of the Colorado River Basin. *Hydrology and Earth System Sciences, 11,* 1417–1434.

Fischbach, J. R., Lempert, R. J., Molina-Perez, E., Tariq, A. A., Finucane, M. L., Hoss, F. (2015). Managing water quality in the face of uncertainty. RAND Corporation, Santa Monica, CA. Retrieved March 8, 2018 from https://www.rand.org/pubs/research_reports/RR720.html.

Groves, D. G., 2006. New methods for identifying robust long-term water resources management strategies for California. Pardee RAND Graduate School, Santa Monica, CA. Retrieved July 08, 2018 from http://www.rand.org/pubs/rgs_dissertations/RGSD196.

Groves, D. G., Bloom, E., Johnson, D. R., Yates, D., Mehta V. K. (2013). Addressing Climate change in local water agency plans. RAND Corporation. Retrieved July 01, 2018 from http://www.rand.org/pubs/research_reports/RR491.html.

Groves, D. G., Davis, M., Wilkinson, R., & Lempert, R. J. (2008). Planning for climate change in the Inland Empire. *Water Resources IMPACT, 10,* 14–17.

Groves, D. G., Fischbach, J. R., Bloom, E., Knopman, D., & Keefe, R. (2013). Adapting to a changing Colorado River. RAND Corporation, Santa Monica, CA. Retrieved July 02, 2018. http://www.rand.org/pubs/research_reports/RR242.html.

Groves, D. G., R. J. Lempert, D. W. May, J. R. Leek, & J. Syme (2016). Using High-performance computing to support water resource planning: a workshop demonstration of real-time analytic facilitation for the Colorado River Basin. Santa Monica, CA. https://www.rand.org/pubs/conf_proceedings/CF339.html.

IPCC (2014a). Climate Change 2014: synthesis report. Contribution of working groups I, II and III to the fifth assessment report of the intergovernmental panel on climate change core writing team. In R. K. Pachauri & L. A. Meyer (Eds.). IPCC, Geneva, Switzerland.

IPCC (2014b). Impacts, adaptation, and vulnerability. part b: regional aspects. contribution of working group ii to the fifth assessment report of the intergovernmental panel on climate change. Barros, V. R., Field, C. B., Dokken, D. J., Mastrandrea, M. D., Mach, K. J., Bilir, T. E., Chatterjee, M., Ebi, K. L., Estrada, Y. O., Genova, R. C., Girma, B., Kissel, E. S., Levy, A. N., MacCracken,

S., Mastrandrea, P. R. & White, L. L. (Eds.). Cambridge University Press, Cambridge UK, New York, NY.

Kalra, N. R., Groves, D. G., Bonzanigo, L, Perez, E. M., Ramos Taipe, C. L., Cabanillas, I. R., et al. (2015). Robust Decision-Making in the water sector: a strategy for implementing Lima's long-term water resources master plan. Retrieved July 03, 2018 from https://openknowledge.worldbank.org/handle/10986/22861.

LeBlanc, J., Ward, M. O., & Wittels, N. (1990). Exploring n-dimensional databases. In *Proceedings of the 1st conference on Visualization '90* (pp. 230–237).

Lempert, R J., Popper, S. W., Groves, D. G., Kalra, N., Fischbach, J. R. Bankes, S. C., et al. (2013b). Making Good Decisions Without Predictions. *RAND Corporation*. Retrieved July 01, 2018 from http://www.rand.org/pubs/research_briefs/RB9701.html.

Lempert, R. J., Kalra, N., Peyraud, S., Mao, Z., Tan, S. B., Cira, D., & Lotsch, A. (2013). Ensuring robust flood risk management in Ho Chi Minh City. Retrieved July 01, 2018 from http://www.rand.org/pubs/external_publications/EP50282.html.

Lindley, D. V. (1972). Bayesian statistics, a review. *Society for industrial and applied mathematics* (Vol. 2) https://doi.org/10.1137/1.9781611970654.

Loucks, D. P., Stedinger, J. R., Haith, D. A. (1981). Water resource systems planning and analysis. Prentice-Hall.

Melillo, J. M., Richmond, T. T., Yohe, G. W. (2014). Climate change impacts in the United States: the third national climate assessment, Vol. 52. https://doi.org/10.7930/j0z31wj2.

Molina-Perez, E. (2016). Directed international technological change and climate policy: new methods for identifying robust policies under conditions of deep uncertainty. Pardee RAND Graduate School. Retrieved July 02, 2018 from http://www.rand.org/pubs/rgs_dissertations/RGSD369.html.

Nash, L. L., & Gleick, P. H. (1993). *The Colorado River Basin and climate change: the sensitivity of streamflow and water supply to variations in temperature and precipitation*. Washington, D.C.: United States Environmental Protection Agency.

Ngo, H. H., Guo, W., Surampalli, R. Y., Zhang, T. C. (2016). Green technologies for sustainable water management. American Society of Civil Engineers, Reston, VA. https://doi.org/10.1061/9780784414422.

Raucher, K., & Raucher, R. (2015). Embracing uncertainty: a case study examination of how climate change is shifting water utility planning. Retrieved July 08, 2018 from https://www.csuohio.edu/urban/sites/csuohio.edu.urban/files/Climate_Change_and_Water_Utility_Planning.pdf.

Retrieved March 8, 2018 from https://openknowledge.worldbank.org/handle/10986/21875.

Seager, R., Mingfang, T., Held, I., Kushnir, V., Jian, L., Vecchi, G., et al. (2007). Model projections of an imminent transition to a more arid climate in Southwestern North America. *Science, 316*(5825), 1181–1184.

Suzuki, S., Stern, D., & Manzocchi, T. (2006). Using association rule mining and highdimensional visualization to explore the impact of geological features on dynamic flow behavior. *SPE annual technical conference and exhibition*.

Swanson, D., Barg, S., Tyler, S., Venema, H., Tomar, S., Bhadwal, S., et al. (2010). Seven tools for creating adaptive policies. *Technological Forecasting and Social Change, 77*, 924–939.

Taylor, A. L., Hickey, T. J., Prinz, A. A., & Marder, E. (2006). Structure and visualization of high-dimensional conductance spaces. *Journal of Neurophysiology, 96*(2), 891–905.

Taylor, K. E., Stouffer, R. J., & Meehl, G. A. (2012). An overview of CMIP5 and the experiment design. *Bulletin of the American Meteorological Society, 93*(4), 485–498.

United Nations (2016). Paris Agreement. Paris: United Nations, pp.1-27.

Working Group on Coupled Modelling. (2015). Coupled Model Intercomparison Project 5 (CMIP5), edited by Programme, World Climate Research. Retrieved March 8, 2018 from https://pcmdi9.llnl.gov/projects/cmip5/.

World Bank Group (2016). Lesotho water security and climate change assessment. World Bank, Washington, DC. Retrieved July 10, 2018 from https://openknowledge.worldbank.org/handle/10986/24905.

Dr. David G. Groves (RAND Corporation) is a Senior Policy Researcher at the RAND Corporation, Co-Director of the Center for Decision Making Under Uncertainty, Co-Director of the Climate Resilience Center, and Core Faculty Member at the Pardee RAND Graduate School. He is a key developer of new methods for decisionmaking under deep uncertainty, and he works directly with natural resources managers worldwide to improve planning for the uncertain future. His primary practice areas include water resources management and coastal resilience planning, with an emphasis on climate adaptation and resilience. He has led numerous projects in which his teams used DMDU methods to improve an agency's long-term plans. For example, he led the development of the decision framework and planning support tool used to develop the State of Louisiana's 2012 and 2017 coastal master plans, and he led the decision analysis for the Government of Puerto Rico's 2018 Economic and Disaster Recovery Plan.

Dr. Edmundo Molina-Perez (School of Government and Public Transformation, Instituto Tecnológico y de Estudios Superiores de Monterrey) is Research Associate Professor at the School of Government and Public Transformation of Tecnológico de Monterrey. His work focusses on the development of new computational methods for the study of socio-technological systems, and the use of Data Science methods for decision analysis under conditions of deep uncertainty. He currently leads applied research work on Mexico's water and energy sectors and directs Conacyt Energy Knowledge Network's Data-Driven Decision Analysis Environment (3D-AE). He teaches courses on systems modeling, advanced simulation techniques, econometrics, and microeconomics.

Dr. Evan Bloom (Netflix Science and Analytics) is a Senior Data Scientist at Netflix where he designs and analyzes experiments to inform key strategic and tactical marketing decisions. He also builds software solutions to automate the analysis workflow. Prior to working at Netflix, he was a Manager of Data Science at Capital One, where he focused on predictive machine learning use-cases ranging from transaction fraud to consumer financial health. Dr. Bloom earned his Ph.D. from the Pardee RAND Graduate School, where his research covered methods and applications of Robust Decision Making to provide stakeholder decision support under deep uncertainty, with a particular focus on climate adaptation in water systems.

Dr. Jordan R. Fischbach (RAND Corporation) is a Senior Policy Researcher at the RAND Corporation, Co-Director of the Climate Resilience Center, and an Affiliate Faculty Member at the Pardee RAND Graduate School. Dr. Fischbach leads RAND research focused on climate adaptation, water resources management, urban resilience, and coastal planning, and has expertise in risk analysis, exploratory simulation modeling, and Robust Decision Making. He currently serves as a Co-Investigator for the NOAA Mid-Atlantic Regional Integrated Sciences and Assessments (MARISA) center and has led multiple projects supporting resilience planning and improved stormwater management in Pittsburgh. He also served as the principal investigator for the flood risk and damage assessment for the State of Louisiana's coastal master planning, and recently led an assessment of damage and needs to support Puerto Rico's recovery after Hurricane Maria.

Open Access This chapter is licensed under the terms of the Creative Commons Attribution 4.0 International License (http://creativecommons.org/licenses/by/4.0/), which permits use, sharing, adaptation, distribution and reproduction in any medium or format, as long as you give appropriate credit to the original author(s) and the source, provide a link to the Creative Commons licence and indicate if changes were made.

The images or other third party material in this chapter are included in the chapter's Creative Commons licence, unless indicated otherwise in a credit line to the material. If material is not included in the chapter's Creative Commons licence and your intended use is not permitted by statutory regulation or exceeds the permitted use, you will need to obtain permission directly from the copyright holder.

Chapter 8
Dynamic Adaptive Planning (DAP): The Case of Intelligent Speed Adaptation

Vincent A. W. J. Marchau, Warren E. Walker and Jan-Willem G. M. van der Pas

Abstract This chapter aims to close the gap between the theory and practice of Dynamic Adaptive Planning (DAP).

- It (1) presents an operationalization of DAP using experts in a workshop setting, (2) applies this operationalization to a real-world policy problem involving a traffic safety technology called intelligent speed adaptation (ISA), and (3) presents both the resulting dynamic adaptive plan and the experiences of the participating stakeholders in designing the plan.
- The workshop was conducted with stakeholders who were likely to actually participate in the planning process for ISA in the Netherlands.
- The workshop was held in a computer-supported group decision room, an interactive, computer-based environment that helps a team of decisionmakers solve problems and make choices, and began with an initial promising plan. The participants then were guided through the process of (1) assessing the strengths, weaknesses, opportunities, and threats (vulnerabilities) of the initial plan, (2) selecting the most uncertain and most important opportunities and threats, (3) defining actions aimed at increasing the robustness of the initial plan, and specifying signposts and trigger values for contingent actions to be taken over time, and (4) testing the proposed plan's performance in the face of 'wildcard scenarios.'
- In an assessment of the workshop, the participants concluded that the adaptive plan that was developed is a promising step toward the large-scale implementation of ISA in the Netherlands, and that DAP is a useful approach for dealing with deep uncertainty.

V. A. W. J. Marchau (✉)
Radboud University (RU) Nijmegen School of Management, Nijmegen, The Netherlands
e-mail: v.marchau@fm.ru.nl

W. E. Walker
Delft University of Technology, Delft, The Netherlands

J.-W. G. M. van der Pas
Municipality of Eindhoven, Eindhoven, The Netherlands

© The Author(s) 2019
V. A. W. J. Marchau et al. (eds.), *Decision Making under Deep Uncertainty*,
https://doi.org/10.1007/978-3-030-05252-2_8

8.1 Introduction to the Approach

This chapter focuses on how to bring Dynamic Adaptive Planning (DAP) into practice. Although DAP was proposed over seventeen years ago (in 2001), and many cases of its use in designing plans for various policy domains have been published (see references in Chap. 3), DAP is still considered a theoretical approach. More particularly:

1. DAP lacks examples of adaptive planning developed by policymakers or domain experts. (Until now, DAP has been carried out almost exclusively by researchers, not by real-world policymakers or domain experts.)
2. DAP lacks realistic examples of real-world policy problems. (Most cases that are published were developed as hypothetical illustrations of DAP).
3. DAP can be defined as a 'high-level concept, captured in a flowchart' (Kwakkel 2010). There is limited insight into the tools and methods needed to design an adaptive plan in practice that can be used in the various steps in the flowchart. A first indication of these is given by Swanson et al. (2010). But this overview is still very broad and needs to be operationalized.

In order to develop and test DAP, a real-world decisionmaking problem involving deep uncertainty was needed. The case we selected was the implementation of a type of innovative traffic safety technology in the Netherlands. For decades, technical systems have been available that make sure a driver cannot exceed the legal speed limit. If these 'intelligent speed adaptation' (ISA) devices would have been implemented years ago, hundreds of thousands of lives worldwide could have been saved. (For example, the Swedish National Road Administration has suggested that ISA could reduce crash injuries by 20–30% in urban areas (see https://trid.trb.org/view.aspx?id=685632).) One barrier to the implementation of ISA is the uncertainty that still exists regarding the effects of (large-scale) ISA implementation (e.g., uncertainty about the acceptance of ISA, and liability in case of the system malfunctions) (van der Pas et al. 2012a). Although policymakers recognize the uncertainties that are involved in implementing ISA, their reaction is usually to initiate more research. Around 2010, Dutch policymakers developed an ISA implementation plan that was focused on implementing the appropriate ISA for the appropriate type of driver. This implementation plan was redesigned as an adaptive plan at a workshop by policymakers, domain experts, and stakeholder representatives, using DAP concepts. In this chapter, we discuss the design and results of this DAP workshop. In addition, we present an evaluation of the workshop by the participants.

8.2 Introduction to the Case

Every day, people in Europe and other parts of the world are confronted with the grim reality of losing loved ones due to traffic accidents. The World Health Organization

(WHO) estimates that every year over 1.2 million people die in traffic accidents and up to 50 million suffer non-fatal injuries (WHO 2015). This means that each day over 3000 people die, which comes down to more than 2 every minute.

Research has shown that 'excessive and inappropriate speed is the number one road safety problem in many countries, often contributing as much as one third to the total number of fatal accidents' (OECD 2006). Speeding not only influences the risk of getting involved in a traffic accident, it also affects the outcomes of an accident. For the Netherlands, it has been estimated that if all motorists were to comply with the legal speed limit using ISA, it would reduce the number of accidents resulting in injury by 25–30% (Oei 2001).

To address speeding behavior, a wide range of policy alternatives have been considered in the past. These measures (speed management measures) are often categorized using the three E's: engineering (related to both vehicle and infrastructure), education, and enforcement. Examples of infrastructure engineering to reduce speeding are speed bumps and roundabouts. Replacing crossings with roundabouts has been shown to reduce the number of crashes with injuries or fatalities significantly (Elvik 2003). In driver education, novice motorists are familiarized with the effects of speed. In the Netherlands, a mandatory educational program for speed offenders is a possible penalty. Enforcement has also proven to be an effective measure against speeding. Stationary speed enforcement alone is estimated to have reduced the number of accidents by 17%, and speed cameras are estimated to have led to a reduction of 39% in fatal accidents (Elvik and Vaa 2009). In addition, a series of effective enforcement measures have been applied in the past, such as trajectory control and undercover surveillance.

History, however, shows that the engineering category of measures in vehicle design is underused: Vehicle design is usually focused on making the vehicle faster instead of making speeding more difficult. For example, research from Sweden shows that the average top speed of all newly sold passenger vehicles in Sweden has increased significantly over the past decades, increasing from 153 km/h in 1975, to 172 km/h in 1985, to 194 km/h in 1995, and to over 200 km/h in 2002 (Sprei et al. 2008). So, the trend in vehicle engineering is not so much to reduce the possibility of exceeding the speed limit, but to enable the driver to drive faster. In-vehicle systems that assist the driver in the task of driving the vehicle are called 'advanced driver assistance systems' (ADAS). An example of an ADAS that is designed to assist the driver in choosing the appropriate speed is 'intelligent speed adaptation' (ISA).

ISA is an in-vehicle system that helps the driver to comply with the legal speed limit at a certain location. ISA technology is relatively straightforward. It uses the functionality of systems that are already available in most vehicles (e.g., a GPS device, digital maps, and engine management systems). Most ISA devices can be assigned to one of three categories depending on how intervening (or permissive) they are (Carsten and Tate 2005):

- Advisory—Display the speed limit and remind the driver of changes in the speed limit.

- Voluntary ("driver-select")—Allow the driver to enable and disable control by the vehicle of maximum speed.
- Mandatory—The vehicle is limited to travel at or below the speed limit at all times.

In addition to categorization by the level of intervention the system gives, ISA devices can be categorized by the type of speed limit information used:

- Fixed: The vehicle is informed of the posted speed limits.
- Variable: The vehicle is additionally informed of certain locations in the network where a lower speed limit is implemented. Examples could include around pedestrian crossings or the approach to sharp horizontal curves. With a variable system, the speed limits are current spatially.
- Dynamic: Additional lower speed limits are implemented because of network or weather conditions, to slow traffic in fog, on slippery roads, around major incidents, etc. With a dynamic system, speed limits are current temporarily.

Since the early 1980s, the effects of ISA have increasingly been studied using different methodologies and data collection techniques, including traffic simulation, driving simulators, and instrumented vehicles. ISA has also been demonstrated in different trials around the world (e.g., Sweden, the Netherlands, the UK, and Australia). The conclusions from all these trials and research are unambiguous regarding the positive effect of ISA on driving speed and the calculated effects on traffic safety. For instance, Carsten and Tate (2005) estimate that fitting all vehicles with a simple mandatory ISA system, making it impossible for vehicles to exceed the speed limit, would reduce injury accidents by 20% and fatal accidents by 37% (see Table 8.1). Adding a capability to respond to real-time network and weather conditions would result in a reduction of 36% in injury accidents and 59% in fatal accidents. In general, it can be concluded that the more permissive the ISA, the less it affects speed choice behavior.

Other studies have confirmed that ISA can significantly reduce speeding and improve traffic safety (Lai et al. 2012). Although speeding is a major internationally recognized policy problem, and ISA seems a proven technology that has the potential to significantly contribute to traffic safety, the obvious question that remains is: Why is it that ISA has not yet been implemented except for some advisory systems? Part of the reason is related to the various (deep) uncertainties surrounding ISA implementation, including ISA technology development (in terms of reliability and accuracy to function properly under all circumstances), the way motorists will comply with ISA support, the impact of large-scale ISA penetration on congestion and travel times, the liability in case of ISA malfunctioning, and the preferences different stakeholders have regarding ISA implementation (van der Pas et al. 2012a).

So, policymakers are aware of the potential of ISA, but policymaking for ISA seems limited to supporting further research and development on ISA implementation. Policymakers are having trouble designing policies that deal appropriately with the uncertainties surrounding ISA implementation.

8 Dynamic Adaptive Planning (DAP): The Case of Intelligent …

Table 8.1 Best estimates of accident reductions by ISA type and severity

System type	Speed limit type	Best estimate of injury accident reduction (%)	Best estimate of fatal and serious accident reduction (%)	Best estimate of fatal accident reduction (%)
Advisory	Fixed	10	14	18
	Variable	10	14	19
	Dynamic	13	18	24
Voluntary	Fixed	10	15	19
	Variable	11	16	20
	Dynamic	18	26	32
Mandatory	Fixed	20	29	37
	Variable	22	31	39
	Dynamic	36	48	59

Source Carsten and Tate (2005)

8.3 Reason for Choosing the DAP Approach

There are many uncertainties that surround ISA implementation. Many of these are not Level 4 (deep) uncertainties (e.g., how the main mechanisms work is understood and the range of things that can go wrong is known). However, there is deep uncertainty about issues such as liability if the system fails, driver acceptance (willingness to use), the behavioral adaptation of ISA drivers and road users who do not have ISA, the effects of different implementation strategies, and the effects of a large-scale real-world implementation of ISA.

One way of obtaining insights into these uncertainties would have been to initiate pilot projects and monitor their results. But, our research was begun shortly after the literature on planning for adaptation to handle deep uncertainty had begun to appear (Walker et al. 2001; Lempert et al. 2003). It was clear that traditional approaches, designed for lower levels of uncertainty, were not entirely appropriate for the higher levels. Given the urgency of the road safety problem, and the availability of new approaches for dealing with deep uncertainty, we decided to get insights into the risks associated with the implementation of ISA by using experts to apply one of the approaches to designing a plan in the face of deep uncertainty—Dynamic Adaptive Planning.

8.4 Methods for Applying DAP

The design of dynamic adaptive plans requires a new and innovative way of using policy analysis methods. These methods are not only used to select an initial promising plan; the methods are also used to probe for the vulnerabilities (and opportunities) of the initial plan and ways to protect it from failing (and to seize opportunities). Thus, decisionmakers can be prepared for the future and can be informed in advance when (under what conditions) and how their plan can be adapted. As described in Chap. 3, the process of designing a dynamic adaptive plan consists of five steps:

I. Setting the stage;
II. Assembling the initial plan;
III. Increasing the robustness of the initial plan;
IV. Setting up the monitoring system;
V. Preparing the contingent actions.

Below, we elaborate on methods that could be used to facilitate each step in practice.

In Step I and Step II, the activities carried out to determine an initial plan are essentially the same as those carried out in choosing a policy in a traditional policy analysis study (see Walker 2000). Step I is the stage-setting step. It involves the specification of objectives, constraints, and available policy alternatives. This specification should lead to a definition of success, in terms of the specification of desirable outcomes.

In Step II, a promising initial plan is identified based on a traditional, *ex-ante* evaluation of the alternatives identified in Step I. There are many well-established methods that can be used to screen the alternatives, including cost-benefit analysis (Sassone and Schaffer 1978), multi-criteria analysis (French et al. 2009), and balanced scorecards (Kaplan and Norton 1993), combined with the results from forecasts, scenarios, models, etc.

Step III focuses on (1) the identification and assessment of the vulnerabilities and opportunities of this initial plan and (2) the design of actions to increase the robustness of the initial plan. Five types of actions can be taken immediately upon implementation of the initial plan to address these vulnerabilities (and opportunities). These five types of actions (which are explained in more detail in Chap. 3) are:

- Mitigating actions (M)—Actions that reduce adverse impacts on a plan stemming from certain (or very likely) vulnerabilities;
- Hedging actions (H)—Actions that reduce adverse impacts on a plan, or spread or reduce risks that stem from uncertain vulnerabilities (much like buying car insurance);
- Seizing actions (SZ)—Actions that take advantage of certain (or very likely) opportunities that may prove beneficial to the plan.
- Exploiting actions (E)—Actions that take advantage of (uncertain) new developments that can make the plan more successful, or succeed sooner;
- Shaping actions (SH)—Actions taken proactively to affect external events or conditions that could either reduce the plan's chance of failure or increase its chance of success.

There are a variety of tools and methods that can be used to identify the vulnerabilities and opportunities of an initial plan. They can be divided into two broad categories:

- Techniques that use (computational) models. Examples include sensitivity analysis (Saltelli et al. 2001), Scenario Discovery (SD) (Bryant and Lempert 2010), and

Exploratory Modeling (EM) (Bankes 1993; Agusdinata 2008). These techniques can be used to identify vulnerabilities and opportunities by varying model inputs across the range of plausible parameter values (Kwakkel et al. 2012) or by exploring outcomes across alternative models of the system of interest (Bankes 1993; Agusdinata 2008).

- Techniques that support experts in the process of identifying assumptions, vulnerabilities, and opportunities. Several examples are included in Sect. 3.2. In this case, we used strengths, weaknesses, opportunities, and threats (SWOT) analysis (Osita et al. 2014), and threats, opportunities, weaknesses, and strengths (TOWS) analysis (Weihrich 1982).

Once the vulnerabilities and opportunities related to the initial plan are identified, these should be assessed in terms of their level of uncertainty (is the vulnerability certain or uncertain), in order specify the type of appropriate action (mitigating or hedging) to take. (If the uncertainty is certain or very likely, take it right away; if it is uncertain, make it a contingent action.) Typically, this 'uncertainty rating or ranking' is very subjective.

The second part in Step III (increasing robustness) requires that specific actions are specified. Dewar et al. (1993) mention five methods for identifying these actions (all of which can be used in a workshop setting): (a) using relevant theories of causation, (b) using historical and comparative experiences, (c) using creativity, (d) using scenarios, and (e) using insurance or regulatory requirements. However, their report does not show how to identify the appropriate actions in a structured and integrated way. SWOT analysis and TOWS analysis can be used to fill this gap. The SWOT analysis reveals the strengths, weaknesses, opportunities, and threats; TOWS then uses the SWOT analysis as input to identify suitable actions in light of the SWOT results.

In Step IV, a monitoring system is designed, and conditions suggesting that a change in the plan is needed (in order to save it from failure or increase the chances for its success) are specified (these are called 'triggers'). The monitoring system tracks the development of the uncertain vulnerabilities of the initial plan (the developments that are being tracked are called 'signposts'). Levels of the signposts beyond which the objectives of the plan will not be reached (the triggers, also known in the DMDU literature as adaptation tipping points) are predefined, and appropriate responsive actions are specified (these are designed in Step V). Several techniques can be used to design a monitoring system. Recently developed techniques that can be used for defining trigger values are based on Exploratory Modeling and Analysis (Agusdinata 2008), Scenario Discovery (Bryant and Lempert 2010), and a technique described by Botterhuis et al. (2010), which combines the detection of weak signals with scenarios.

Once the adaptive plan has been designed, the results from Steps I–IV are implemented and the contingent actions (Step V) are prepared. The implementation process is suspended until a trigger event occurs.

8.5 Setting up a DAP Workshop on ISA Implementation

We operationalized the steps of DAP through means of a structured workshop, using a group decision room (GDR) to support the workshop process. The GDR is a group decision support tool that supports quick and efficient teamwork and the generation of information. In our GDR, the participants provided input using a laptop computer that was connected to a server. The (anonymous) results are directly visible to the participants, so participants are confronted with their own input and that of other participants. Also, there is the opportunity to react to each other's input or to add information. Because the information is anonymous, nobody can dominate the discussion. Figure 8.1 summarizes the workshop process (which is fully described in van der Pas (2011)). The relationship between this process and the DAP steps is the following:

DAP Steps I and II: Setting the stage and assembling an initial plan

Steps I and II used the actual ISA implementation strategy of the Dutch Ministry of Transport (as specified in interviews with policymakers and internal policy documents of the Ministry). This made it possible for the Dutch policymakers to use the generated information in their everyday jobs, and it allowed for an *ex post* assessment comparison (in a few years) of the actual plan with the adaptive plan that they had designed.

DAP Steps III, IV, and V: Increasing the robustness of the initial plan, setting up the monitoring system, and preparing the contingent actions

Steps III, IV, and V were supported through a combined SWOT-TOWS analysis, as follows:

i. List the various strengths, weaknesses, opportunities, and threats for the initial ISA implementation plan (structured brainstorm in the GDR). (Note: Both weaknesses and threats were considered to be vulnerabilities of the plan.)
ii. Identify a 'top 10' for each SWOT category (ranking using the GDR).
iii. Score the items in these top 10's on their uncertainty and importance for the outcomes of the initial plan (scoring items on a five-point scale).
iv. Define actions, signposts, and trigger values for the vulnerabilities and the opportunities for the high uncertain/high-impact items (TOWS uses the SWOT analysis as input and translates the outcomes of the SWOT into actions).

Figure 8.2 is the flowchart for vulnerabilities, which was developed to support these activities. (There was also a flowchart for opportunities. For reasons of space, only the flowchart for vulnerabilities is shown.)

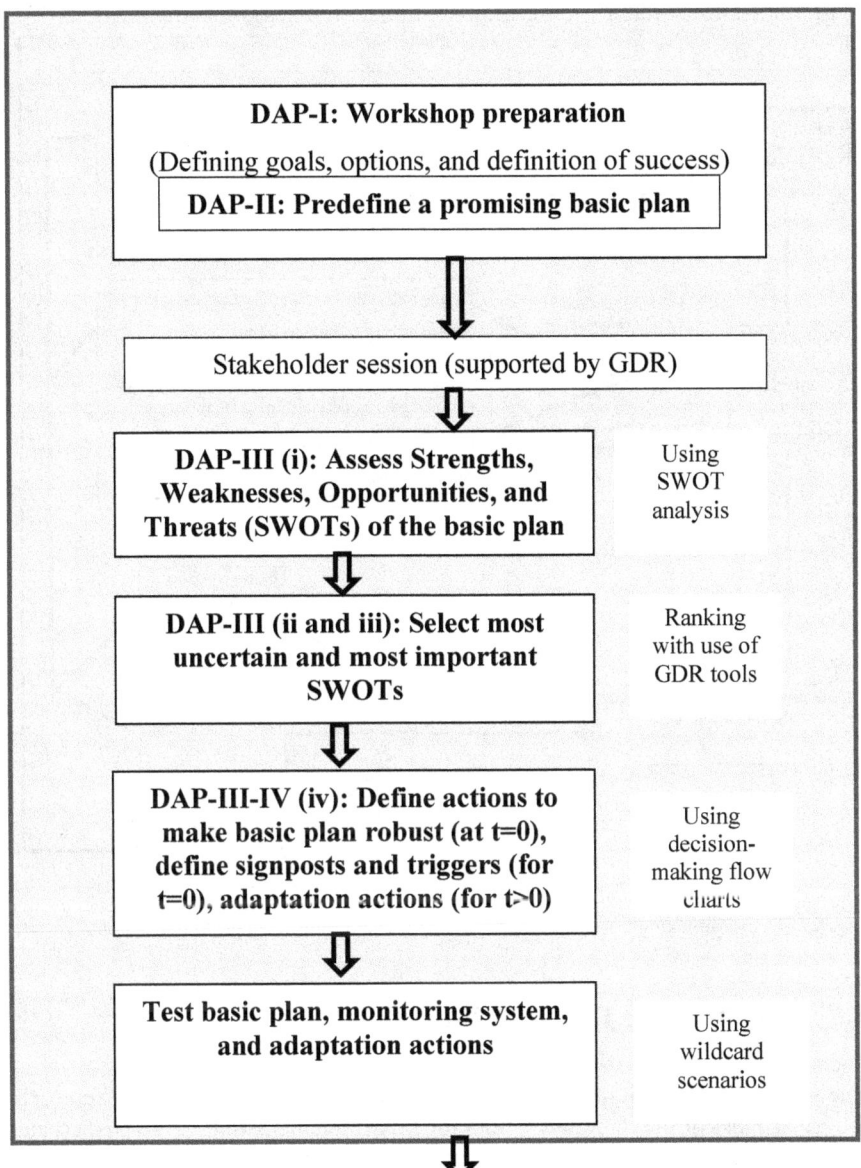

Fig. 8.1 Workshop process

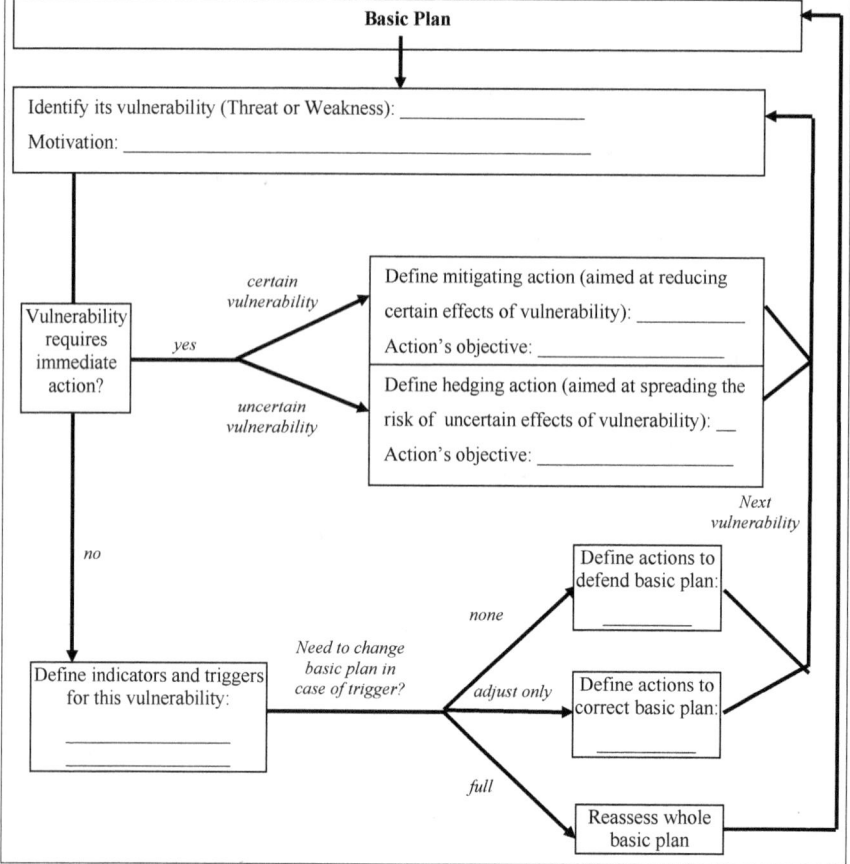

Fig. 8.2 Decisionmaking flowchart for identifying vulnerabilities and protective actions

8.6 Results of the DAP-ISA Workshop

The workshop was held in December 2010. For the workshop, we invited representatives of most of the important actors for ISA implementation in the Netherlands (Walta 2011), resulting in 18 participants. The participants included policymakers (4), scientists (7), consultants (5), an ISA system developer (1), and an insurance specialist (1). So, most of the important actors for ISA implementation were represented (at least one representative of each important group was included). Each of the respondents had over six years of work experience (the majority had over 10 years of experience).

Steps I and II: Setting the stage and assembling an initial plan

An initial plan was presented to the participants. Table 8.2 presents an overview of the initial plan. The initial plan was formulated as follows: The Dutch Ministry of Transport wants to implement the most appropriate ISA for the most appropriate driver. Three types of vehicle drivers are distinguished:

- The compliant driver: This type of driver has the intrinsic motivation to stick to the speed limit. Their ISA would be only a speed alert. Implementation would primarily involve an education campaign.
- The less compliant driver: This type of driver lacks the intrinsic motivation to stick to the speed limit. They would get to choose the type of ISA to be used. Incentives would be offered by insurance and lease companies.
- The notorious speed offender: Under the current regime, this type of driver would lose his or her driver's license (and would be obliged to follow a traffic behavior course). In place of these draconian measures, they would be forced to use restricting ISA. A pilot test would be undertaken to provide information on how well the plan might be expected to work.

The implementation of ISA would consist of two phases. Phase I would run up to the end of 2012. In 2013, a still undefined Phase II would start. During the workshop, the participants were asked to reflect upon this initial plan.

Step III: Increasing the robustness of the basic plan

The participants identified several vulnerabilities and a few opportunities related to the initial ISA plan shown in Table 8.2. Some of these vulnerabilities and opportunities could be addressed at the time the plan was implemented—i.e., the initial plan could be made more robust. Table 8.3 presents an overview of these vulnerabilities and opportunities and the actions that could be taken to improve the robustness of the initial plan.

For example, 'the availability of an accurate speed limit database' was considered a certain high-impact weakness. Speed limit data must be correct for the right time (dynamic), the right location, and the right vehicle. The participants discussed whether they should deal with this vulnerability right away, or whether they could wait until a predefined situation occurs (i.e., a trigger event). They decided that it is important to immediately deal with this uncertainty (so they followed the arrow down in the decisionmaking flowchart in Fig. 8.2).

The availability of an up-to-date database was considered to be either a certain or uncertain vulnerability. The participants next discussed its uncertainty level and decided that while the effects of incorrect speed limit data are very uncertain, it is fairly certain that this vulnerability will occur. They therefore filled in the box at the bottom of Fig. 8.2 (indicating the need for mitigating actions).

Steps IV and V: Setting up the monitoring system and preparing the contingent actions

Using the same decisionmaking flowcharts shown in Fig. 8.2, the participants defined the signposts, triggers, and contingent actions. A subset of these is shown in Table 8.4.

Table 8.2 Initial plan, as input to the workshop (from Steps I and II)

Initial plan

Type of driver	Type of ISA	Action	Definition of success	Constraints
Phase I (2009–2012)				
Compliant driver	Warning ISA (speed alert)	• Start a campaign aimed at persuading people to turn on the speed alert functionality on their navigation device • Make agreements with companies that develop navigation devices	• Before 2013: 50% of the people that own and use a navigation device actively use the speed alert functionality	• Budget for a campaign
Less compliant and compliant driver	Free to be selected	• Develop a business case with insurance companies and lease companies	• Before 2013: 50% of the car owners and 50% of lease drivers can choose insurance or lease a product that involves ISA	
Notorious speed offender	Restricting ISA	• Perform a pilot test aimed at assessing the effects of implementing a restricting ISA for notorious speed offenders • Make an evidence-based decision regarding implementation of such a system for notorious speed offenders	• Before 2013: A decision has to be made on implementation of ISA for notorious speed offenders, based on, among others, the outcomes of the pilot test	• Budget/time

Phase II (beginning in 2013) will be dependent on the results of Phase I. For this phase, more restricting types of ISA will be considered

Table 8.3 Increasing the robustness of the initial plan

Vulnerabilities (V) and Opportunities (O)	Hedging (H), Mitigating (M), Seizing (S), and Exploiting (E) actions
V: Implementing a restricting ISA for notorious speed offenders will damage the image of the less intervening ISA systems. ISA will be associated with punishment not with assistance (as it is now)	H: Decouple the pilot from the rest of the initial plan, and avoid the term ISA (currently done by calling it speed-lock)
V: The availability of an accurate speed limit database. Speed limit data have to be correct for right time (dynamic), the right location, and the right vehicle	This is a certain vulnerability, so: M: Define and apportion responsibilities before starting with implementation M: Issue a request for bids for the development of a speed limit database (this should be arranged by public authorities) M: Guarantee quality through a third party that is under the supervision of the public authorities M: Develop a system based on beacons that overrule the static speed limit information (failsafe design)
V: Automotive lobby will oppose the large-scale implementation of ISA	H: Include automobile manufacturers in the implementation strategy
V: Speed limit data change more frequently than expected (by time and location)	H: Implement ISA systems that are robust against this situation (i.e., systems that allow for communication with the infrastructure, to transmit temporary speed limits—e.g., radio, Bluetooth)
O: Cars and ISA draw lots of attention and appeal to people's emotions. Instead of seeing this as a threat, this can be used as an opportunity	S: Invite stakeholders that have positive feelings about ISA to participate in improving and implementing ISA (e.g., the presenters of Top Gear, race drivers)
O: People/companies are more willing to adopt a technology if they can see the technology in practice. Creating a pool of cars that are equipped can result in an uptake of the technology	S: Practice what you preach. Let the Ministry equip its fleet with ISA to set an example. Prove that ISA can significantly reduce the number of accidents and can result in fewer insurance claims

Chapter 3 describes the four types of contingent (trigger event) actions that can be taken:

- **Defensive actions (DA)**—Responsive actions taken *after initial implementation* to clarify the plan, preserve its benefits, or meet outside challenges in response to specific triggers, but that leave the initial plan unchanged;
- **Corrective actions (CR)**—Adjustments to the plan in response to specific triggers;
- **Capitalizing actions (CP)**—Responsive actions taken *after initial implementation* to take advantage of opportunities that further improve the performance of the initial plan;

Table 8.4 Monitoring system and contingent actions

Vulnerabilities (V) and Opportunities (O)	Monitoring and triggering system	Actions: Reassessment (R), Corrective (CR), Defensive (D), and Capitalizing (CP)
V: Implementing a restricting ISA for notorious speed offenders will damage the image of the less obtrusive ISA systems. ISA will be associated with punishment not with assistance (as it is now)	Monitor the: • Number of negative press publications • Level of acceptance of different ISA systems • Number and type of ISA-related questions asked of politicians in the lower house	D: Media campaigns to manage the perception of people regarding ISA (and the speed-lock); explain the difference and the need for implementing such an ISA for this type of driver
V: The availability of an accurate speed limit database. Speed limit data have to be correct for the right time (dynamic), the right location, and the right vehicle	Monitor the: • Level of accuracy/reliability of speed limit database	D: Initiate database accuracy enhancement CR: Stop implementation of certain types of ISA or combine with on/off switch and overruling possibilities CR: Design the system in such a way that it only warns/intervenes in areas with certain accuracy levels
V: Technology can fail: • Location determination can be inaccurate (e.g., in tunnels, in cities with high buildings) • Systems can stop functioning (sensors fail, etc.)	Monitor the: • Cause of accidents (relationship ISA—cause of accident) • Press releases on ISA and accidents	D: Make sure the market improves the systems (adjust implemented rules and regulations regarding system functioning) R: When large-scale failure occurs or the effects are drastic (ISA implementation leads to fatalities)
V: Speed limit data become more and more dynamic	Monitor the: • Availability of dynamic speed limits	D: Make sure road authorities equip new dynamic speed limit infrastructure with infra-to-vehicle communication (so in-vehicle systems can be easily adjusted) D: Standardize communication protocols and communication standards

(continued)

Table 8.4 (continued)

Vulnerabilities (V) and Opportunities (O)	Monitoring and triggering system	Actions: Reassessment (R), Corrective (CR), Defensive (D), and Capitalizing (CP)
O: ISA implementation can result in larger cost savings than expected, lower and more homogeneous speeds, lower consumption costs (fuel savings + lower maintenance), resulting in higher levels of acceptance	Monitor additional effects of implementation on: • Emissions • Fuel use • Throughput/congestion	CP: Increase the number of participating insurance companies CP: Use this information in the business case for new insurance and lease companies

- **Reassessment (RE)**—A process initiated when the analysis and assumptions critical to the plan's success have lost validity (i.e., when unforeseen events cause a shift in the fundamental goals, objectives, and assumptions underlying the plan).

Table 8.4 shows the direct results from the workshop. Further operationalization of the signposts and related trigger values would be required to develop a monitoring system. For instance, consider the case of the second vulnerability mentioned: 'the availability of an accurate speed limit database.' The signpost suggested by the participants is: 'level of accuracy/reliability of speed limit database.' The next step would be to assign specific trigger values to the actions. For instance, if the accuracy of the speed limits in the database drops under a prespecified level of accuracy (e.g., 97%), a defensive action should be triggered, and additional effort should be taken to make the speed limit database more accurate. If the accuracy drops below an even lower level (e.g., 60%), a corrective action should be taken (i.e., the initial plan should be changed) as follows: 'Stop implementation of certain types of ISA or combine with on/off switch and overruling possibilities.' The specific trigger values would have to be determined (e.g., by using literature or additional modeling efforts).

After the dynamic adaptive plan was designed (as described above), the participants were asked to 'test' the extent to which they considered the designed plan to be 'future proof.' This process was supported by the use of wildcard scenarios (van Notten 2004). Examples of wildcard scenarios that were used are:

- After ISA is implemented, hackers develop ways to mislead the ISA systems, allowing people to speed without the system noticing.
- Current ISA systems use the USA satellite system to determine their position. The Americans 'play' with the accuracy of the system. In times of war, the system is more accurate than in times of peace and less accessible to civilians. In 2013, the USA is no longer at war and the accuracy is reduced. After 2013, the system becomes so inaccurate that safety issues arise.

The participants were asked to think about 'what if' the wildcard scenarios were to occur. In particular, they were asked to answer the following questions for each:

- What would happen to the (road) transport system?
- What would happen to your adaptive plan, and how would the outcomes of the plan be influenced if this scenario were to occur?
- Is your adaptive plan capable of dealing with this scenario?

These wildcard scenarios led to interesting (and lengthy) discussions. A total of nine wildcard scenarios were assessed. In six cases, the groups indicated that their plan was capable of dealing with the wildcard scenario. Open questions asked of the participants in a follow-up questionnaire revealed that they appreciated the wildcard scenario portion of the workshop, and they stressed the added value of these scenarios.

8.7 Evaluation of the DAP Approach

In addition to the above assessment of workshop outcomes, we also used a (Web-based) questionnaire to elicit the participants' opinions on the DAP approach. The main results are displayed in Table 8.5.

As can be seen from the table, the participants were very positive about the suitability and usefulness of adaptive plans (#1, 2, and 5). They thought that the elements of the adaptive plan (actions, monitoring system, etc.) and the process of designing and implementing a plan would be useful for ISA implementation in the Netherlands (#3, 7) and that policymakers were capable of using the process (#6). However, they agreed that the process of developing and implementing such a plan was more time-consuming than that required for a traditional static plan (#4). For an in-depth analysis of the results from the questionnaire, and an overview of the complete evaluation, see van der Pas et al. (2012b).

8.8 Lessons Learned About the Process of Developing Dynamic Adaptive Plans

By using a structured, participatory approach to developing a dynamic adaptive plan, we addressed the issues and challenges mentioned in Sect. 8.1. We used SWOT, TOWS, a group decision room, wildcard scenarios, decisionmaking flowcharts, etc. In addition, we tested this approach in a workshop with experts, stakeholders, and policymakers and evaluated it by canvassing the opinions of the workshop participants. We found that the workshop approach we used is promising and can produce a usable adaptive robust plan. However, we found that better ways are needed to identify the signposts and trigger values. This information could come from the use of other methods. A promising technique to do this might be Scenario Discovery (SD) (Bryant and Lempert 2010). Future applications of SD in the context of DAP can prove useful in identifying signposts and trigger values. Hermans et al. (2017) also provide some useful suggestions for designing monitoring arrangements.

8 Dynamic Adaptive Planning (DAP): The Case of Intelligent ...

Table 8.5 Participants' evaluation of the DAP approach ($n = 18$)

#	Workshop	Median[a] [interquartile range]	Max	Min
1	DAP is an appropriate way to develop a plan for the implementation of ISA	4 [3–4.25]	5	2
2	ISA implementation using an adaptive plan increases the chance of reaching the ISA-related policy goals	4 [3.5–4]	5	3
3	The generated strengths, weaknesses, opportunities, and threats, and the defined actions, signposts, and trigger values can be used in the ongoing effort of developing ISA implementation plans for the Netherlands	4 [4–5]	4	3
4	Developing and implementing adaptive plans are more time-consuming than developing and implementing traditional static plans for ISA implementation	3 [2–4]	5	1
5	The expected benefits of developing adaptive plans are bigger than the expected costs (for problems that are surrounded with deep uncertainty)	4 [3–4]	5	2
6	Policymakers in general are capable of identifying the strengths, weaknesses, opportunities, and threats, and to think of actions to counter the weaknesses and threats and profit from the strengths and opportunities	4 [2–4]	5	2
7	Designing and implementing adaptive plans fits the current practice of policymaking in the Netherlands	3 [3–4]	5	2

[a] 1 = strongly disagree, 2 = disagree, 3 = neither disagree, nor agree, 4 = agree, 5 = strongly agree

Several further lessons were learned from the workshop and subsequent feedback. First, some decisions need to be taken before starting to design an adaptive plan. These are often related to the political process (e.g., the decision to use an analytical approach) and to educating the policymakers (e.g., the design of an adaptive plan might be costly and time-consuming, and the concepts are new). During the workshop, the experts had trouble dealing with these issues. This problem can be prevented by increasing the involvement of the participants over time. Instead of one workshop plus assessment and questionnaire, a dedicated task force could be formed with the responsibility to formulate an advice to the decisionmakers. The major stakeholders would be represented in the task force. The task force would meet 3 times. The first meeting would focus on making an inventory of the level of knowledge and the questions the members have. The second meeting would be a workshop in which the largest knowledge gaps and questions are dealt with. The third meeting would be to design the adaptive plan.

Second, the participants indicated that a vulnerability or opportunity is often neither 100% certain or uncertain. Although implicit in the DAP framework, it is not necessarily true in practice. An action can be assessed to be both fairly certain in some respects and uncertain in others. Consider, for instance, the vulnerability 'the ISA technology can fail; location determination can be inaccurate.' It is fairly certain that this will occur. However, one stakeholder judged this as uncertain, because the magnitude of the effects if it occurs is uncertain. So, one might want to define both a mitigating action (e.g., provide a warning to the driver when the system fails) and a reassessment action (in case fatalities with the system occur). Distinguishing between the uncertainty of occurrence and the uncertainty of impact when it occurs is an important distinction that should be made when developing adaptive plans.

Third, after the assessment of vulnerabilities and opportunities, a choice can be made whether to handle a specific vulnerability or opportunity through actions to be taken immediately or in the future in response to a trigger event. An assessment of the costs of both approaches is required to make a reasoned choice. No clear guidance on how to choose between taking immediate action or preparing for the future currently exists. Related to this, it proved to be impossible to specify trigger values during the workshop. This was not due only to practical time constraints, but also because defining trigger values involves very specific expertise and knowledge. A possible solution to this problem could be to introduce a fast simple transport model (FSM) into the workshop. This would allow for running simulations during the workshop in order to determine trigger values (FSMs are used in SD and EM). This solution was prototyped in a project related to the Colorado River, which was documented by Groves et al. (2016). They did this by reconvening a panel of experts at Lawrence Livermore Laboratory and going once more through the deliberative process, this time with the underlying exploratory models being operated and producing the required analyses in real time.

Fourth, although the initial plan distinguished three types of motorists (compliant, less compliant, and notorious speeders), the analysis showed that the vulnerabilities and opportunities mostly address either the notorious speeders or the overall initial plan (without distinguishing among the other types of motorists). This indicates that experts may find it difficult to assess a plan that consists of multiple alternatives (they might not address each alternative consistently). In the workshop, we had the impression that the experts focused on the underlying assumptions and tried to find vulnerabilities and opportunities for these (e.g., ISA should be a reliable technology, and for a GPS-based ISA, an accurate speed limit database is required). As a result, they came up with more generic vulnerabilities. Later, these would have to be translated into specific vulnerabilities for each of the alternatives for each of the target groups for which the plan is to be applied.

8.9 Conclusions

Previous literature suggested that, conceptually, DAP is a promising policy design approach that is able to handle deep uncertainties. However, it also pointed out some major remaining challenges. Three of these challenges were addressed in this case. First, *DAP lacked examples of adaptive plans developed by policymakers or domain experts*. By designing an adaptive plan with policymakers, domain experts, and other stakeholders, we showed that policymakers are capable of designing dynamic adaptive plans. Second, *DAP lacked well worked out examples of real-world policy problems*. By using DAP to design an actual ISA implementation plan with policymakers, domain experts, and other stakeholders, we showed it can be used to address a real-world policy problem. In addition, the experts indicated that the plan could be used to encourage and speed up real-world implementation of ISA in the Netherlands. Third, *DAP was previously a high-level concept, captured only in a flowchart, yielding only limited insight into its operationalization*. By designing, applying, and evaluating a workshop, we showed how DAP can be operationalized by integrating existing research methods into an integrated design approach. Moreover, the process revealed problems that can be expected in doing so.

This case showed that the workshop approach is a promising way to design a robust, adaptive plan in the face of deep uncertainty. However, it also showed that there are still some issues that need to be addressed in future research in order to design an adaptive plan using DAP:

- The framework seems to be ambiguous regarding the moment of implementation. In particular, the participants had trouble defining when some of the actions should take place. (For example, certain required actions had to be taken before the adaptive plan could even begin to be implemented. These mainly related to the political process and to the perceived time and cost involved in designing and implementing a plan using DAP).
- Participants had trouble dealing with the fact that a vulnerability or opportunity is not either certain or uncertain (which influences the type of action that needs to be defined). There is a scale between certain and uncertain, so an action can be assessed by participants to be both fairly certain in some respects and uncertain in others. In addition, although DAP is selected because the decision problem at hand is considered deeply uncertain, this does not mean that every aspect of the decision problem is uncertain. It is important for the participants to have a clear picture of which dimension of deep uncertainty is being addressed. In terms of the definition given by Lempert et al. (2006), we distinguished three dimensions: deep uncertainty regarding the appropriate model, (2) deep uncertainty regarding the prior probability distributions for inputs to the model and their interdependencies, and (3) deep uncertainty regarding the stakeholder preferences that can be used to rank policy alternatives. Participants need to have a clear picture of what aspect of these dimensions is being addressed by each of the vulnerabilities and opportunities. Methods to indicate the dimensions of deep uncertainty could be useful here.

- For each of the vulnerabilities and opportunities, a choice can be made whether to handle these through actions to be taken immediately (an initial action) or in the future in response to a trigger event (a contingent action). Participants struggled to find criteria that could be used to resolve the timing of the actions.
- Experts had trouble dealing with the fact that the initial plan was a package of actions. Going through the process of defining vulnerabilities and opportunities, signposts and trigger values, and contingent actions for each of the actions in the initial plan proved a challenge. As a result, they did not address each of these actions consistently. This is likely due to the setup of the workshop. Future workshops should include mechanisms to deal with policy packages more effectively.

All of these issues should be addressed in work aimed at improving the DAP approach.

With regard to ISA, the work confirmed the hypothesis that DAP is a useful approach for dealing with the uncertainties related to its implementation. Traditional/current ISA policies involve either 'do nothing' (no ISA) or 'do the wrong thing' (e.g., ISA for everybody). DAP enables policymakers to begin implementation, to monitor developments, and to adjust the plan based on real-world developments that cannot be predicted.

References

Agusdinata, D. B. (2008). *Exploratory Modeling and Analysis to Deal with Deep Uncertainty*. Ph.D. Thesis, Delft University of Technology, Delft.

Bankes, S. (1993). Exploratory modeling for policy analysis. *Operations Research, 4*(3), 435–449.

Botterhuis, L., van der Duin, P., de Ruijter, P., & van Wijck, P. (2010). Monitoring the future: Building an early warning system for the Dutch ministry of justice. *Futures, 42*(5), 454–465.

Bryant, B. P., & Lempert, R. J. (2010). Thinking inside the box: A participatory, computer-assisted approach to scenario discovery. *Technological Forecasting and Social Change, 77*(1), 34–49.

Carsten, O. M. J., & Tate, F. N. (2005). Intelligent speed adaptation: Accident savings and cost-benefit analysis. *Accident Analysis and Prevention, 37*(3), 407–416.

Dewar, J. A., Builder, C. H., Hix, W. M., & Levin, M. (1993). *Assumption-based planning: A planning tool for very uncertain times*. Santa Monica, CA: MR114-A, RAND.

Elvik, R. (2003). Effects on road safety of converting intersections to roundabouts: Review of evidence from non-U.S. studies. *Transportation Research Record: Journal of the Transportation Research Board, 1847*, 1–10.

Elvik, R., & Vaa, T. (2009). *The handbook of road safety measures* (2nd ed.). Oxford: Elsevier Ltd.

French, S., Maule, J., & Papamichail, N. (2009). *Decision behaviour analysis and support*. Cambridge (UK): Cambridge University Press.

Groves, D. G., R. J. Lempert, D. W. May, J. R. Leek, and J. Syme (2016). *Using high-performance computing to support water resource planning: A workshop demonstration of real-time analytic facilitation for the Colorado River Basin,* RAND Corporation and Lawrence Livermore National Laboratory, CF-339-RC, 2016. Retrieved December 30, 2017 from https://www.rand.org/pubs/conf_proceedings/CF339.html.

Hermans, L. M., Haasnoot, M., ter Maat, J., & Kwakkel, J. H. (2017). Designing monitoring arrangements for collaborative learning about adaptation pathways. *Environmental Science & Policy, 69,* 29–38.

Kaplan, R. S., & Norton, D. P. (1993). Putting the balanced scorecard to work. *Harvard Business Review, 71*(5), 134–147.

Kwakkel, J. H. (2010). *The treatment of uncertainty in airport strategic planning*. Ph.D. Thesis, Delft University of Technology, Delft.

Kwakkel, J. H., Walker, W. E., & Marchau, V. A. W. J. (2012). Assessing the efficacy of adaptive airport strategic planning: Results from computational experiments. *Environment and Planning B: Planning and Design, 39,* 533–550.

Lai, F., Carsten, O., & Tate, F. (2012). How much benefit does intelligent speed adaptation deliver: An analysis of its potential contribution to safety and environment. *Accident Analysis and Prevention, 48,* 63–72.

Lempert, R. J., Popper, S. W., & Bankes, S. C. (2003). *Shaping the next one hundred years: New methods for quantitative long-term policy analysis*. Santa Monica: RAND.

Lempert, R. J., Groves, J. D. G., Popper, S. W., & Bankes, S. C. (2006). A general, analytic method for generating robust strategies and narrative scenarios. *Management Science, 52*(4), 514–528.

OECD—Organization for Economic Co-Operation and Development (2006). *Speed Management.* Paris: OECD.

Oei, H. L. (2001). *Safety Consequences of Intelligent Speed Adaptation (ISA)*. Report R-2001-11. Institute for Road Safety Research (SWOV), Leidschendam.

Osita, H. C., Idoko O. R., & Nzwkwe, J. (2014). Organization's stability and productivity: The role of SWOT analysis. *International Journal of Innovative and Applied Research, 2*(9), 23–32.

Saltelli, A., Chan, K., & Scott, E. M. (2001). *Sensitivity analysis*. New York: Wiley.

Sassone, P. G., & Shaffer, W. A. (1978). *Cost-benefit analysis—A handbook*. San Diego: Academic Press.

Sprei, F., Karlsson, S., & Holmberg, J. (2008). Better performance or lower fuel consumption: Technological development in the Swedish new car fleet 1975–2002. *Transportation Research Part D, 13*(2), 75–85.

Swanson, D., Barg, S., Tyler, S., Venema, H., Tomar, S., & Bahdwal, S. (2010). Seven tools for creating adaptive policies. *Technological Forecasting and Social Change, 77*(6), 924–939.

van der Pas, J. W. G. M. (2011). *Clearing the road for ISA implementation? applying adaptive policymaking for the implementation of intelligent speed adaptation*. Delft, the Netherlands: TRAIL Thesis Series, T2011/13.

van der Pas, J. W. G. M., Marchau, V. A. W. J., Walker, W. E., van Wee, G. P., & Vlassenroot, S. H. (2012a). ISA implementation and uncertainty: A literature review and expert elicitation study. *Accident Analysis and Prevention, 48,* 83–96.

van der Pas, J. W. G. M., Kwakkel, J. H., & van Wee, B. (2012b). Evaluating adaptive policymaking using expert opinions. *Technological Forecasting and Social Change, 79,* 311–325.

van Notten, P. (2004). *Writing on the wall: Scenario development in times of discontinuity*. Ph.D. thesis, University of Maastricht, Maastricht.

Walker, W. E. (2000). Policy analysis: A systematic approach to supporting policymaking in the public sector. *Journal of Multicriteria Decision Analysis, 9*(1–3), 11–27.

Walker, W. E., Rahman, S. A., & Cave, J. (2001). Adaptive policies, policy analysis, and policymaking. *European Journal of Operational Research, 128,* 282–289.

Walta, L. (2011). *Getting ADAS on the Road—Actors' interactions in advanced driver assistance systems deployment*. Ph.D. Thesis, Delft University of Technology, Delft.

Weihrich, H. (1982). The TOWS matrix: A tool for situational analysis. *Long Range Planning, 15*(2), 54–66.

WHO—World Health Organization. (2015). *Global status report on road safety 2015*. Geneva: WHO Press.

Prof. Vincent A. W. J. Marchau (Radboud University (RU), Nijmegen School of Management) holds a chair on Uncertainty and Adaptivity of Societal Systems. This chair is supported by The

Netherlands Study Centre for Technology Trends (STT). His research focuses on long-term planning under uncertainty in transportation, logistics, spatial planning, energy, water, and security. Marchau is also Managing Director of the Dutch Research School for Transport, Infrastructure and Logistics (TRAIL) at Delft University of Technology (DUT), with 100 Ph.D. students and 50 staff members across 6 Dutch universities.

Prof. Warren E. Walker (Emeritus Professor of Policy Analysis, Delft University of Technology). He has a Ph.D. in Operations Research from Cornell University, and more than 40 years of experience as an analyst and project leader at the RAND Corporation, applying quantitative analysis to public policy problems. His recent research has focused on methods for dealing with deep uncertainty in making public policies (especially with respect to climate change), improving the freight transport system in the Netherlands, and the design of decision support systems for airport strategic planning. He is the recipient of the 1997 President's Award from the Institute for Operations Research and the Management Sciences (INFORMS) for his 'contributions to the welfare of society through quantitative analysis of governmental policy problems.'

Dr. Jan-Willem G. M. van der Pas (Eindhoven municipality) has a Ph.D. from Delft University of Technology. His research focused on policymaking under deep uncertainty for Intelligent Transport Systems. After receiving his Ph.D., he worked, respectively, as a researcher at the Delft University of Technology and as team manager, Smart Mobility, for a research and consultancy firm (DTV Consultants). Currently, Jan-Willem works as a strategist/coordinator, Smart Mobility, at Eindhoven municipality where he continues to improve the theory and practice of Dynamic Adaptive Planning.

Open Access This chapter is licensed under the terms of the Creative Commons Attribution 4.0 International License (http://creativecommons.org/licenses/by/4.0/), which permits use, sharing, adaptation, distribution and reproduction in any medium or format, as long as you give appropriate credit to the original author(s) and the source, provide a link to the Creative Commons licence and indicate if changes were made.

The images or other third party material in this chapter are included in the chapter's Creative Commons licence, unless indicated otherwise in a credit line to the material. If material is not included in the chapter's Creative Commons licence and your intended use is not permitted by statutory regulation or exceeds the permitted use, you will need to obtain permission directly from the copyright holder.

Chapter 9
Dynamic Adaptive Policy Pathways (DAPP): From Theory to Practice

Judy Lawrence, Marjolijn Haasnoot, Laura McKim, Dayasiri Atapattu, Graeme Campbell and Adolf Stroombergen

Abstract

- Decision making by flood risk managers is challenged by uncertainty and changing climate risk profiles that have elements of deep uncertainty.
- Flood risk managers at a regional level in New Zealand requested better understanding and tools for decision making under conditions of uncertainty and changing conditions.
- A game was used to catalyze a process of new knowledge transfer and its uptake in technical assessments and decision making processes.
- The understanding enabled DAPP to be used to develop a long-term plan that can accommodate changes in flood frequency from climate change (as projected in three climate change scenarios) over at least 100 years.
- Use of "new" economic tools with DAPP facilitated decision making to consider the sensitivity of alternative policies to a range of climate change scenarios, to discount rate, decision review date, and costs and losses, thus addressing deep uncertainty by considering the long-term effects of initial decisions to changing conditions.
- The risk of path-dependent decisions and the role that current frameworks and practices play in blinding actors to the range of possible outcomes that could evolve in the future, was reduced.

Daya Atapattu is deceased (October 2018).

J. Lawrence (✉)
Victoria University of Wellington, Wellington, New Zealand
e-mail: judy.lawrence@vuw.ac.nz

M. Haasnoot
Water Resources and Delta Management, Deltares, Delft, The Netherlands

Faculty of Physical Geography, Utrecht University, Utrecht, The Netherlands

L. McKim · D. Atapattu
Greater Wellington Regional Council, Wellington, New Zealand

G. Campbell · A. Stroombergen
Infometrics Consulting, Wellington, New Zealand

- A series of interventions to raise and increase awareness through new information and its framing, experimentation, and leadership enabled the uptake of DAPP.

9.1 Introduction to the Case

The adequacy of the tools typically used by natural resource managers is being questioned in the face of climate change and increasing human exposure to the natural hazards, because of future uncertainties and increasing risk. Large investments have been made worldwide on structural flood protection and for coastal flooding. These are increasingly being found to be inadequate as risk profiles increase with changing climate impacts. This chapter describes how flood risk managers at a regional level in New Zealand applied the Dynamic Adaptive Policy Pathways (DAPP) approach (see Chap. 4) to managing deep uncertainty around flood frequency associated with changing climate, and examines the lessons learned from taking DAPP theory into practice. New Zealand is a country that is geologically young, seismically active, and highly pluvial. It is located in the South Pacific Ocean, where the dynamics of changing climate are readily apparent and have a disproportionate effect on human habitation. For example, flooding is the most frequent natural hazard experienced in New Zealand, with a major damaging flood occurring on average every eight months (Ministry for the Environment 2008). This means that there are long-established institutional arrangements for governing flood risk and civil emergencies arising from floods, and there are professional standards and practices used commonly by flood risk managers. These have however developed under past climate conditions, which makes them less well suited to addressing climate change induced risk, and particularly to deep uncertainty (Lawrence et al. 2013a, b).

In the past in New Zealand, flood managers built large structural protection works to manage flood risk in river catchments. Structural protection, such as levees, has been the dominant measure historically used by the regional councils responsible for flood management, largely designed on the (implicit) assumption that the future will be like the past. In the last few decades, flood risk management has also been augmented with statutory planning measures that can accommodate specific flood levels in existing areas of development, and reduce the consequences of floods by restricting new development in known flood-prone areas. There have been only a few instances of retreat from flood-prone areas, and only one anticipatory one, achieved through multiple objectives defined by the community with the council responsible for the project and funding of it (Vandenbeld and MacDonald 2013).

The specter of increasing frequency of extreme precipitation events as a consequence of global climate change (Reisinger et al. 2014), has stimulated New Zealand's interest in what this might mean for the integrity of current flood protection measures. There is well-documented New Zealand-based research (Ericksen 1986) highlighting the role that structural protection has played in creating disasters, through a "safety paradox" based on the "levee effect" (Burby and French 1981; Burby et al. 2001; Tobin 1995). The ensuing damages, in event of the protection levels being exceeded, will be exacerbated by the impact of climate change on flood flows.

With this background, the Greater Wellington Regional Council (GWRC) sought to understand the "effects of climate change" (as required under the Resource Management Act 1991 as amended in 2004) for the completion of the Hutt river protection scheme (one of the largest such schemes in New Zealand, funded by regional rates (property taxes) on communities) and for the surrounding development and its proposed intensification. The Council started by better understanding the nature of the changes in climate and their effects on flood flows, based on research at the New Zealand Climate Change Research Institute and Victoria University of Wellington (Lawrence et al. 2013b). This raised awareness by showing the effect of climate change on the current protection levels, and therefore when the scheme would fail to reach its objective of protecting the surrounding city (Hutt City) from a 1:440 year flood event. This protection level had been decided following extensive community engagement as the scheme was being reviewed (Wellington Regional Council 2001). While previously the Council had identified the need for complementary planning measures, such as rules for flood proofing new developments and identifying flood paths (Wellington Regional Council 1996), these were never fully implemented in land-use plans. This was in large part because the Regional Council had no planning rules in place in its Regional Plan that were required to be implemented by the district council (Hutt City) responsible for land-use planning; this was assumed to be the preserve of the city council. The protection level was perceived as a high level of protection and thus conveyed "safety." This had the perverse effect of diminishing the need for planning rules to address the "residual risk," should there be a breach or overtopping of the levee. Consequently, implementation of such measures has been limited to the floodway and some unprotected areas. Previous studies had identified the effect of climate change on flood risk (Pearson and McKerchar 1999; Tait et al. 2002), which was considered when setting the design standard for the Hutt River Flood Management Plan (HRFMP, Sect. 3.4, p. 36). However, climate change had not featured in long-term land-use development planning. Thus, the legacy of assets and people exposed to increasing flood frequency has persisted and increased.

9.2 Reason for Choosing DAPP

An opportunity arose in 2011, when the GWRC sought to complete the Hutt City corridor part of the flood scheme, to further examine whether the 1:440 Average Recurrence Interval (ARI) objective could be maintained over the next 100 years (the nominal life of the flood scheme). The Council staff were aware that they needed to update their knowledge and capability to address changing climate risk, due to the IPCC Fifth Assessment Report (AR5) review that was underway, which is used by government agencies as the benchmark for monitoring climate change knowledge.[1]

---

[1] It should be noted that by the time the IPCC review reports are published, the knowledge base can already be several years out of date, due to the cut off dates for peer review of the papers.

The Council sought assistance from the New Zealand Climate Change Research Institute, Victoria University of Wellington, on how to incorporate new knowledge of climate change effects using new tools, into the completion stage of the flood scheme. This coincided with the application of the Dynamic Adaptive Policy Pathways (DAPP) planning approach for flood risk management elsewhere. DAPP is an exploratory model-based planning tool that helps in the design of strategies that are adaptive and robust over different scenarios of the future, which was evolving in the Netherlands (Haasnoot et al. 2013), and a similar application undertaken in the Thames catchment (Ranger et al. 2010). The Council sought guidance on such approaches. They were found to be successful in communicating uncertainty and dynamic change in both climate and non-climate conditions over time frames of "at least 100 years" (the statutory benchmark used in planning for climate change in New Zealand).[2]

9.3 Setup of Approach for Case Study in Practice

With an enabling environment setup and the expertise and facilitation in place, interest was first stimulated using presentations of the problem and tools used to address the problem at a generic level. The framing of the problem proved crucial in getting Council staff to understand the changing risk. This was demonstrated by their organizing several further staff workshops/meetings, and the ultimate inclusion of the theoretical DAPP approach into their final work on the Hutt River Flood Scheme. This occurred first with the technical advisory staff within the Council, then with the elected Council members at several workshops/meetings. Further awareness was raised by introducing the Sustainable Delta game [a serious game to experience and thus learn how to make decisions under deep uncertainty over long timeframes (Valkering et al. 2013; http://deltagame.deltares.nl)] to the Council staff. Their interest was stimulated to the extent that they helped fund further development of the game and its tailoring for New Zealand settings (visually, and the models/scenarios that drive it). The latter activity had two purposes: satisfying the demand from the Council, and enabling the University and Deltares to experiment with the use of the game, undertake refinements, and develop further modules for river and coastal settings. This subsequently catalyzed demand for the game and the associated pathways generator (for drawing alternative pathways and stress testing them against different scenarios of the future: http://pathways.deltares.nl) to be used elsewhere in New Zealand for national guidance on coastal hazards and climate change (Ministry for the Environment 2017), for strategy development (e.g., Hawkes Bay Clifton to Tangoio Coastal Hazards Strategy 2120[3]), and for other planning purposes. Staff were trained by

---

[2] The timeframe set out in the New Zealand Coastal Policy Statement, the Ministry for the Environment guidance documents, and the practice for flood risk management.

[3] See: https://www.hbcoast.co.nz/.

9 Dynamic Adaptive Policy Pathways (DAPP): From Theory to Practice

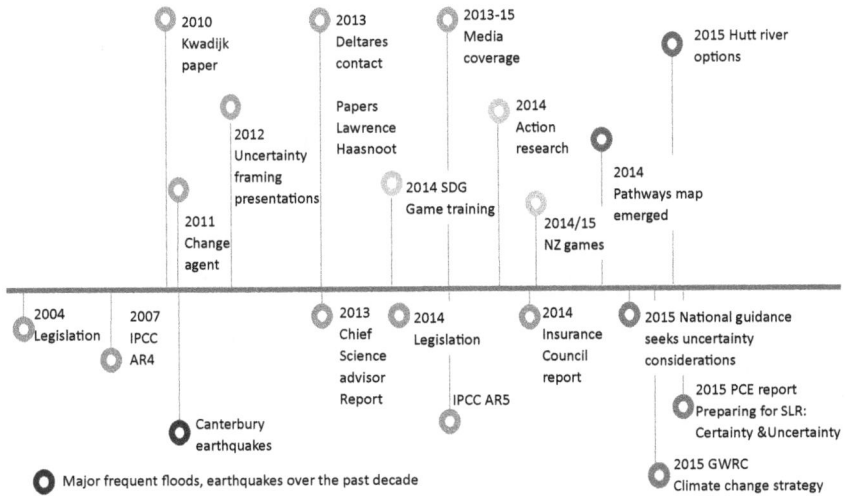

Blue=Creating interest, Yellow=Increasing awareness, Purple=Experiment Hutt River, Red=DAPP uptake, Dark Blue=Major hazards, Green=context.

Fig. 9.1 Timeline of interventions. *Source* Lawrence and Haasnoot (2017)

Deltares and the University-based facilitator to use the approach (DAPP) and the tools (the Sustainable Delta game).

The steps undertaken to introduce and implement the approach are shown in Fig. 9.1 (the events above the timeline). In addition, relevant environmental, social, and political events are shown (below the timeline) that created a receptive space for use of the approach. These enabling conditions in the context (shown below the line in Fig. 9.1) included the growing number of national and international reports and hazard events signaling changing risk; the flood manager's wish to increase the capability of the staff; the ability to do embedded action research; the opportunity provided by the interest of the Council; a trusted facilitator; an implementable theoretical approach; and support from the top of the Regional Council and by the functional managers (Lawrence and Haasnoot 2017).

9.4 Applying DAPP in Practice

Within DAPP, an adaptive plan is conceptualized as a series of actions over time (pathways). The essence of the approach is the proactive planning for flexible adaptation over time in response to how the future actually unfolds. The approach starts from the premise that policies/decisions have a design life, and might fail as the operating conditions change or underlying assumptions turn out to be wrong (Kwadijk et al. 2010). A set of questions are asked that facilitate consideration of the risk over a long time frame and enable different strategies that can meet the long- and short-term objectives to be met under different scenarios of the future and consideration of their consequences: What are the first issues that we will face as a result of climate change? Under what conditions will current strategies become ineffective in meeting objectives, and thus reach a threshold?[4] When should the various actions be triggered given that their implementation has a lead time?[5] What are alternative policy pathways that can be taken? How robust are the pathways over a range of plausible future climate scenarios? Are we able to change paths easily and with little disruption and cost? The formulation of these questions, based on the triggering issues, are part of the first element of the generalized framework for dealing with deep uncertainty in decision making. The framework is described in more detail in Sect. 4.2 of Chap. 4.

The next step involves the specification of objectives in terms of goals. The objective was set by the Council: to protect the city for at least 100 years for the 1:440-year flood event. The system structure was modeled using three components (Lawrence et al. 2013a):

- Twelve Global Climate Models (statistically downscaled) and 4 emission scenarios, to produce 48 alternative climates (i.e., changes in monthly average rainfall and temperature) over the twenty-first century for the Hutt River catchment;
- A simple procedure (algorithm) to estimate changes in extreme rainfall for the catchment;
- Hourly rainfall data (historical and adjusted for future climate changes in means and extremes) using a hydrological model (Topnet) to derive flood frequencies under both historical and the 48 alternative future climates.

Subsequently, alternative sets of actions are proposed (which were called "options"). In this specific case, the options, alternative pathways, and thresholds were identified using iterative processes involving technical advisors, decision makers, and community stakeholders, as input to the decision-making process.

Futures were generated on the basis of IPCC scenarios. The discussions produced three sets of actions to be studied in detail (which were labelled Option 1, Option 2C, and Option 4). The options were then tested against the scenarios and a pathways map was generated, as shown in Fig. 9.2. If an initial action fails to meet the stated objective, additional actions or other actions are needed, and a several pathways emerge. At predetermined thresholds, the course can change while still achieving the objectives. By exploring different pathways, and considering whether actions will lock in those actions and prevent adjustments in the future and thus create path dependency, an adaptive plan can be designed that includes initial actions and long-term options.

The pathways map resulted in a set of six feasible pathways. Figure 9.2 presents a scorecard that shows the effects of each of the pathways (both the direct effects and

---

[4] What is called a 'threshold' in this chapter is called an 'Adaptation Tipping Point (ATP)' in Chap. 4 (DAPP theory). They mean the same thing (see also the Glossary).

[5] Since this case study predated the explicit identification of adaptation signals and trigger points (as described in Chap. 4), they are not mentioned further here, or shown on the pathways map, although consideration was given to lead times.

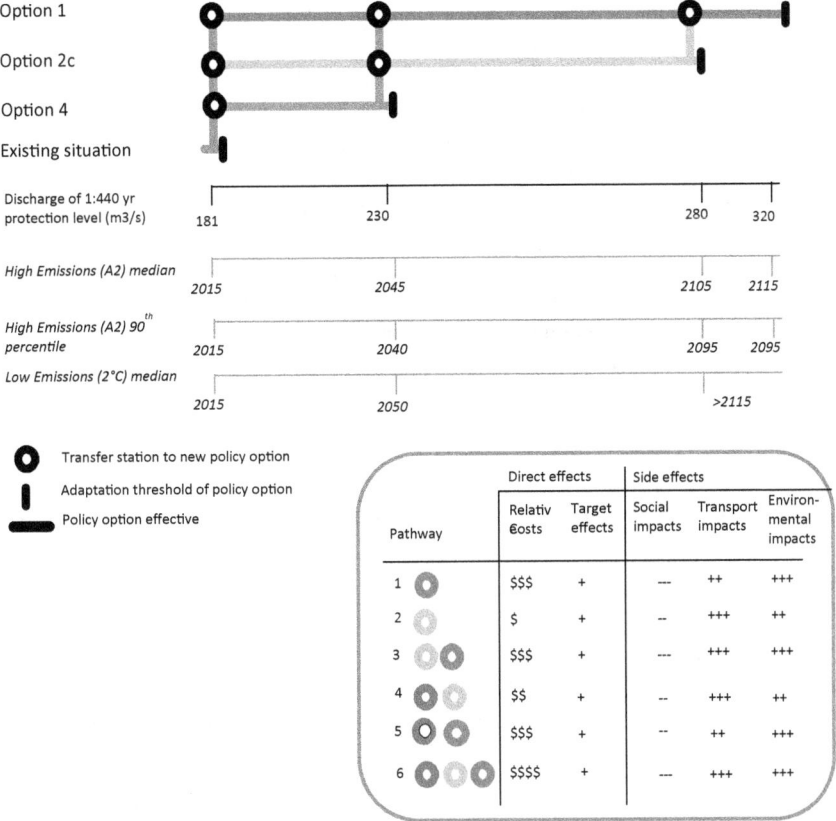

Relative impacts are indicated with – and +; - is negative impact and + positive impact. All pathways have negative social impacts, as land has to be purchased. (Option 1: A 90m river channel and 50 m berm; right and left stopbanks meets the standard over 100 years in all scenarios; cost $267m. Option 2C: A 90m river channel 25 m berm; properties to be purchased; cost of $143 million. Option 4: 70 m river channel; 30 years of flood protection; lower level of protection (2300 m3/s); properties purchased after 20 years; cost $114m until 2035. Staged Option 4 to Option 2C will cost an additional $68 million; total cost $182.m million).

Fig. 9.2 Hutt River City Centre Upgrade Project: adaptation pathways map showing options, pathways, scenarios, and adaptation thresholds; scorecard showing relative costs and direct effects of pathways, and potential side effects requiring consideration. *Source* Generated by the Pathways Generator (http://pathways.deltares.nl/), based on Greater Wellington Regional Council (2015)

the side effects). Because we do not know the probabilities of different climate/flood damage scenarios but we do know the net cost of each pathway, we used a variation of Real Options Analysis to ascertain what occurrence probability threshold a particular climate/flood scenario would have to attain, that would justify choosing one pathway over another. The net cost of each pathway was also tested for robustness to different assumptions about the discount rate, implementation costs, residual loss, decision review dates, and different climate/flood damage scenarios. The decision makers discussed the alternatives, and decided to consult the community on Options 2C and 4 prior to making their final decision on a preferred pathway. A brochure setting out the two alternative options was circulated within the community, and a survey of public perceptions and preferences for the alternatives was undertaken. The community's preference was for Option 2C as the initial set of actions. The Council then agreed to Option 2C in its final decision, which was then communicated to the affected properties directly, through the local media, and via the Council's online decisions record. The plan is being implemented and will be monitored using signals that indicate when the next step of the pathway should be triggered or whether reassessment of the plan is needed. The signals will be socially defined, reflecting the tolerability of the adverse consequences by the community, or at the 10-year review period for the Hutt River scheme, whichever comes earlier, before thresholds in the physical processes are reached.

9.5 Results of Applying the Approach

The use of DAPP enabled a systematic analysis of alternatives for achieving the objective set by the Council: to protect Hutt City for at least 100 years for the 1:440-year flood event. It also enabled the use of Real Options Analysis (ROA) to be undertaken on the pathways developed by the Council. This in turn enabled the Council to consider the "effects of climate change" on the flood flows, and to design a number of pathways that had choices in the future for transferring to other actions when the initial actions in the chosen pathway no longer met the objective. This gave flexibility to change course in the future, whatever the future might bring—i.e., "being prepared."

Staff capability increased by providing a tool that could motivate futures consideration in the analysis of options (the Sustainable Delta game) and that could motivate the use of ROA analysis alongside DAPP, which also increased the capability in the use of DMDU tools. The pathways in this case study were developed manually, and then redrawn using a pathways generator[6] developed by Deltares. The theoretical pathways approach was embedded in the staff knowledge and operations of the flood group. This was demonstrated by requests to repeat the process in other catchments, which was subsequently done. The use of DMDU tools also brought together a wide range of local and central government officers and elected representatives, which enabled a multi-disciplinary environment to develop for problem-solving.

Importantly, the Council reached a decision on which packages of initial actions (i.e., options) to discuss with the community. Two alternatives were chosen—one starting now and continuing quite far into the future (Option 2C), and a staged one that was more cost-effective in the short term, but would have to be changed fairly soon (Option 4). Option 1 was not consulted on, because the Council decided it was too expensive to implement now (although it was likely to be needed in the long term).

---

[6]The pathways generator can be found here: http://pathways.deltares.nl/.

The Council made the decision to begin by implementing Option 2C after receiving overwhelming support for it from the whole community. Cost-effectiveness is thus not necessarily the primary motivator of decisions in such contexts.

The Council's decision to commit to land purchase under Option 2C created a certainty of funding for those in the urban area affected by this option. While this was open for both options 2C and 4, property owners preferred near-term certainty over waiting 10 years for the land purchase to take place, when another Council might have different priorities. Thus, the certainty created was a stronger motivator for the decision taken, compared with the more cost-effective staged decision (Greater Wellington Regional Council 2015). This was supported by both directly affected property owners and the wider community.

9.6 Reflections (Lessons Learned) for Practice and Theory

How far does DAPP take us in addressing uncertainty? This case study was in part experimental in the New Zealand context, providing an opportunity to identify where further theoretical and methodological development might take place.

We found that for new approaches to be adopted, a "knowledge broker" was necessary to enable the transfer of new knowledge and to facilitate learning. The request to Victoria University of Wellington to assist in this regard, to improve the capability of technical staff at the Council to address "the effects of climate change," is evidence of that. The Council was aware of the IPCC 2013 and 2014 reports and their statutory mandate to address the impacts of climate change. Using the University as a broker of new knowledge and practice through participation in workshops and conferences internationally opened up wider research networks to the councils, bringing in new understandings of the effects of climate change and new tools for practice.

The DAPP approach was attractive to the Council because it deals with uncertainty by enabling a full range of plausible futures to be explored, and thus also addresses tail-end risk (surprise or "Black Swans"). The relationship between the Council and the University facilitator enabled the Council access to tools and techniques that otherwise were not readily accessible to them. Their focus was on solving problems on the ground within a political setting that requires certainty and progress on projects that deliver community services. It thus also had utility for the political decision maker who wanted a systematic, yet flexible tool, to assess changing risk profiles within their term of office (three years), while also achieving flood protection over the long term.

Communities may prefer to pay more for certainty of outcome than a pure cost-effectiveness analysis might suggest, as evidenced in the Hutt River case study. Understanding such preferences will change the way the pathways are quantitatively assessed. It will affect how to weigh certainty against cost (the decision taken was more costly to the Council over time than the alternative). This factor needs to be understood better in terms of its effect on how the different actions and pathways are framed.

The need to reframe the problem to increase understanding and implications of the uncertainties was recognized early on in the process. This recognition became catalytic in the uptake of DAPP within the work of the council flood group. In this context, leadership inside and at different levels in the organization contributed to its mainstreaming, as it did across the region in other cities (e.g., Wellington City Council). The political leadership of the joint councils flood committee, and its chair in particular, also contributed to the uptake of DAPP in flood management assessments and decision making. Funding support from the national government helped to get other councils involved in using the tools, and this is ongoing now through the Resilience and the Deep South National Science Challenges for theoretical and methodological developments, and through testing their application in other case study areas (e.g., the "Living at the Edge" researchers working with the Hawkes Bay joint councils developing the Clifton to Tangoio Coastal Hazard Strategy 2120). Applying the DAPP approach through serious game workshops across New Zealand has helped bring together a wide range of local and national government officers and elected representatives, enabling a multi-disciplinary environment for problem-solving.

This particular case study was constrained by the boundary condition of a single river, which enabled a focused application of the theoretical approach. On the other hand, further applications will enhance learning about the method, if they are focused on wider regional strategies and multiple stressors, and are applied with communities that have a wider stake in the outcomes of the decision making process. Embedding DAPP within New Zealand's national coastal hazards and climate change guidance (Lawrence et al. 2018), centered around community engagement processes, reinforces wider use of DAPP in decision making for other problems that involve deep uncertainty about the future. Further research is needed for implementing DAPP in spatial planning contexts that typically use static consideration of future risk within legal frameworks that emphasize "certainty." Overcoming the certainty/change paradox is necessary to address the path dependency of current approaches for dealing with the future. The risk of path-dependent decisions and the role that current frameworks and practices play in blinding actors to the range of possible outcomes that could evolve in the future, was therefore reduced.

In this case study, the adaptation thresholds and decision trigger values (trigger points) for deriving signals of change were not fully developed. The economic analysis of the pathways was undertaken as a sensitivity analysis of the factors noted in Sect. 9.4. Further work in the New Zealand context is underway to define trigger points for future thresholds for changing pathways. Some work has started on this (Barnett et al. 2014; Stephens et al. 2018; Werners et al. 2013), addressing physical impacts, socially defined triggers that are decision relevant and meaningful to the affected communities, and the relationship between the trigger points and decisions that can be taken, and when.

However, whether signals and trigger points can be monitored and acted upon in practice is unclear in a river management context. This is because funding allocations by councils are made on a 10-year basis with three-yearly reviews [through statutory Long-Term Plans (LTPs)]. This funding system, with its focus on structural investments for "protecting" communities from floods, as opposed to reducing future risk, will affect how reviews at trigger points can be implemented. For example, it is never certain that funds will be allocated at any time in the future through the LTP, because of other competing projects, the priority for which will also be affected by future values, social and economic preferences, and vulnerabilities.

The ability to implement pathway choices made today is influenced by the dominant political setting, governance arrangements, and economic conditions. Addressing the following questions is therefore critical for whether policy choices can be implemented: How can the planning approach, and the decisions made, persist? How can the planning objectives be revisited when the operating conditions and enablers change? Does the theoretical approach hold under "fire" from communities directly affected by the risk, and when "surprises" or disasters happen? Can DAPP facilitate the identification of enablers and entry points for a robust and flexible implementation pathway? Better understanding of the influence of these questions and their answers on the planning and implementation process, and how to make their constraining characteristics into opportunities for adaptive decision making, will be essential for deep uncertainty to be addressed where it matters.

References

Barnett, J., Graham, S., Mortreux, C., Fincher, R., Waters, E., & Hurlimann, A. (2014). A local coastal adaptation pathway. *Nature Climate Change, 4,* 1103–1108.

Burby, R. J., & French, S. P. (1981). Coping with floods: the land use management paradox. *Journal of the American Planning Association, 47,* 289–300.

Burby, R. J., Nelson, A. C., Parker, D., & Handmer, J. (2001). Urban containment policy and exposure to natural hazards: Is there a connection? *Journal of Environmental Planning and Management, 44,* 475–490.

Erickscn, N. J. (1986). *Creating flood disasters* (p. 323). Water and Soil Miscellaneous Publication 77. Water and Soil Division, Ministry of Works and Development, Wellington.

Greater Wellington Regional Council, (2015). *Flood protection: Option flexibility and its value Hutt River City Centre Upgrade River Corridor Options Report* (p. 31). Prepared for GWRC by Infometrics & PS Consulting Ltd. Greater Wellington Regional Council, Wellington.

Haasnoot, M., Kwakkel, J., Walker, W., & ter Maat, J. (2013). Dynamic adaptive policy pathways: A method for crafting robust decisions for a deeply uncertain world. *Global Environmental Change, 23,* 485–498.

Kwadijk, J., Haasnoot, M., Mulder, J., Hoogvliet, M., Jeuken, A., Van der Krogt, R., et al. (2010). Using adaptation tipping points to prepare for climate change and sea level rise: A case study in the Netherlands. *Wiley Interdisciplinary Reviews: Climate Change, 1,* 729–740.

Lawrence, J., Bell, R., Blackett, P., Stephens, S., & Allan, S. (2018). National guidance for adapting to coastal hazards and sea level rise: Anticipating change, when and how to change pathway. *Environmental Science & Policy, 82,* 100–107.

Lawrence, J., & Haasnoot, M. (2017). What it took to catalyze uptake of dynamic adaptive pathways planning to address climate change uncertainty. *Environmental Science & Policy, 68,* 47–57.

Lawrence, J., Reisinger, A., Mullan, B., & Jackson, B. (2013a). Exploring climate change uncertainties to support adaptive management of changing flood-risk. *Environmental Science & Policy, 33,* 133–142.

Lawrence, J., Sullivan, F., Lash, A., Ide, G., Cameron, C., & McGlinchey, L. (2013b). Adapting to changing climate risk by local government in New Zealand: institutional practice barriers and enablers. *Local Environment*, 1–23.
Ministry for the Environment. (2008). *Meeting the challenges of future flooding in New Zealand*. Wellington: Ministry for the Environment.
Ministry for the Environment. (2017). *Coastal hazards and climate change: Guidance for local government.* In R. G. Bell, J. Lawrence, S. Allan, P. Blackett, & S. A. Stephens (Eds.). Ministry for the Environment Publication ME-1341. Ministry for the Environment, Wellington.
Pearson, C., & McKerchar, A. (1999). *An update of Hutt river flood frequency* (p. 24). NIWA Client Report CHC99/11, Wellington.
Ranger, N., Millner, A., Dietz, S., Fankhauser, S., Lopez, A., & Ruta, G. (2010). *Adaptation in the UK: A decision-making process. A policy brief*. Grantham Institute on Climate Change and the Environment and Centre for Climate Change Economics and Policy.
Reisinger, A., Kitching, R. L., Chiew, F., Hughes, L., Newton, P. C. D., Schuster, S., Tait, A., & Whetton, P. (2014). *Australasia climate change 2014: Impacts, adaptation and vulnerability. Part B: Regional aspects. Contribution of working group II to the fifth assessment report of the intergovernmental panel on climate change* In V. R. Barros, C. B. Field, D. J. Dokken, M. D. Mastrandrea, K. J. Mach, T. E. Bilir, M. Chatterjee, K. L. Ebi, Y. O. Estrada, R. C. Genova, B. Girma, E. S. Kissel, A. N. Levy, S. MacCracken, P. R. Mastrandrea, & L. L. White (Eds.) (pp. 1371–1438). Cambridge, UK and New York, USA: Cambridge University Press.
Stephens, S., Bell, R., & Lawrence, J. (2018). Developing signals to trigger adaptation to sea-level rise. *Environmental Research Letters, 13*(10). https://doi.org/10.1088/1748-9326/aadf96.
Tait, A., Bell, R., & Burgess, S. (2002). *Meteorological hazards and the potential impacts of climate change in Wellington region—A scoping study*. Report for NIWA & Wellington Regional Council WLG2002/19. 152.
Tobin, G. A. (1995). The levee love affair: A stormy relationship. *Journal of the American Water Resources Association, 31*, 359–367.
Valkering, P., van der Brugge, R., Offermans, A., Haasnoot, M., & Vreugdenhil, H. (2013). Perspective-based simulation game to explore future pathways of a water-society system under climate change. *Simulation and Gaming, 44*, 366–390.
Vandenbeld, A., & MacDonald, J. (2013). Fostering community acceptance of managed retreat in New Zealand. In J. Palutikof, S. Boulter, A. Ash, M. Stafford Smith, M. Parry, M. Waschka, & D. Guitart (Eds.), *Climate adaptation futures* (pp. 161–166). UK: Wiley-Blackwell.
Wellington Regional Council. (1996). *Living with the river: Hutt River floodplain management plan phase one summary report* (p. 113).
Wellington Regional Council. (2001). *Hutt River floodplain management plan: For the Hutt river and its environment*. Wellington.
Werners, S., Pfenninger, S., van Slobbe, E., Haasnoot, M., Kwakkel, J., & Swart, R. (2013). Thresholds, tipping and turning points for sustainability under climate change. *Current Opinion in Environmental Sustainability, 5,* 334–340.

Dr. Judy Lawrence is a Senior Research Fellow (Climate Change Research Institute, Victoria University of Wellington). Her research focus is on climate change adaptation, uncertainty, and institutional policy design. Judy co-chaired the New Zealand government's Climate Change Adaptation Technical Working Group, was co-author of the new Zealand Coastal hazards and Climate Change Guidance and is a Coordinating Lead Author for the IPCC Sixth Assessment Review, Working Group 2. Judy leads the development and application of Dynamic Adaptive Policy Pathways (DAPP) planning in New Zealand.

Dr. Marjolijn Haasnoot is a Senior Researcher and Advisor at Deltares and researcher in adaptive delta management at Utrecht University, the Netherlands. Her research focus is on water management, integrated assessment modeling, and decision making under deep uncertainty. She is working and consulting internationally on projects assessing impacts of climate change and socioeconomic developments and alternative management options to develop robust and adaptive plans. Marjolijn developed the Dynamic Adaptive Policy Pathways planning approach. She has a Ph.D. in engineering and a Masters in environmental science.

Laura McKim, M.Sc. was a Strategic Planner at Greater Wellington Regional Council (GWRC) developing the GWRC Climate Change Strategy, and worked on the dissemination and facilitation of the river and coastal serious games with council staff. She has a Masters in environmental studies.

Dayasiri Atapattu, M.Sc. was in the Flood Protection Group of the Greater Wellington Regional Council as Team Leader for the Hutt River Floodplain Management Plan (HRFMP). His work focus was investigating, designing, and constructing hydraulic structures and has a Bachelor of Science (Engineering) from the University of Sri Lanka and a Master of Science (Engineering Hydrology) from the Imperial College, University of London.

Graeme Campbell, B.Sc. is currently Manager Flood Protection at the Greater Wellington Regional Council. He works with communities, individuals, asset managers, local and regional councils, and politicians to mitigate the effects of natural hazards related to flooding, cyclones, and tsunami. Graeme was previously Manager of AC Consulting Group Ltd, (River, Marine and Civil Works), undertaking pre- and post-disaster assessment and hazard mapping for structural and non-structural mitigation options. He has a Bachelor of Civil Engineering.

Dr. Adolf Stroombergen is Chief Economist and Director at Infometrics Consulting, Wellington, New Zealand. His work focus is on economic modeling, public and social economics, transport, energy, econometrics, efficiency, risk analysis, and benchmarking. He undertook the Real Options Analysis of DAPP for the Hutt River flood and Hawkes Bay coastal case studies. He has a Bachelor's degree in Economics and Mathematics, and a Ph.D. in Economics.

Open Access This chapter is licensed under the terms of the Creative Commons Attribution 4.0 International License (http://creativecommons.org/licenses/by/4.0/), which permits use, sharing, adaptation, distribution and reproduction in any medium or format, as long as you give appropriate credit to the original author(s) and the source, provide a link to the Creative Commons licence and indicate if changes were made.

The images or other third party material in this chapter are included in the chapter's Creative Commons licence, unless indicated otherwise in a credit line to the material. If material is not included in the chapter's Creative Commons licence and your intended use is not permitted by statutory regulation or exceeds the permitted use, you will need to obtain permission directly from the copyright holder.

Chapter 10
Info-Gap (IG): Robust Design of a Mechanical Latch

François M. Hemez and Kendra L. Van Buren

Abstract

- Info-Gap (IG) Decision Theory, introduced in Chap. 5, is used to formulate, and solve, the analysis of robustness in the early design of a mechanical latch.
- The three components necessary to assess robustness of the latch design (a system model, performance requirement, and representation of uncertainty) are discussed.
- The robustness analysis indicates that the nominal design can accommodate significant uncertainty before failing to deliver the required performance.
- The discussion concludes with the assessment of a variant design to show how a decision (*"which design should be chosen?"*) can be confidently reached despite the presence of significant gaps in knowledge.

10.1 Introduction

The role of numerical simulation to aid in decisionmaking has grown significantly in the past three decades for a diverse number of applications, such as financial modeling, weather forecasting, and design prototyping (Oden et al. 2006). Despite its widespread use, numerical simulations suffer from unavoidable sources of uncertainty, such as making simplifying assumptions to represent non-idealized behavior, unknown initial conditions, and variability in environmental conditions. The question posed herein is: How can simulations support a confident decision, given their inherent sources of uncertainty? This chapter gives an answer to this question using Info-Gap (IG) Decision Theory, the philosophy of which is presented in Chap. 5.

The main factors in selecting an approach to evaluate the effect of uncertainty on a decision are (1) the nature and severity of the uncertainty, and (2) computational feasibility. In cases of deep uncertainty, the development of a probability distribution

---

F. M. Hemez (✉)
Lawrence Livermore National Laboratory, Livermore, CA, USA
e-mail: hemez1@llnl.gov

K. L. Van Buren
Los Alamos National Laboratory, Los Alamos, NM, USA

cannot confidently be made. Further, uncertainty can often be unbounded and decisionmakers will need to understand how much uncertainty they are accepting when they proceed with a decision. A few examples of comparing methods can be found in Hall et al. (2012) and Ben-Haim et al. (2009). Hall et al. (2012) compare IG with RDM (Chap. 2). Ben-Haim et al. (2009) compare IG with robust Bayes (a min–max approach with probability bounds known as P-boxes, and coherent lower previsions). In all cases, the major distinction is the prior knowledge, or how much information (theoretical understanding, physical observations, numerical models, expert judgment, etc.) is available, and what the analyst is willing to assume if this information does not sufficiently constrain the formulation of the decisionmaking problem. One can generally characterize IG as a methodology that demands less prior information than other methods in most situations. Furthermore, P-boxes can be viewed as a special case of IG analysis (Ferson and Tucker 2008), which suggests that these two approaches may be integrated to support probabilistic reasoning. The flexibility promoted by IG, however, also makes it possible to conveniently analyze a problem for which probability distributions would be unknown, or at least uncertain, and when a realistic and meaningful worst case cannot be reliably identified (Ben-Haim 2006).

Applying IG to support confident decisionmaking using simulations hinges on the ability to establish the robustness of the forecast (or predicted) performance to modeling assumptions and sources of uncertainty. Robustness, in this context, means that the performance requirement is met even if some of the modeling assumptions happen to be incorrect. In practice, this is achieved by exercising the simulation model to explore uncertainty spaces that represent gaps in knowledge—that is, the "difference" between best-known assumptions and how reality could potentially deviate from them. An analysis of robustness then seeks to establish that performance remains acceptable within these uncertainty spaces. Some requirement-satisfying decisions will tolerate more deviation from best-known assumptions than others. Given two decisions offering similar attributes (feasibility, safety, cost, etc.), preference should always be given to the more robust one—that is, the solution that tolerates more uncertainty without endangering performance.

As discussed in Chap. 5, three components are needed to carry out an IG analysis: A system model, a performance requirement, and a representation of uncertainty. This chapter shows how to develop these components to assess robustness for the performance of a mechanical latch in an early phase of design prototyping. IG is particularly suitable for design prototyping, because it offers the advantages of accommodating minimal assumptions while communicating the results of an analysis efficiently through a robustness function.

Our discussion starts in Sect. 10.2 by addressing how the three components of an info-gap analysis (system model, performance requirement, and uncertainty model) can be formulated for applications that involve policy topics. The purpose is to emphasize that info-gap applies to a wide range of contexts, not just those from computational physics and engineering grounded in first-principle equations. Section 10.3 introduces our simple mechanical example—the design of a latch, and its desired performance requirements given that the geometry, material properties, and loading conditions are partially unknown. The section also discusses the system model defined to represent the design problem. Details of the simulation (how the geometry is simplified, how some of the modeling assumptions are justified, how truncation errors are controlled, etc.) are omitted, since they are not essential to understand how an analysis of robustness is carried out.

Section 10.4 discusses the main sources of uncertainty in the latch design problem, how they are represented with an info-gap model of uncertainty, and the implementation of robustness analysis. Two competing latch designs are evaluated in Sect. 10.5 to illustrate how confident decisions can be reached despite the presence of significant gaps in knowledge. A summary of the main points made is provided in Sect. 10.6.

10.2 Application of Info-Gap Robustness for Policymaking

Our application suggests how info-gap robustness (see Chap. 5) can be used to manage uncertainty in the early-stage design of a latch mechanism (Sects. 10.3 and 10.4) and how robustness functions may be exploited to support decisionmaking (Sect. 10.5). The reader should not be misled, however, in believing that this methodology applies only to contexts described by first-principle equations such as the conservation of momentum (also known as Newton's 2nd law, "Force = Mass × Acceleration") solved for the latch example. The discussion presented here emphasizes the versatility of IG robustness for other contexts, particularly those involving policy topics for which well-accepted scientific models might be lacking.

Consider two high-consequence policymaking applications:

- *Climate change*: Policy decisions to address the impact on human activity of changes in the global climate (and vice versa) tend to follow either the precautionary principle or the scientific design of intervention. In the first case (precautionary principle), decisionmakers would err on the side of early intervention to mitigate the potentially adverse consequences of climate change, even if the scientific understanding of what causes these changes, and what the consequences might be, is lacking. In the second case (scientific design of intervention), longer-range planning is implied while a stronger emphasis would be placed on gaining a better scientific understanding and reducing the sources of uncertainty before policy is enacted. In the presence of incomplete understanding of the phenomena that drive changes in the global climate, effects on the planet's eco-system, and potential consequences for human activity, it is unclear which early intervention strategies to adopt and how effective they might be. Given scientific uncertainty, however, policymakers are inclined to adopt early precautionary intervention.
- *Long-range infrastructure planning*: The world population is increasingly concentrated in urban areas. The ten largest urban areas, such as Tokyo (Japan), Jakarta (Indonesia), Delhi (India), and New York City (USA), feature population densities between 2,000 and 12,000 individuals per km^2, thus exceeding densities in rural areas by more than two to three orders-of-magnitude. Managing these population centers offers serious challenges in terms of housing, transportation, water and

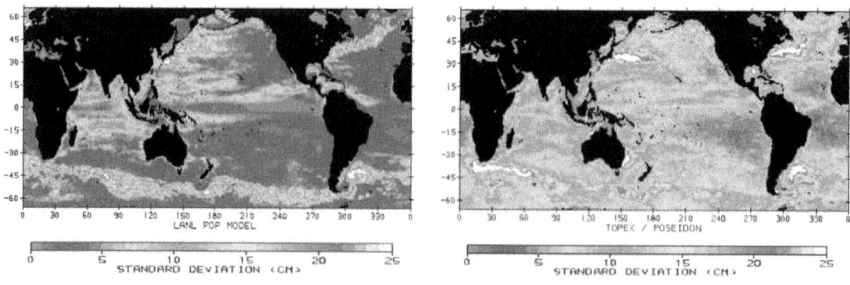

(a) Left: Satellite measurement of sea variability   (b) Right: Model prediction of sea variability

Fig. 10.1 Measurement and prediction of sea surface variability (Malone et al. 2003)

power supplies, access to nutrition, waste management, and many other critical systems. Future infrastructure needs to be planned for many decades in the presence of significant uncertainty regarding population and economic growth, urbanization laws, and the adoption of future technologies. The development, for example, of peer-to-peer transportation systems might render it necessary to rethink how conventional public transportation networks and taxi services are organized. The challenge of infrastructure planning is to design sufficient flexibility in these interconnected engineered systems when some of the factors influencing them, together with the performance requirements themselves, might be partially unknown.

The challenge of policymaking for these and similar problems is, of course, how to manage the uncertainty. These applications often involve incomplete scientific understanding of the processes involved, elements of extrapolation or forecasting beyond known or tested conditions, and aspects of the decisionmaking practice that are not amenable to being formulated with mathematical models. Info-gap robustness, nevertheless, makes it possible to assess whether a policy decision would deliver the desired outcomes even if the definition of the problem features significant uncertainty and some of the assumptions formulated in the analysis are incorrect.

Consider, for example, climate change. Developing a framework to support policymaking might start with a scientific description of how the oceans, atmosphere, and ice caps interact. Figure 10.1 illustrates a satellite measurement of sea-level variability (left) compared to the prediction obtained with a global circulation model (right). The latter is based on historical data and observations made in the recent past that are extrapolated to portray the conditions observed by the satellite over a similar period. Smith and Gent (2002) describe the physics-based models solved numerically to describe this phenomenology.

Even though simulations such as Fig. 10.1 are grounded in first-principle descriptions, they are not immune to uncertainty. Executing this calculation with a one-degree resolution (*i.e.*, 360 grid points around the Earth), for example, implies that some of the computational zones are as large as 314 km^2 near the equator, which is nearly five times the surface area of Washington, D.C. It raises the question of whether localized eddies that contribute to phenomena such as the Atlantic Ocean's

Gulf Stream are appropriately represented. Beyond the question of adequacy, settings such as resolution, fidelity with which various processes are described, and convergence of numerical solvers generate numerical uncertainty. These imply that code predictions could differ, maybe significantly, from the "true-but-unknown" conditions that analysts seek to know.

Other commonly encountered sources of uncertainty in first-principle simulations include the variability or incomplete knowledge of initial conditions, boundary conditions, constitutive properties (material properties, reactive chemistry, etc.), and source functions (*e.g.*, how much greenhouse gas is introduced into the atmosphere?). This is in addition to not always understanding how different processes might be coupled (*e.g.*, how does the chemistry of the ocean change due to increased acidity of the atmosphere?). Model-form uncertainty, which refers to the fact that the functional form of a model might be unknown, is also pervasive in computational sciences. An example would be to select a mathematical equation to represent the behavior of a chemical at conditions that extrapolate beyond what can be experimentally tested in a laboratory. Finally, large-scale simulation endeavors often require passing information across different code platforms. Such linkages can introduce additional uncertainty, depending on how the variables solved for in one code are mapped to initialize the calculation in another code.

The aforementioned sources of uncertainty, while they are multifaceted in nature and can be severe, are handled appropriately by a number of well-established methods, such as statistical sampling (Metropolis and Ulam 1949), probabilistic reliability (Wu 1994), worst-case analysis, and IG robustness. In the last case, the system model is the simulation flow that converts input settings to performance outcomes. The performance requirement defines a single criterion or multiple criteria that separate success from failure. The uncertainty model describes the sources of variability and lack-of-knowledge introduced by the simulation flow. Once the three components are defined, a solution procedure is implemented to estimate the robustness function of a given decision. Competing decisions can then be assessed by their ability to meet the performance requirement. Likewise, the confidence placed in a decision is indicated by the degree to which its forecasted performance is robust, or insensitive, to increasing levels of uncertainty in the formulation of the problem. Regardless of how sophisticated a simulation flow might be, IG analysis always follows this generic procedure, as is discussed in Sects. 10.3 and 10.4 for the latch application.

Eventually, information generated from first-principle models yields indicators that need to be combined with "soft" data to support policy decisions. Figure 10.2 (Bamber et al. 2009) is a notional illustration where sea levels (left) predicted over Western Europe would be combined with population densities (right) to assess how to mitigate the potentially adverse consequences of rising waters. In this example, one source of information (sea levels) comes from a physics-based global circulation model, while the other one (projected population levels) represents "softer" data, since the future population levels need to be extrapolated from spatially and temporally sparse census data. As one moves away from science-based modeling and simulation, the data, opinions, and other considerations integrated to support policy decisions contribute sources of uncertainty of their own. This uncertainty might

(a) Left: Prediction of costal sea level   *(b) Right: Prediction of population*

Fig. 10.2 Predictions of coastal sea-level rise and population growth in Western Europe

Fig. 10.3 Extrapolation of populations beyond 2012, by continent

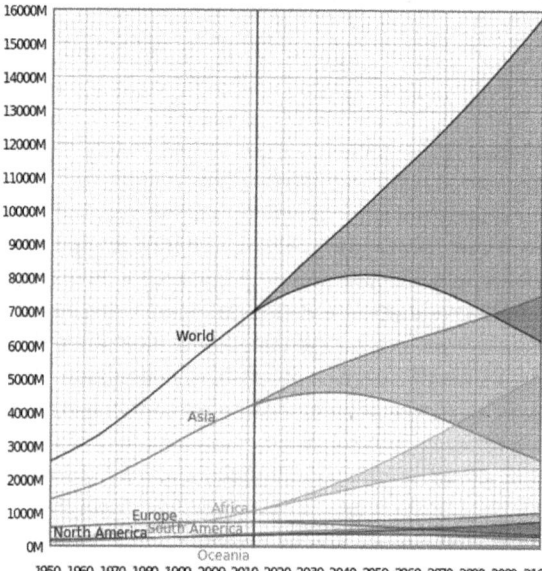

also get amplified from extrapolating to conditions that have not been previously observed, or when forecasts are projected into the future. Figure 10.3 suggests what happens, for example, to forecasts of the world's population (United Nations 2014).

IG robustness makes it possible to assess the extent to which a policy decision is affected by what may be unknown, even in the presence of sources of uncertainty that do not lend themselves to parametric representations such as probability distributions, polynomial chaos expansions, or intervals. Accounting for an uncertainty such as the gray region of Fig. 10.3 is challenging if a functional form is lacking. One might not know if the world's population can be modeled as increasing or decreasing, or even if the trend can be portrayed as monotonic.

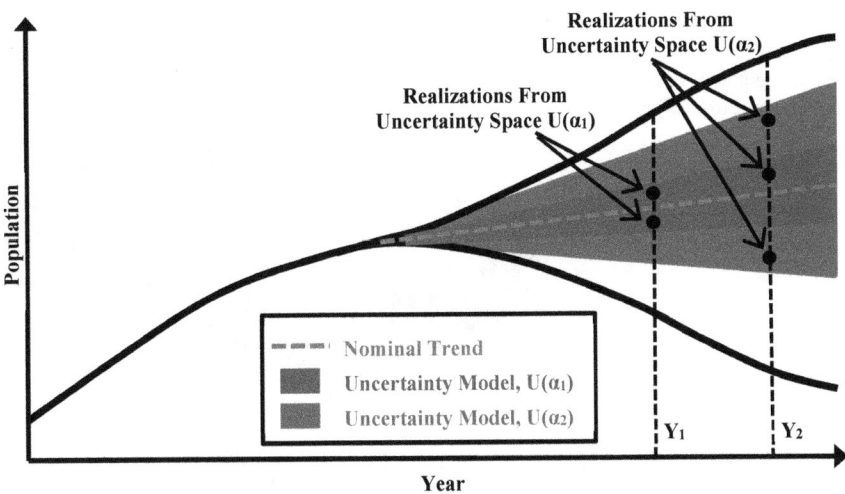

Fig. 10.4 Representing increasing levels of uncertainty for the world's population

Figure 10.4 suggests one possibility to handle this challenge, whereby increasing levels of uncertainty, as indicated by the uncertainty spaces $U(\alpha_1)$ (blue region) and $U(\alpha_2)$ (green region), are defined around a nominal trend (red-dashed line). The uncertainty can be explored by selecting population values within these sets and without necessarily having to formulate a parametric representation (*e.g.*, "population growth is exponential") if policymakers are not willing to postulate such an assumption. The figure illustrates two values chosen in set $U(\alpha_1)$ at year Y_1, and three values selected in the larger-uncertainty set $U(\alpha_2)$ at year Y_2. This procedure would typically be implemented to assess if the policy objective is met as future populations deviate from the nominal trend in unknown ways.

Another type of uncertainty, often encountered in the formulation of policy problems and which lends itself naturally to info-gap analysis, is qualitative information or expert opinions that introduce vagueness or non-specificity. For example, one might state from Fig. 10.3 that "World population is growing", without characterizing this trend with a mathematical equation. Policymakers might seek to explore if decisions they consider can accommodate this type of uncertainty while delivering the desired outcomes. The components of such an analysis would be similar to those previously discussed. A system model is needed to analyze the consequences of specific conditions, such as "the population is growing" or "the population is receding", and a performance requirement is formulated to separate success from failure. The uncertainty model would, in this case, include alternative statements (*e.g.*, "the population is growing" or "the population is growing faster") to assess if the decision meets the performance requirement given such an uncertainty.

Fig. 10.5 Conceptual illustration of the compartment door of a consumer electronics product

10.3 Formulation for the Design of a Mechanical Latch

To provide a simple example of applying IG, we illustrate its use in the design of a latch for the compartment door of a consumer electronics product conceptually illustrated in Fig. 10.5. The objective of the design is to ensure proper opening and closing of the door. The challenge is that the geometry of the door, material properties, and loading conditions are not precisely known, which is common in an early design phase. Establishing that the performance of a given design is robust to these gaps in knowledge, as discussed in Sect. 10.4, demonstrates that the requirement can be met in the presence of potentially deep uncertainty. The decisionmaker or customer can rely on this information to appreciate the relative merits of different design decisions.

The first step is to define the problem, its loading scenario, and decision criterion. Expert judgment suggests that the analysis be focused on stresses generated in the latch when opening and closing the door. Figure 10.6 indicates the latch component considered for analysis. For simplicity, the rest of the compartment is ignored, the contact condition is idealized, and severe loading conditions, such as those produced when dropping the device on the ground, are not modeled. Likewise, non-elastic deformation, plasticity, damage, and failure mechanics are not considered. A linear, isotropic, and homogeneous material model is specified. Such a model is useful to characterize the performance of the nominal design as long as one understands that the final material selected for manufacturing could exhibit characteristics that deviate significantly from this baseline. Assessing performance solely based on these nominal properties is, therefore, not a sound design strategy. It is necessary to assess the robustness of that performance to changes in the material (and other) properties.

The geometry of the latch shown on the right of Fig. 10.6 is simplified by converting the round corners to straight edges. This results in dimensions of 3.9 mm (length) by 4.0 mm (width) by 0.8 mm (thickness). The latch's head, to which the contact displacement is applied, protrudes 0.4 mm above the surface. A perfectly rigid attachment to the compartment door is assumed, which makes it possible to neglect the door altogether and greatly simplifies the implementation.

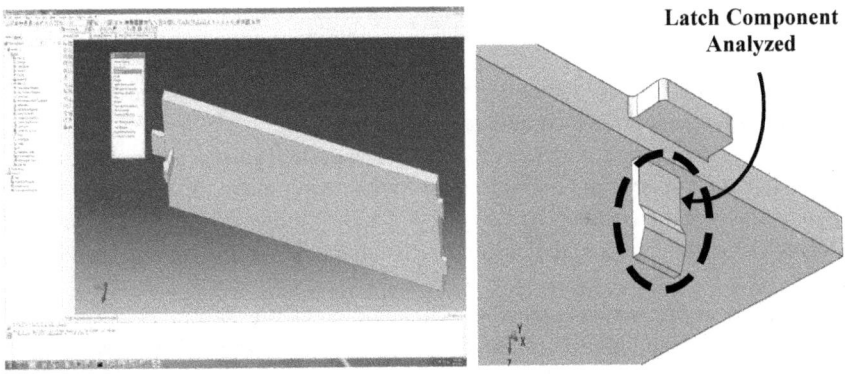

(a) Left: Compartment door geometry   *(b) Right: Latch component analyzed*

Fig. 10.6 Geometry of the compartment door and detail of the latch analyzed

A loading scenario also needs to be specified to carry out the analysis. It is assumed that the latch locks securely in place in the compartment's receptacle to close the door after deflecting by a specified amount. Given the simplification of the geometry considered, this condition can be analyzed by applying a displacement whose nominal value is:

$$U_{\text{Contact}} = 0.50 \text{ mm.} \tag{10.1}$$

The nominal value, however, is only a best estimate obtained from the manufacturing of similar devices, and it is desirable to guarantee that the latch can meet its performance requirement given the application of contact displacements different from 0.50 mm (either smaller or greater).

Finally, a performance requirement must be defined for simulation-based decisionmaking. Multiple criteria are often considered in structural design, sometimes conflicting with each other. One wishes, for example, to reduce the weight of a component while increasing its stiffness. For clarity, a single requirement is proposed here. Applying the displacement U_{Contact} generates a bending deformation, which produces a force that varies in time as the latch relaxes from its initial deflection. The dynamically applied load that results produces stresses in the material. A common performance criterion is to ensure that the material can withstand these stresses without breaking. The design is said to be requirement-compliant if the maximum stress anywhere in the latch does not exceed a critical stress value:

$$\sigma_{\text{Max}} \leq \sigma_{\text{Critial}} \tag{10.2}$$

The upper bound is defined as a fraction of the yield stress that indicates material failure:

$$\sigma_{\text{Critical}} = (1 - f_S)\sigma_{\text{Yield}} \quad (10.3)$$

where f_S denotes a safety factor defined in $0 \leq f_S \leq 1$. The yield stress is $\sigma_{\text{Yield}} = 55$ MPa for the generic polycarbonate material analyzed.

With this formulation, the analyst can select a desired safety factor and ascertain how much uncertainty can be tolerated given this requirement. A well-known trade-off, which is observed in Sect. 10.4, is that demanding more performance by selecting a larger value of f_S renders the design more vulnerable (or less robust) to modeling uncertainty.

The analysis of mechanical latches is a mature field after several decades of their common use in many industries. An example is given in BASF (2001), where peak strain values obtained for latches of different geometries are estimated using a combination of closed-form formulae and empirical factors. While these simplifications enable derivations "by hand," they also introduce assumptions that are thought to be inappropriate for this design problem. The decision is therefore made to rely on a finite element representation (Zienkiewicz and Taylor 2000) to estimate stresses that result from imposing the displacement (10.1) and assess whether the requirement-compliant condition (10.2) is met. The finite element model discretizes the latch's geometry into elemental volumes within which the equations-of-motion are solved. The Newmark algorithm is implemented to integrate the equations in time (Newmark 1959). This procedure also introduces assumptions, such as the type of interpolation function selected for the elemental volumes. These choices add to the discretization of the geometry to generate truncation errors. Even though they are important, these mesh-size and run-time considerations are not discussed in order to keep our focus on the info-gap analysis of robustness.

To briefly motivate the modeling choices made, it suffices to state that the three-dimensional representation of the latch's geometry yields a more realistic prediction of the deformation shape by capturing the curvature caused by the applied load. This is illustrated in Fig. 10.7, which depicts the computational mesh and deformation pattern resulting from applying the nominal displacement of 0.50 mm. This simulation is performed with standalone, MATLAB®-based, finite element software developed by the authors. The predicted deformation, *i.e.*, displacements such as those depicted in Fig. 10.7, and corresponding forces are extracted and provided to a one-dimensional approximation based on linear beam theory to estimate the peak stress, σ_{Max}. Predicting σ_{Max} depends, naturally, on choices made to setup the simulation such as values of the applied displacement ("Is the displacement equal to 0.50 mm or something else?") and material properties ("Are the stiffness and density properties prescribed using the nominal values or something else?"). Next, an analysis of robustness is carried out to assess the extent to which the design will remain requirement-compliant even if some of these assumptions are changed.

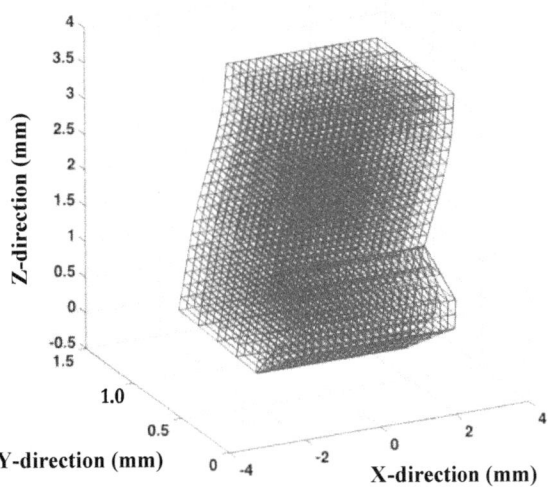

Fig. 10.7 Computational mesh and deformation of the latch due to a 0.50-mm displacement

10.4 The Info-Gap Robust Design Methodology

This section discusses the methodology applied to achieve an info-gap robust design. Three issues need to be discussed before illustrating how the robustness function is calculated and utilized to support decisionmaking. The first issue is to define the design space. The second issue is to determine the sources of uncertainty against which the design must be robust. The third question is how to represent this uncertainty mathematically without imposing unwarranted assumptions. These choices are discussed before showing how the robustness function is derived.

Several parameters of the geometry are available to define the design space, including the length, width, thickness, and overall configuration of the latch's geometry. For computational efficiency, it is desirable to explore an as-small-as-possible design space while ensuring that the parameters selected for design optimization exercise an as-significant-as-possible influence on performance, which here is the peak stress, σ_{Max}, of Eq. (10.2).

The first issue is to define the design space by judiciously selecting parameters that describe the geometry of the latch. This is achieved using global sensitivity analysis to identify the most influential parameters (Saltelli et al. 2000). Five sizing parameters are considered. They are the total length (L), width (W_C), and depth (D_C) of the latch; and geometry (length, L_M, and depth, D_H) of the surface where the displacement $U_{Contact}$ is applied. An analysis-of-variance is performed based on a three-level, full-factorial design of computer experiments that requires $3^5 = 243$ finite element simulations. The parameters L (total length) and W_C (width) are found to account for approximately 76% of the total variability of peak-stress predictions when the five dimensions ($L; W_C; D_C; D_H; L_M$) are varied between their lower and upper bounds. Design exploration is therefore restricted to the pair ($L; W_C$).

Table 10.1 Definition of sources of uncertainty in the simulation model

Variable	Description	Nominal value	Typical range
E	Modulus of elasticity	2.0 GPa	2.0–3.0 GPa
G	Shear modulus	0.73 GPa	0.71–1.11 GPa
ν	Poisson's ratio	0.37	0.35–0.40
ρ	Mass density	$1.20 \times 10^{+3}$ kg/m^3	1.20–$1.25 \times 10^{+3}$ kg/m^3
U_{Contact}	Applied contact displacement	0.50 mm	0.20–0.80 mm
F_{OS}	Dynamic overshoot factor	1.0	0.5–1.5

The second issue is to define the sources of modeling uncertainty against which the design must be robust. This uncertainty represents the fact that real-world conditions might deviate from what is assumed in the simulation model. To make matters more complicated, the magnitude of these deviations, which indicates by how much the model could be incorrect, is unknown. Furthermore, precise probability distributions are lacking. It is essential that the representation of uncertainty can account for these attributes of the problem without imposing unwarranted assumptions.

Table 10.1 defines the sources of modeling uncertainty considered in the analysis. The first four variables (E; G; ν; ρ) represent the variability of polycarbonate plastics. The nominal values (third column) and typical ranges (fourth column) originate from surveying material properties published by various manufacturers. We stress, however, that the actual values may fall outside of these ranges. The fifth variable, U_{Contact}, accounts for uncertainty of the actual displacement to which the latch might be subjected when opening and closing the compartment door. The dynamic load overshoot factor (sixth variable), F_{OS}, is purely numerical. It expresses that the actual loading may differ from how the displacement condition is specified in the simulation. Variable F_{OS} is used as a scaling factor that changes the dynamic overshoot resulting from the application of a short-duration transient load.

The modulus of elasticity (E) is an uncertain material property. It is estimated at 2.0 GPa, and it is confidently known that it will not be less than this value, though it could be greater by one GPa or more. The most that can be said about this variable is that it falls in a one-sided range of unknown size, which can be represented by an unbounded family of nested intervals:

$$E^{(0)} \leq E \leq E^{(0)} + h W_E^{(\text{Upper})}, \tag{10.4}$$

where $E^{(0)} = 2.0$ GPa and $W_E^{(\text{Upper})} = 1.0$ GPa. In this formulation, the quantity h represents the unknown horizon-of-uncertainty ($h \geq 0$). Likewise, the nominal value of the shear modulus (G) is 0.73 GPa, an estimate that could err as low as 0.71 GPa (or less), and as high as 1.11 GPa (or more). Thus, a family of nested asymmetric intervals captures the uncertainty in G:

$$G^{(0)} - hW_G^{(\text{Lower})} \leq G \leq G^{(0)} + hW_G^{(\text{Upper})}, \quad (10.5)$$

where $G^{(0)} = 0.73$ GPa, $W_G^{(\text{Lower})} = 0.02$ GPa, and $W_G^{(\text{Upper})} = 0.38$ GPa. Uncertainty in v and ρ is represented by uncertain intervals similar to Eq. (10.5).

The fifth variable of Table 10.1 is the displacement U_{Contact}. The latch must allow reliable opening and closing of the compartment door for a nominal 0.50-mm displacement. This value, however, is only an estimate and the range defined in the table (0.20 mm $\leq U_{\text{Contact}} \leq$ 0.80 mm) expresses that the applied displacement is unknown. There is no fundamental reason that U_{Contact} cannot be less than 0.20 mm or cannot exceed 0.80 mm. These are estimates based on extreme events, which are typically poorly known because they tend to be postulated rather than being observed. A formulation with nested intervals acknowledges that this range is uncertain:

$$U_{\text{Contact}}^{(0)} - hW_U^{(\text{Lower})} \leq U_{\text{Contact}} \leq U_{\text{Contact}}^{(0)} + hW_U^{(\text{Upper})}, \quad (10.6)$$

where $U_{\text{Contact}}^{(0)} = 0.50$ mm and $W_U^{(\text{Lower})} = W_U^{(\text{Upper})} = 0.30$ mm.

The above description of intervals for variables $(E; G; v; \rho; U_{\text{Contact}}; F_{\text{OS}})$ addresses the third and final question, which is how to mathematically represent the model uncertainty. Little is typically known about sources of uncertainty such as these in the early stage of a design. For this reason, and to avoid injecting unsubstantiated assumptions in the analysis, no probability law or membership function is assumed. Even the ranges listed in Table 10.1 are questionable, as collecting more information or choosing a different material for manufacturing could yield values outside of these assumed intervals. For these reasons, the uncertainty of each variable is represented as a range of unknown size, which defines a family of six-dimensional hypercubes:

$$U(h) = \left\{ \theta = (\theta_k)_{1 \leq k \leq 6} \text{ such that } hW_k^{(\text{Lower})} \leq \theta_k - \theta_k^{(0)} \leq hW_k^{(\text{Upper})} \right\}, \quad (10.7)$$

with $h \geq 0$. The vector $\theta = (\theta_k)_{1 \leq k \leq 6}$ collects the six variables $(E; G; v; \rho; U_{\text{Contact}}; F_{\text{OS}})$, and $\theta_k^{(0)}$ denotes a nominal value (third column of Table 10.1). The IG model of uncertainty, $U(h)$, is not a single set (hypercube in this case), but rather an unbounded family of nested sets (hypercubes). The hypercubes grow as h gets larger, endowing h with its meaning as a horizon-of-uncertainty. The scaling coefficients $W_k^{(\text{Lower})}$ and $W_k^{(\text{Upper})}$ are set such that the assumed ranges (fourth column of Table 10.1) are recovered when $h = 1$. It is emphasized that this definition is arbitrary. What is essential in this formalism is that the horizon-of-uncertainty, h, is unknown, which expresses our ignorance of the extent to which modeling assumptions might deviate from reality. The definition of Eq. (10.7) makes it explicit that there is no worst case, since the horizon-of-uncertainty can increase indefinitely.

Table 10.2 summarizes the components of info-gap analysis for the latch problem, as defined in Chap. 5. For a given horizon-of-uncertainty, h, numerical values of the six variables are selected from the IG model of uncertainty, $U(h)$, of Eq. (10.7).

Table 10.2 The three components of IG analysis applied to the latch design problem

IG Component	Application to the Latch Design Problem
System model	Finite element simulation, as illustrated in Fig. 10.7
Performance requirement	Equations (10.2) and (10.3)
Uncertainty model	Equation (10.7), informed from values listed in Table 10.1

These variables define a single realization of the system model analyzed to evaluate the performance of the latch, which is defined herein as a peak stress. This is repeated with newly selected values from $U(h)$ until the uncertainty model at this horizon-of-uncertainty has been thoroughly explored and the maximal (worst) stress, $\max_{\{\theta \in U(h)\}} \sigma_{Max}(\theta)$, has been found. The maximal stress can be compared to the compliance requirement of Eq. (10.2). Equation (10.8) shows how searching for the maximal stress within the uncertainty model, $U(h)$, relates to the robustness of the design.

Note that the IG uncertainty model (10.7) does not introduce any correlation between variables, because such information is usually unknown in an early design stage. A correlation structure that would be only partially known can easily be included. An example of info-gapping the unknown correlation of a financial security model is given in Ben-Haim (2010).

At this point of the problem formulation, a two-dimensional design space $\mathbf{p} = (L; W_C)$ is defined together with the performance requirement (10.2). Modeling uncertainty is identified in Table 10.1 and represented mathematically in Eq. (10.7). The finite element simulation indicates that the peak stress experienced by the nominal design is $\sigma_{Max} = 28.07$ MPa, which does not exceed the yield stress of 55 MPa and provides a safety factor of $f_S = 49\%$. Even accounting for truncation error introduced by the lack of resolution in the discretization (see the discussion of Fig. 10.9), the conclusion is that the nominal design is requirement-compliant.

The question we wish to answer is whether the design remains requirement-compliant if real-world conditions to which the latch might be subjected deviate from those assumed in the simulation. More explicitly, we ask: What is the greatest horizon-of-uncertainty, $\hat{h}$, up to which the predicted peak stress, σ_{Max}, does not violate the requirement (10.2) for all realizations of the uncertain variables in the info-gap model (10.7)? The question is stated mathematically as:

$$\hat{h}(\sigma_{Critical}) = \max_{\{h \geq 0\}} \left(\max_{\{\theta \in U(h)\}} \sigma_{Max}(\theta) \leq \sigma_{Critical} \right) \quad (10.8)$$

where $\hat{h}$ is the robustness of the design given a performance requirement, $\sigma_{Critical}$.

Answering this question amounts to assessing how performance, such as the peak stress σ_{Max} here, evolves as increasingly more uncertainty is explored using the simulation model (Ben-Haim 2006). Section 10.5 shows how robustness functions of

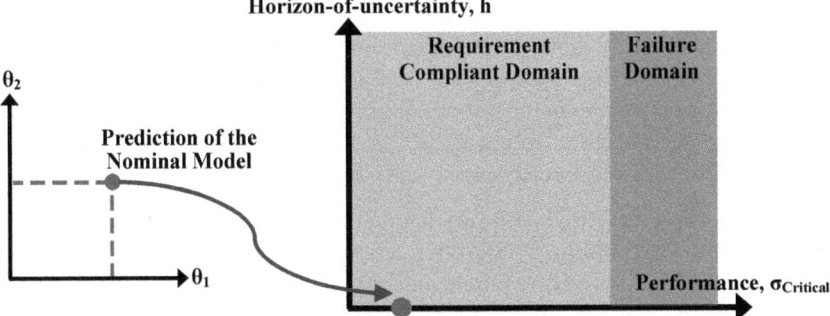

(a) Top: Prediction of the nominal model used to initialize the robustness function

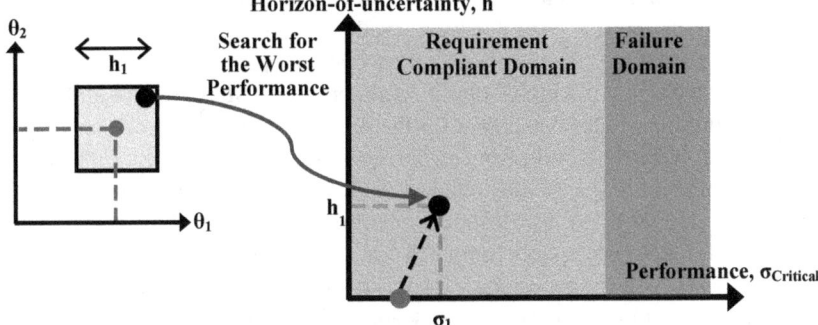

(b) Middle: Mapping of the worst performance found within an uncertainty space $U(h_1)$

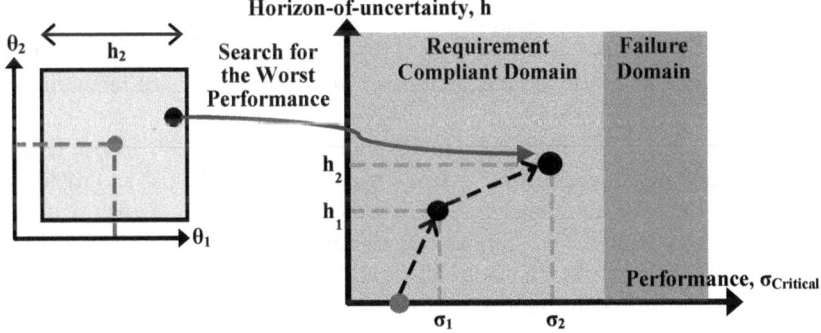

(c) Bottom: Mapping of the worst performance found within an uncertainty space $U(h_2)$

Fig. 10.8 Conceptual illustration of how the robustness function of a design is constructed

competing designs can be utilized to support decisionmaking. Figure 10.8 conceptually illustrates how robustness of the design is evaluated, as we now explain.

The robustness function, which is progressively constructed in Fig. 10.8 by exploring larger-uncertainty spaces, $U(h)$, maps the worst-case performance as a function of horizon-of-uncertainty. Its shape indicates the extent to which performance deteriorates as increasingly more uncertainty is considered. A robust design is one that tolerates as much uncertainty as possible without entering the "failure" domain (red region) for which the requirement is no longer met.

Applying the concept of robustness to the latch design problem is simple. One searches for the maximal (worst) stress, $\max_{\{\theta \in U(h)\}} \sigma_{Max}(\theta)$, obtained from finite element simulations where the variables θ vary within the uncertainty space $U(h)$ defined in Eq. (10.7). As stated in Eq. (10.8), robustness is the greatest size of the uncertainty space such that the design is requirement-compliant irrespective of which model is analyzed within $U(\hat{h})$. Said differently, all system models belonging to the uncertainty space $U(\hat{h})$ are **guaranteed** to meet the performance requirement of Eq. (10.2). The horizon-of-uncertainty, h, is nevertheless unknown and may exceed $\hat{h}$. Not all system models in uncertainty sets $U(h)$, for h greater than $\hat{h}$, are compliant.

The procedure, therefore, searches for the worst-case peak stress within the uncertainty space $U(h)$. This is a global optimization problem (Martins and Lambe 2013) whose resolution provides one datum for the robustness function, such as the point $(\sigma_1; h_1)$ in Fig. 10.8b that results from exploring the uncertainty space $U(h_1)$. For simplicity, the uncertainty spaces illustrated on the left side of the figure are represented with two variables, $\theta = (\theta_1; \theta_2)$. It should not obscure the fact that most applications will deal with larger-size spaces (the latch has six variables θ_k). Figure 10.8c indicates that the procedure is repeated by increasing the horizon-of-uncertainty from h_1 to h_2, hence performing an optimization search over a larger space $U(h_2)$.

The procedure outlined above stops when requirement-compliance is no longer guaranteed—that is, as the worst-case peak stress exceeds the critical stress $\sigma_{Critical}$. The corresponding point $(\sigma_{Critical}; \hat{h})$ is necessarily located on the edge of the (red) failure region. By definition of robustness (10.8), $\hat{h}$ is the maximum level of uncertainty that can be tolerated while guaranteeing that the performance requirement is always met.

Figure 10.9 depicts the robustness function of the nominal latch design. It establishes the mapping between critical stress ($\sigma_{Critical}$, horizontal axis) and robustness ($\hat{h}$, vertical axis). The horizontal gray lines indicate the truncation error that originates from discretizing the continuous equations on a finite-size mesh (Hemez and Kamm 2008). They are upper bounds of truncation error that can be formally derived (Mollineaux et al. 2013) and numerically evaluated through mesh refinement to support sensitivity and calibration studies (van Buren et al. 2013). For the nominal geometry ($L = 3.9$ mm and $W_C = 4.0$ mm), running the simulation with a 200-µm mesh on a dual-core processor of a laptop computer takes four minutes and produces a truncation error of 4.76 MPa. A run with a 100-µm mesh reduces this upper bound of error to 1.63 MPa at the cost of a 54-min solution time. The need to perform several hundred runs to estimate the robustness function motivates the choice of the 200-µm mesh resolution. The resulting level of truncation error (4.76 MPa) is acceptable for decisionmaking, since it represents a numerical uncertainty of only $\approx 10\%$ relative to the critical stress ($\sigma_{Critical} = 55$ MPa).

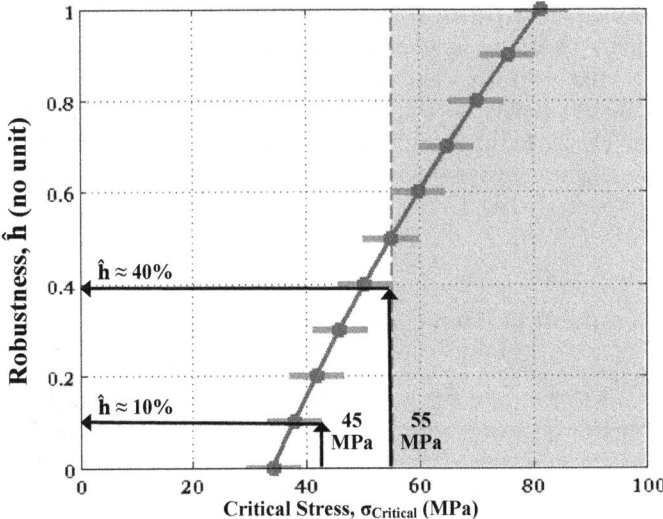

Fig. 10.9 Robustness function of the nominal design

The robustness function is obtained by continuing the process suggested in Fig. 10.8, where more and more points are added by considering greater and greater levels of horizon-of-uncertainty. For any performance limit σ_{Critical} on the horizontal axis, the corresponding point on the vertical axis is the greatest tolerable uncertainty, namely, the robustness $\hat{h}(\sigma_{\text{Critical}})$. The positive slope of the robustness function indicates a trade-off between performance requirement and robustness, as we now explain. Suppose that the analyst is willing to allow the peak stress to reach 55 MPa, which provides no safety margin ($f_S = 0$). From Fig. 10.9, it can be observed that the greatest tolerable horizon-of-uncertainty is $\hat{h} = 0.40$. (Note that this value accounts for truncation error, which effectively "shifts" the robustness function by 4.76 MPa to the right.) It means that the design satisfies the performance requirement (10.2) as long as none of the model variables $\theta = (E; G; \nu; \rho; U_{\text{Contact}}; F_{\text{OS}})$ deviates from its nominal value by more than 40%. Said differently, the design is guaranteed to satisfy the critical stress criterion as long as real-world conditions do not deviate from nominal settings of the simulation by more than 40%, even accounting for truncation effects.

Suppose, however, that the analyst wishes to be more cautious—for instance, by requiring that the peak stress not exceed 45 MPa. Now the safety factor is $f_S = 18\%$. From Fig. 10.9, not exceeding this peak stress is satisfied if model variables do not deviate from their nominal values by more than approximately 10%. In other words, the more demanding requirement ($f_S = 18\%$) is less robust to uncertainty ($\hat{h} = 0.10$) than the less demanding requirement ($f_S = 0$ and $\hat{h} = 0.40$). More generally, the positive slope of the robustness function expresses the trade-off between greater caution in the mechanical performance (increasing f_S) and greater robustness against uncertainty in the modeling assumptions (increasing $\hat{h}$).

Choices offered to the decisionmaker are clear. If it can be shown that real-world conditions cannot possibly deviate from those assumed in the simulation model by more than 40%, then the nominal design is guaranteed requirement-compliant. Otherwise an alternate design offering greater robustness should be pursued. This second path is addressed next.

10.5 Assessment of Two Competing Designs

This section illustrates how robustness functions, such as the one discussed in Sect. 10.4, can be exploited to select between two competing designs. Figure 10.10 depicts a comparison between geometries that differ in their choices of design variables, $\mathbf{p} = (L; W_C)$. The left side is the nominal geometry ($L = 3.9$ mm, $W_C = 4.0$ mm), and the right side shows a 20% larger design ($L = 4.68$ mm, $W_C = 4.80$ mm). Given that the thickness is kept the same in both geometries, the volume of the variant design increases by $\approx$44%. This consideration is important because selecting the variant design would imply higher manufacturing costs. The decisionmaker, therefore, would want to establish that the performance of the variant geometry is significantly more robust to the modeling uncertainty than what the nominal design achieves.

Figure 10.11 compares the robustness functions of the nominal and 20% larger geometries. The blue-solid line identifies the nominal design, and the variant is shown with a green-dashed line. Horizontal gray lines quantify the upper bounds of truncation error that originates from mesh discretization. The results are meaningful precisely because the prediction uncertainty due to truncation effects is sufficiently small with the mesh discretization used.

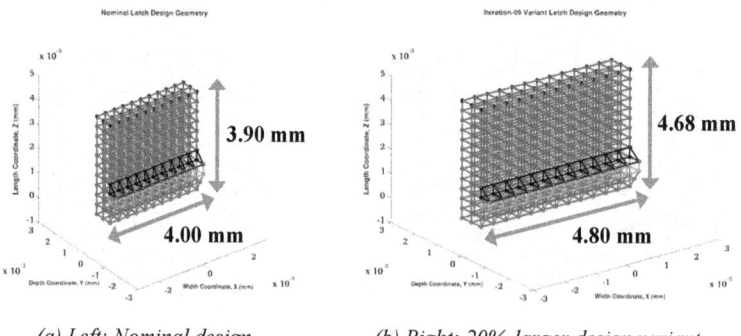

(a) Left: Nominal design   (b) Right: 20%-larger design variant

Fig. 10.10 Meshes of the competing nominal and variant latch designs

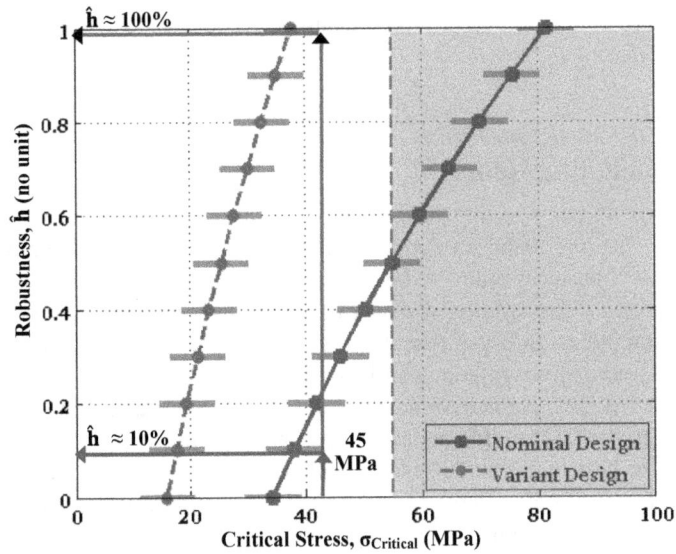

Fig. 10.11 Robustness functions of the nominal (blue) and variant (green) designs

Figure 10.11 illustrates that, when no modeling uncertainty is considered, the variant design clearly predicts a better performance. This is observed on the horizontal axis (at $\hat{h} = 0$) where the peak stress of the variant geometry ($\sigma_{Max} = 16$ MPa) is less than half the value for the nominal design ($\sigma_{Max} = 34$ MPa). This result is consistent with the fact that in the variant design the applied force is spread over a larger surface area, which reduces stresses generated in the latch.

Suppose that the analyst requires a safety factor of $f_S = 18\%$, implying that the stress must be no greater than 45 MPa. As observed in Fig. 10.9 (reproduced in the blue curve of Fig. 10.11), the nominal geometry tolerates up to 10% change in any or all of the model variables without violating the performance requirement. The larger-size geometry, however, can tolerate up to 100% change without violating the same requirement. In other words, the variant design is more robust ($\hat{h} = 1.0$ instead of $\hat{h} = 0.10$ nominally) at this level of stress ($\sigma_{Critical} = 45$ MPa).

The slopes of the two robustness functions can also be compared in Fig. 10.11. The slope represents the trade-off between robustness and performance requirement. A steep slope implies a low cost of robustness that can be increased by a relatively small relaxation of the required performance. The figure suggests that the cost of robustness for the nominal design (blue curve) is higher than for the variant geometry (green curve). Selecting the 20% larger design is undoubtedly a better decision, given that it delivers better predicted performance (lower predicted value of σ_{Max}) and is less vulnerable to potentially incorrect modeling assumptions. In fact, the variant design offers an 18% safety margin ($f_S = 18\%$) even if model variables deviate from their nominal values by up to 100% ($\hat{h} = 1.0$). The only drawback of the variant design is the ≈44% larger volume that increases manufacturing costs relative to those of the nominal design.

10.6 Concluding Remarks

This chapter has presented an application of simulation-based IG robust design. The need for robustness stems from recognizing that an effective design should guarantee performance even if real-world conditions deviate from modeling and analysis assumptions. Info-gap robustness is versatile, easy to implement, and does not require assuming information that is not available.

IG robust design is applied to the analysis of a mechanical latch for a consumer electronics product to provide a simple, mechanical illustration. The performance criterion is the peak stress at the base of the latch resulting from displacements that are applied to open or close the compartment. The geometry, simulation model, and loading scenario are simplified for clarity. Round corners, for example, that mitigate stress concentrations, are altered to straight edges. Likewise, severe impact loads experienced when dropping the device on a hard surface are not considered. The description of the analysis, however, is comprehensive and can easily be translated to other, more realistic, applications.

The robustness of the nominal design is studied to assess the extent to which performance is immune to sources of uncertainty in the problem. This uncertainty expresses the fact that real-world conditions could differ from what is assumed in the simulation without postulating either probability distributions or knowledge of worst cases. One example of how real-world conditions can vary from modeling assumptions is the variability of material properties. Uncertainty also originates from assumptions embodied in the simulation model that could be incorrect. One example is the dynamic overshoot factor used to mitigate the ignorance of how materials behave when subjected to fast-transient loads. The analysis of the mechanical latch pursued in this chapter indicates that the design can tolerate up to 40% uncertainty without exceeding the peak-stress performance requirement.

The performance of an alternate design, which proposes to spread the contact force over a larger surface area, is assessed for its ability to provide more robustness than the nominal design. The analysis indicates that the variant latch is predicted to perform better, while its robustness to modeling uncertainty is greater at all performance requirements. The variant geometry features, however, a 44% larger volume, which would imply higher manufacturing costs. The discussion presented in this chapter illustrates how an analysis of robustness helps the decisionmaker answer the question of whether an improvement in performance, or the ability to withstand more uncertainty about real-world conditions, warrants the cost associated with a design change.

The simplicity of the example discussed here should not obscure the fact that searching for a robust design might come at a significant computational expense if the simulation is expensive or the uncertainty space is large-dimensional. This is nevertheless what automation is for and what software is good at. Developing the technology to perform large-scale explorations frees the analyst to apply his/her creativity to more challenging aspects of the design.

References

Bamber, J. L., Riva, R. E. M., Vermeersen, B. L. A., & Le Brocq, A. M. (2009). Reassessment of the potential sea-level rise from a collapse of the West Antarctic Ice Sheet. *Science, 324*(5929), 901–903.

Ben-Haim, Y., Dasco, C. C., Carrasco, J., & Rajan, N. (2009). Heterogeneous uncertainties in cholesterol management. *International Journal of Approximate Reasoning, 50,* 1046–1065.

Ben-Haim, Y. (2006). *Info-gap decision theory: Decisions under severe uncertainty* (2nd edn.). Oxford: Academic Press Publisher.

Ben-Haim, Y. (2010). *Info-gap economics: An operational introduction* (pp. 87–95). Palgrave-Macmillan.

Ferson, S., & Tucker, W. T. (2008). Probability boxes as info-gap models. In *Annual Conference of the North American Fuzzy Information Processing Society—NAFIPS 2008.* Article number 4531314.

Hall, J. W., Lempert, R. J., Keller, K., Hackbarth, A., Mijere, C., & McInerney, D. J. (2012). Robust climate policies under uncertainty: A comparison of robust decision making and info-gap methods. *Risk Analysis, 32*(10), 1657–1672.

Hemez, F. M., & Kamm, J. R. (2008). A brief overview of the state-of-the-practice and current challenges of solution verification. In *Computational methods in transport: Verification and validation* (pp. 229–250). Springer Publisher.

Malone, R. C., Smith, R. D., Maltrud, M. E., & Hecht, M. W. (2003). Eddy-resolving ocean modeling. *Los Alamos Science, 28,* 223–231.

Martins, J. R. R. A., & Lambe, A. B. (2013). Multidisciplinary design optimization: A survey of architectures. *AIAA Journal, 51,* 2049–2075.

Metropolis, N., & Ulam, S. (1949). The Monte Carlo method. *Journal of the American Statistical Association, 44,* 335–341.

Mollineaux, M. G., Van Buren, K. L., Hemez, F. M., & Atamturktur, S. (2013). Simulating the dynamics of wind turbine blades: Part I. *Model Development and Verification, Wind Energy, 16,* 694–710.

Newmark, N. M. (1959). A method of computation for structural dynamics. *ASCE Journal of Engineering Mechanics, 85,* 67–94.

Oden, J. T., Belytschko, T., Fish, J., Hughes, T. J. R., Johnson, C., Keyes, D., Laub, A., Petzold, L., Srolovitz, D., & Yip, S. (2006) Revolutionizing engineering science through simulation. In *National science foundation blue ribbon panel on simulation-based engineering.*

Saltelli, A., Chan K., & Scott, M. (2000). *Sensitivity analysis.* Wiley.

Smith, R., & Gent, P. (2002). Reference manual for the parallel ocean program (POP), Ocean Component of the Community Climate System Model, *National Center for Atmospheric Research,* Boulder, CO. (Also, Technical Report LA-UR-02-2484 of the Los Alamos National Laboratory, Los Alamos, NM.).

United Nations. (2014). United Nations Department of Economic and Social Affairs (DESA) Continent Population 1950 to 2100, *Wikimedia Commons, The Free Media Repository*, Retrieved October 27, 2017, From: commons.wikimedia.org/w/index.php?title=File:UN_DESA_continent_population_1950_to_2100.svg&oldid=130209070.

van Buren, K. L., Mollineaux, M. G., Hemez, F. M., & Atamturktur, S. (2013). Simulating the dynamics of wind turbine blades: Part II, model validation and uncertainty quantification. *Wind Energy, 16,* 741–758.

Wu, Y.-T. (1994). Computational method for efficient structural reliability and reliability sensitivity analysis. *AIAA Journal, 32,* 1717–1723.

Zienkiewicz, O. C., & Taylor, R. L. (2000). *The finite element method, volume 1: The basis.* Butterworth-Heinemann Publisher.

Dr. François M. Hemez is a scientist in the Design Physics Division at Lawrence Livermore National Laboratory, where he is contributing to nuclear non-proliferation efforts. He spent twenty years at Los Alamos National Laboratory, with responsibilities in the nuclear weapon simulation and certification programs. François is recognized for his expertise in model validation, uncertainty quantification, and decisionmaking. He graduated from Ecole Centrale Paris and earned a doctoral degree in aerospace engineering (1993) from the University of Colorado Boulder.

Dr. Kendra L. Van Buren is a scientist in the Computational Physics Division at Los Alamos National Laboratory. She is an expert in the validation and uncertainty quantification of multi-physics simulation models. Kendra's interests include the theoretical and experimental understanding of energetic materials, with applications to high explosive systems and nuclear non-proliferation. She graduated from the University of California Los Angeles, and earned a doctoral degree in civil engineering (2012) from Clemson University.

Open Access This chapter is licensed under the terms of the Creative Commons Attribution 4.0 International License (http://creativecommons.org/licenses/by/4.0/), which permits use, sharing, adaptation, distribution and reproduction in any medium or format, as long as you give appropriate credit to the original author(s) and the source, provide a link to the Creative Commons licence and indicate if changes were made.

The images or other third party material in this chapter are included in the chapter's Creative Commons licence, unless indicated otherwise in a credit line to the material. If material is not included in the chapter's Creative Commons licence and your intended use is not permitted by statutory regulation or exceeds the permitted use, you will need to obtain permission directly from the copyright holder.

Chapter 11
Engineering Options Analysis (EOA): Applications

Richard de Neufville, Kim Smet, Michel-Alexandre Cardin and Mehdi Ranjbar-Bourani

Abstract This chapter illustrates the use and value of Engineering Options Analysis (EOA) using two case studies. Each describes the analysis in detail. Each entails the need for plans to monitor projects so that managers know when to exercise options and adapt projects to the future as it develops. The Liquid Natural Gas case (Case Study 1) concerns the development of a liquid natural gas plant in Australia. It:

- Provides a generic prototype for the analysis of projected innovative developments.
- Demonstrates the kind of insights that EOA can provide.
- Highlights the potential advantage of flexibility in size, time, and location of facilities. In particular, it develops the important insight that modular designs may be more profitable than monolithic designs because they enable managers to reduce the significant risk of overdesigned plants, and they increase opportunities by taking advantage of the time and location of increases in demand.

The IJmuiden case (Case Study 2) concerns water management and flood control facilities in the Netherlands. It:

- Demonstrates the application of EOA to cope with uncertainty in natural processes, in contrast to the more traditional context of market uncertainties.
- Uses diverse scenarios to identify conditions for which a strategy is valid across significant ranges of future conditions, and contrary situations in which a choice depends on belief about the level of risks.
- Documents how EOA shows which forms of flexibility in design justify their cost (in this case, flexibility in pumping facilities adds significant value, but flexibility in the flood defense height of the structure does not).

R. de Neufville (✉)
Massachusetts Institute of Technology, Cambridge, MA, USA
e-mail: ardent@mit.edu

K. Smet
University of Ottawa, Ottawa, Canada

M.-A. Cardin
Imperial College London, London, UK
e-mail: m.cardin@imperial.ac.uk

M. Ranjbar-Bourani
University of Science and Technology of Mazandaran, Behshahr, Iran

© The Author(s) 2019
V. A. W. J. Marchau et al. (eds.), *Decision Making under Deep Uncertainty*,
https://doi.org/10.1007/978-3-030-05252-2_11

- Shows how the calculation of distributions of possible outcomes provides decisionmakers with useful information concerning worst-case outcomes, unavailable from average outcomes alone.

11.1 Case Study 1: Liquid Natural Gas in Victoria State, Australia

This case illustrates how Engineering Options Analysis (EOA) can provide effective, useful guidance for the planning of infrastructure developments surrounded by deep uncertainty. By exploring the performance of alternative strategies under a broad range of uncertainties, the analysis can identify which approaches provide the best opportunities for good design. That is, although we cannot conclusively identify which plan is best (given the uncertainty of our data and the future), EOA makes it possible for us to identify preferable strategies, and thus guide decisionmakers and stakeholders toward better solutions. The case study demonstrates this possibility.

How does EOA resolve the apparent conflict between its capability to justify desirable strategies and its inability to defend detailed calculations (given the incertitude of the data and the futures the data represent)? This is because EOA views any project as a series of possible actions over time and space. In general, we do not, indeed most often cannot, build projects all at once (de Neufville and Scholtes 2011). We do not develop a power grid or a manufacturing system in one go; we build up such enterprises in various steps over time and in different locations. In this context, planning is about choosing first steps wisely, avoiding locking in a fixed solution. The idea is to select initial steps that give us the flexibility (the continuing opportunity) to make further productive steps as the currently deep uncertainty resolves into increasing certainty about future events. Through the selection of wise first steps, EOA can thus lead us to productive, value-enhancing strategies, even though it cannot specify the value in detail.

This case concerns the proposed development of a system to produce Liquid Natural Gas (LNG) as a fuel for road transport by trucks (lorries). The overall concept is to make productive use of this relatively clean resource (compared to diesel fuels rich in particulates) that has generally been wasted (flared off) or otherwise used inefficiently. The idea is to ship gas ashore from deepwater wells, deliver it to a plant for liquefaction, and distribute it subsequently by fuel trucks to distribution points throughout the region. For reference, the proposed project is to be located around Melbourne in Victoria State, Australia.

The question is: *how should we plan the development of this project? Specifically, what is the most advantageous design of the production facilities?* This is clearly a case of decisionmaking under deep uncertainty: Nothing like this has been done in the region, and the market potential over the life of the project is highly uncertain. Not only are the relative prices of the fuels uncertain over the long run of the project, but also the rate at which road transporters will shift from diesel to LNG is totally unknown in this context. This case thus provides a good opportunity to explore the use of EOA in the context of deep uncertainty.

11.2 Setup of the EOA Approach for the LNG Case Study

11.2.1 Design Alternatives

The analysis compares the flexible design strategies resulting from the EOA to a traditional engineering design, taken as the base case. This comparison documents the value of EOA analysis.

The traditional engineering design came from the Keppel Offshore & Marine company, which also computed the relevant design parameters. It represents the conventional engineering approach, which leads to the "optimal" solution to meet predicted future demand for LNG. Its design process identifies a solution that maximizes value, assuming no uncertainty, treating the future as deterministic. Specifically, it creates a single large facility taking advantage of economies of scale (Fig. 11.1a). In fact, this design does not maximize the actual value of the project, since the realized demand for LNG in 10 or 20 years will differ, possibly substantially, from the predicted values used in the design analysis. Sometimes the plant will be too large compared to the actual use, and will incur great losses as the actual average costs of production using an oversized plant will soar. Sometimes the plant will be too small and miss opportunities. Either way, even though the design derives from an optimization process, the design based on a fixed forecast is unlikely to be optimal for the actual conditions that will prevail.

The proposed flexible designs use modularity to achieve their adaptability. The first flexible strategy focuses on when to implement the additional modules. The second expands on this by adding the consideration of where to place the modules. Figure 11.1 illustrates these alternatives:

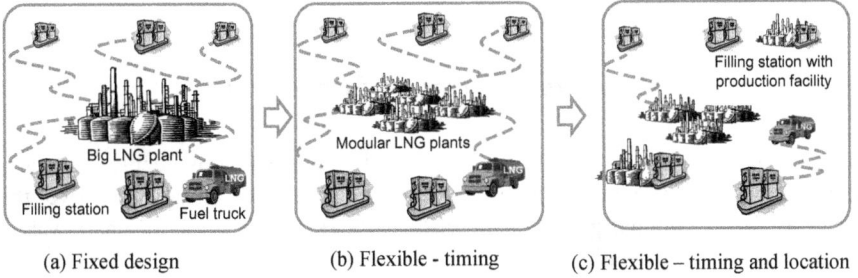

(a) Fixed design   (b) Flexible - timing   (c) Flexible – timing and location

Fig. 11.1 Alternative design configurations of the LNG production and distribution system. Adapted from Cardin et al. (2015)

- *Flexible strategy—timing*: this strategy deploys capacity at the central site according to how demand does or does not grow over time (Fig. 11.1b).
- *Flexible strategy—timing and location*: this strategy allows for gradual deployment of capacity both over time AND geographically at the distribution sites (Fig. 11.1c).

11.2.2 Parameter Values

This case study uses cost data available from previously completed studies about the project, as of around 2013. These data came from domain experts who were associated with the formulation of the traditional fixed design. Salient specifics include:

- Economies of scale drive down the unit cost of capacity for larger plants, according to the standard formula: Capital cost (Capex) of plant $= K$ (Capacity)$^\alpha$ (K being a constant and α being the relevant economies of scale factor. α can range between 0 and 1. When $\alpha = 1$, Capex increases linearly with Capacity, indicating there are no economies of scale. Smaller values of α indicate increasing economies of scale. In practice, usual economies of scale exist up to $\alpha = 0.85$, and $\alpha \sim 0.6$ is the maximum observed.)
- The Capex of a LNG plant with a capacity of 25 tons/day is $25 million.
- The annual operating costs for a LNG plant (Opex) are 5% of its Capex.
- A management-imposed discount rate of 10%, given the risks and the opportunity cost of capital.
- A management-defined project lifetime of 20 years; corporate tax rate of 15%; and depreciation as straight-line over 10 years with zero salvage value.

Interested readers can find additional details in Ranjbar-Bourani (2015) and Cardin et al. (2015). While these specific cost parameters are treated as fixed inputs in this study, an EOA analysis in general could just as easily treat any of these parameters, specifically including the discount rate, as uncertain variables with associated distributions.

11.2.3 Characterization of Sources of Uncertainty

The actual future demand for the LNG product of the system is deeply uncertain, due to the fact that future market acceptance of the product, energy costs over the long term, and eventual government subsidies or restrictions, are all unknowable. Despite this obvious reality, the previously completed consultant analyses conventionally assumed a fixed projection for the demand (red line on Fig. 11.2) when they determined the optimal size of the fixed design. The analysis presented here builds on this previous study by working with available market research at the collaborating

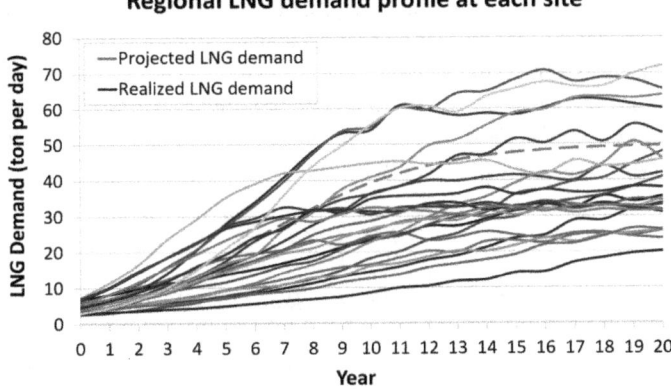

Fig. 11.2 Projected deterministic demand (red) and example projected scenario demands (black). Adapted from Cardin et al. (2015)

consultant firm to generate a broad range of possible variants of the base case fixed demand (gray lines on Fig. 11.2).

To illustrate the EOA process, this case study applies a distribution over the possible initial demand, the subsequent rate of growth, and the volatility around the projected deterministic trend. The choice of which distribution to use, while also a deeply uncertain assumption, should be made based on available data quantifying parameter uncertainty, where available. In our case we apply a uniform distribution over the range as a matter of simplicity and to illustrate the process, without advocating it as a preferred choice. Using Monte Carlo simulations, we develop thousands of possible demand paths over the projected life of the project. Figure 11.2 illustrates some possibilities compared to the deterministic assumption. As we should expect for long-term forecasts for new technologies in new markets, the possibilities deviate substantially from the best deterministic estimate (de Neufville and Scholtes 2011). While this characterization of uncertainty remains imperfect, the point here is simply to point us toward more productive first steps, NOT to give us a perfectly rationalized, complete plan for the future.

There is also deep uncertainty about the system itself. Although we know that petrochemical plants routinely exhibit substantial economies of scale, we do not know what these will actually be for the proposed system, when developed in a new context. The analysis reflects this deep uncertainty by examining several possibilities for the economies of scale factor α: 1, 0.95, 0.9, and 0.85. To our knowledge, these values represent the range of possibilities for such plants.

Further deep uncertainty lies in the learning rate we should apply to the implementation of a series of modular plants in any flexible development. The learning rate refers to the descriptive fact that the cost of modular capacity routinely decreases with the number of units produced. This is a common observation from everyday life: when we do something for the first time we are relatively inefficient; the more we repeat the task, the more we learn to be productive. Hence the appellation of "learning" for the phenomenon, even though in practice the observed cost reductions may also come from design innovations and manufacturing improvements. Such learning has been well documented in the context of infrastructure projects (see Appendix C of de Neufville and Scholtes 2011). Equations 11.1 and 11.2 represent this situation, where U_1 and U_i are the Capex of the first and ith modules, and B is the slope of the learning curve determined empirically from case studies:

$$U_i = U_1 \times i^B \tag{11.1}$$

$$B = \log(100\% - LR\%)/\log(2) \tag{11.2}$$

To account for uncertainty in the learning rate for the modular design, the analysis examines learning rates corresponding to LR = 0, 5, 10, 15, and 20%.

Note in this context that the learning phenomenon encourages the use of modular flexibility. It counterbalances the effect of economies of scale, which is to promote large early investments in infrastructure. The learning effect reduces the cost of small modular increments of capacity compared to large units. In this respect, the learning phenomenon is comparable to the time value of money: they both encourage smaller investments tailored to more immediate demands.

11.3 Results from Applying the EOA Approach to the LNG Case Study

11.3.1 Fixed Design

The conventional engineering analysis (based on a previous consultant's study) calculated the optimal size of the central plant for a fixed forecast of the LNG demand. How could the designers do this one might well ask, given the many uncertainties? Different engineering practices invoke a variety of sources, such as industry forecasts provided by economists, midpoint estimates from geologists, etc. The point here is that designing around specific forecasts is common practice, and this is what the consultants for the fixed design did. The optimum in their case was the maximum net present value (NPV); i.e., the solution that provided the greatest discounted benefits over discounted costs.

To illustrate the process leading to identifying the fixed design, Fig. 11.3 shows the NPV associated with each plant size for different economies of scale factors, from none ($\alpha = 1.0$) to the highest assumed ($\alpha = 0.85$). The intuitive understanding is that:

- There is a design "sweet spot" for the optimal, most profitable, plant size (the stars on the curves) for any level of economies of scale: build too small, and the design

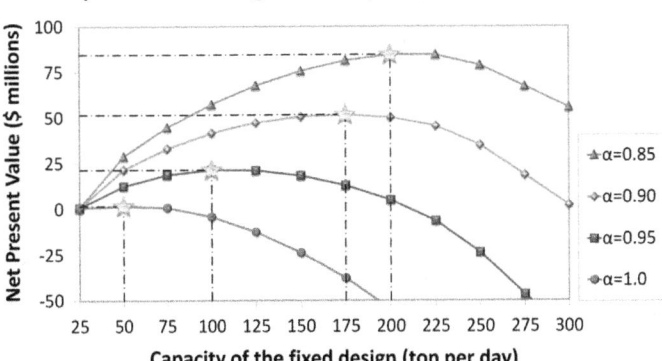

Fig. 11.3 NPV of fixed designs under deterministic demand. Stars show the optimum design for a given economies of scale factor α (Cardin et al. 2015)

loses out on potential profit from higher demands; build too large, and there is overcapacity and attendant lower values.
- The greater the economies of scale (smaller α), the larger the optimal fixed design should be. This is because the economies of scale lower the average unit cost of capacity, and thus favor larger designs.

Given the assumed existence of aggressive economies of scale for petrochemical plants, the previous consultant's fixed design process produced a plan to build a single large plant of 200 tons/day (i.e. corresponding to $\alpha = 0.85$). As Fig. 11.3 indicates, this plan seemed highly profitable, delivering a NPV of about $84.56 million, a substantial return over the cost of capital indicated by the 10% discount rate.

11.3.2 Performance of Fixed Design Under Uncertainty

Before proceeding further, we should explicitly note that—in the realistically uncertain world—the results from a deterministic analysis can be systematically incorrect and misleading. This phenomenon is particularly true for capacity-constrained systems, such as plants with limiting capacities. Such facilities do not benefit from excess demand over their capacity, but do lose when the demand is less than anticipated. For cases where the uncertain demands are equally likely to exceed or be less than expected, the uncertainty systematically leads to lower overall value. Gains are limited and unlikely to counterbalance losses. In general, we must expect that the average overall result of an uncertain environment is not equal to the result associated with the average of the uncertain environment. This is the common fact known as Jensen's Law. To assume incorrectly that system value calculated using average parameters is the true expected value is to fall victim to the "Flaw of Averages" (de

Table 11.1 Optimum fixed designs under deterministic and uncertain LNG demand with different economies of scale factor α

Economies of scale factor, α	Most profitable capacity (tons/day)		Present value (PV) of plant at most profitable capacity ($ millions)	
	Deterministic analysis (ignoring uncertainty)	Stochastic analysis (recognizing uncertainty)	Deterministic net PV (NPV) (ignoring uncertainty)	Expected net PV (ENPV) (recognizing uncertainty)
1.0	50	25	1.75	0.87
0.95	100	75	21.51	14.27
0.90	175	125	51.75	37.18
0.85	200	175	84.56	61.18

Adapted from Cardin et al. (2015)

Neufville and Scholtes 2011). The true expected value may easily differ greatly, as Table 11.1 indicates.

The expected performance of the fixed design in the context of uncertainty analysis differs systematically and significantly from that of the deterministic analysis. Our analyses demonstrate the phenomenon by calculating and comparing the performance of the most profitable fixed design under conditions of deterministic and uncertain demand. Specifically, we use Monte Carlo simulations to subject the optimal plant size to 2000 different futures, comparable to those in Fig. 11.2. The value of the design under uncertainty is then a distribution associated with the possible demand realizations. Because of the non-linearities within the system, inherent in both its cost structure and the variations in demand, the shape of the distribution of values differs from the distribution of demand. Focusing on the plant design set forth by the consultant (200 tons/day for an economies of scale factor $\alpha = 0.85$), the deterministic analysis estimates its net present value (NPV) at $84.56 million. By contrast, the simulation analysis calculates the expected net present value (ENPV) of this same plant design to be only $61.18 million, 20% less when we take uncertainty into account. This discrepancy between the values resulting from the deterministic and uncertainty analyses is across the board for all of the optimal plant sizes associated with different levels of economies of scale. Results obtained from a deterministic analysis systematically fail to recognize the value under uncertainty. Table 11.1 presents the comparisons.

Moreover, analyses recognizing uncertainty lead to systematically different optimal designs than the deterministic analysis. In this example, the most profitable fixed designs for the uncertain demand are all systematically smaller than those suggested by the deterministic analysis. For example, with the greatest economy of scale factor ($\alpha = 0.85$), the most profitable design is 200 tons/days when ignoring uncertainty, but 175 tons/day from our analysis with uncertainty. This is due to the lower project values that occur once the analysis accounts for uncertainty.

11.3.3 Flexible Strategies

The essence of the flexible strategies in the LNG case is to deploy capacity in modules according to proven (realized) demand. The idea is to build less capacity at the start to avoid premature over commitment and overcapacity, and to add capacity modules according to realized demand. In this case, field experts indicated that available standard modules have a capacity of 25 tons/day.

The operational question is: when should we make use of the flexibility to expand? Given the uncertainty in the evolution of demand, there is no a priori absolute answer to this question. In practice, we want to expand capacity when demand has grown sufficiently. Thus the operational issue for the analysis is to determine, for each possible realization of the future, when the growth meets the criterion to justify extra capacity.

The analysis needs to specify the criterion for expanding capacity or, in general, for exercising the option. Analysts can determine this "decision rule" from system managers or some other reasonable representation of how decisionmakers react to what they see actually happening. For example, the analysis might posit a decision rule such as: "Expand capacity once the demand in each of the preceding two years has exceeded some 'threshold value' of existing capacity". The Monte Carlo simulation running period by period then checks in each period to see if the criterion is met for that scenario, and then adds capacity if justified.[1]

11.3.4 Flexible Strategy—Timing (But No Learning)

A decision rule incorporates the answer to the operational question of when to exercise flexibility. Decision rules typically consist of a sequence of logical IF/THEN/ELSE operators. For the simple case of capacity expansion, we used the following decision rule:

- IF "the difference between the observed demand and current capacity is higher than 90% of the capacity of the module in the previous period,"
- THEN "expand current capacity by adding a module,"
- ELSE "do nothing".

The values that trigger action in the decision rules are the threshold values.

Figure 11.4 illustrates a typical result of the flexibility analysis. (To see a complete set of comparative target curves for all the different economies of scale (not just $\alpha = 0.95$ shown in Fig. 11.4), see Cardin et al. 2015). Figure 11.4 compares the performance under uncertainty of:

- Our base case optimal fixed design (specifically with a medium $\alpha = 0.95$), and

---

[1] Note: "Threshold values" in EOA serve a similar function to that of the "triggers" or "Adaptation Tipping Points" used in other DMDU approaches.

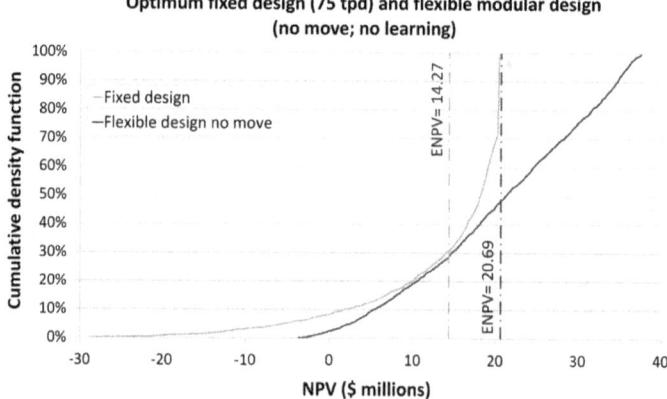

Fig. 11.4 Target curves for optimum fixed design ($\alpha = 0.95$, 75 tons/day) and for flexible strategy—timing (Cardin et al. 2015)

- A flexible strategy that enables capacity expansion at the main production site when actual demand justifies this investment.

It displays the cumulative distribution of the performance of each design (that is, the target curves). A few words explaining cumulative target curves. The lower left side of each target curve indicates the lowest level of performance of each design as observed in the simulation, which is at 0% on the vertical scale of the cumulative distribution. The curve extends to the upper right, where it indicates the maximum performance observed, at the 100% level of the cumulative distribution. As an illustrative example, the curve for the fixed design in Fig. 11.4 could lead to an NPV as low as minus $25 million, with a maximum of a nearly plus $21 million. Overall, the ENPV of this fixed design is $14.27 million.

Notice that this fixed design, which takes advantage of economies of scale to build a larger facility (3 times the size of the minimum module) at the central site, has two unattractive features. It:

- Can lead to large losses (NPV could be as low as−$25M), because the larger plant leads to large losses if sufficient demand does not materialize; and
- Cannot gain more than an NPV = $21M, since it cannot serve demands exceeding its fixed capacity.

The flexible strategy starting with one module (25 tons/day) does significantly better than the fixed design, with the same assumed range of uncertainties. As Fig. 11.4 shows:

- Its ENPV = $20.69M, nearly 44% better than the fixed design ($20.69M vs. $14.27M)!
- Its overall performance in this case stochastically dominates that of the fixed design (that is, its target curve is absolutely to the right of, and thus better than that of the fixed design).

- It reduces exposure to downside risks: the decision to start small puts less investment at risk and lowers maximum losses if demand is low. In this particular example, the flexible design strategy reduces maximum NPV loss from about −$25M to less than −$5M.
- Conversely, it provides the ability to take advantage of upside opportunities: it enables the easy addition of capacity when demand soars, and increases the maximum NPV gain from about $21M to nearly $38M.

In this case, the flexible strategy systematically leads to more valuable results in all respects.

11.3.5 Flexible Strategy—Timing and Location (But No Learning)

The flexibility analysis for the strategy that allows flexibility both as to when and where to add capacity is similar to the previous analysis. However, this analysis uses additional decision rules to explore the locational flexibility. These decision rules address three questions: *When should we build the modular plant for the first time? When should we expand it? And where should we build it?* Essentially, the decision rules result in capacity expansions being undertaken in those areas where observed demand is the greatest.

Figure 11.5 shows the additional advantages of the flexibility to locate capacity away from the main site. As expected, looser constraints on system design increase the maximum potential value. In this case, the ability to distribute capacity across the region (and thus to reduce logistic costs) further increases the system ENPV (from $20.69M to $23.29M) and the maximum NPV (from about $38M to about $60M).

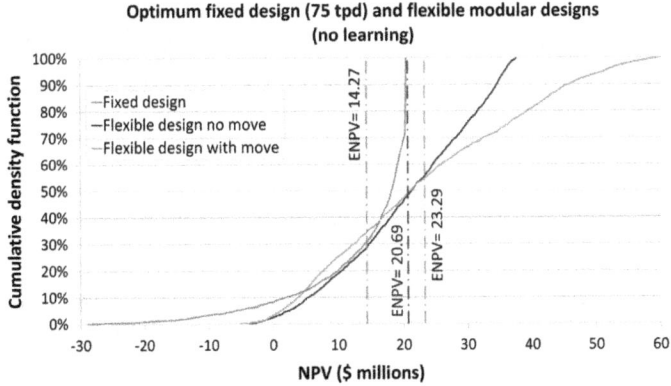

Fig. 11.5 Target curves for optimum fixed design ($\alpha = 0.95$, 75 tons/day) and for flexible strategy—timing and location (Cardin et al. 2015)

Table 11.2 Multi-criteria performance metrics for alternative designs with no learning, economies of scale $\alpha = 0.95$

Criterion	ENPV value ($ millions)			Improvement (%)	
	Optimum fixed design	Flexible timing	Flexible time + place	Flexible timing	Flexible time + place
Expected NPV	14.27	20.69	**23.29**	45	**63**
Value at Risk, 10%	1.82	**5.40**	3.74	**197**	105
Value at Gain, 90%	20.46	34.54	**45.78**	69	**124**

Adapted from Cardin et al. (2015)

However, in this case, the strategy with locational flexibility does not stochastically dominate alternative designs. Visually, its target curve crosses the target curves of the alternative designs. In this and similar cases, designers may not want to choose the solution based upon a single metric such as ENPV. Indeed, no single metric is sufficient to characterize a general distribution. In this context, we need to consider multiple criteria of evaluation.

Table 11.2 provides a multi-criteria display of the performance of the preceding fixed design and the two flexible strategies. It displays the average ENPV and two measures of the extreme values. The best results for any criterion are bold-faced. In terms of extremes, better practice generally focuses on some threshold level of cumulative performance rather than on the absolute maxima and minima values from the Monte Carlo simulation. This is because those highest and lowest values, being very rare, can vary considerably among simulations. The threshold measures are quite stable, however. Standard thresholds of interest are the 10% Value at Risk (the performance at the 10% cumulative probability or percentile), and the 90% Value at Gain. Table 11.2 compares the performance of the fixed and two flexible designs in these terms. In this case, the flexible strategy involving both time and location might seem better, even though it does not dominate the flexible strategy that only considers the timing of additions. Specifically, the more inclusive flexible strategy offers the highest expected NPV—less downside risk than the fixed design, and greater upside potential.

11.3.6 Flexible Strategy—Learning

Learning increases the value of flexibility. Because it reduces the cost of modules as more get implemented, it favors their use and thus increases the value of flexibility. Figure 11.6 shows how this occurs. It compares the target curves for the flexible strategy—timing at various levels of learning—from none (LR = 0%) to 20%. The message is clear: the greater the rate of learning, the more valuable the flexibility using modules.

11 Engineering Options Analysis (EOA): Applications

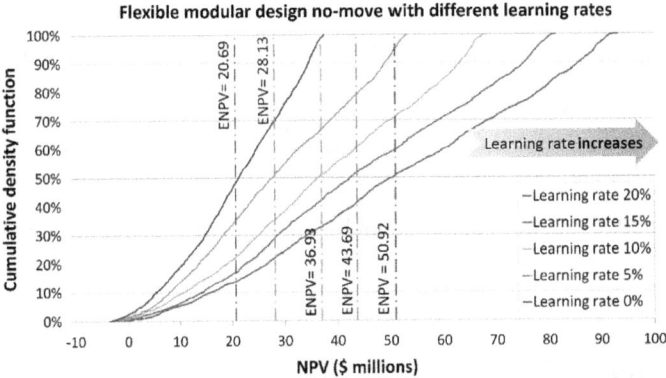

Fig. 11.6 Target curves for flexible strategy—timing with different learning rates ($\alpha = 0.95$, 75 tons/day) (Cardin et al. 2015)

11.3.7 Learning Combined with Economies of Scale

Given the deep uncertainty concerning the intensity of both economies of scale and learning rate, the analysis explores their joint effect on the desirability of possible design strategies. Figure 11.7 displays the results. This diagram brings out important results:

- As expected, lower economies of scale and greater learning rates increase the value of flexibility.
- The value of flexibility in this case ranges up to $60 million. Flexibility thus offers significant potential, which demands exploration.

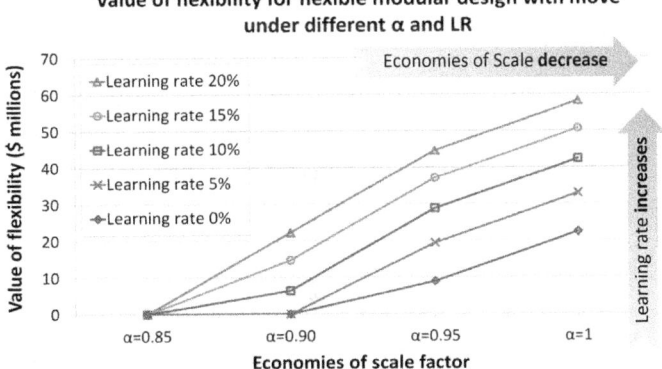

Fig. 11.7 Value of flexibility with different economies of scale and learning rates (Cardin et al. 2015)

- In this case, the flexible strategies are valuable for all but the most extreme cases (that is, cases where the economies of scale are particularly high and there is no learning). For even modest learning rates and economies of scale, the flexible modular design is valuable overall.
- One may thus conclude that, in this case, the modular flexible strategies fare well over a wide range of variations in these parameters.

In general terms, the result is that, even faced with deep overall uncertainty for this system, we can conclude that a flexible strategy is ultimately desirable.

11.3.8 Multi-criteria Comparison of Strategies

EOA generates a large amount of relevant data. It offers many metrics for comparison of the alternatives. This is a great advantage over economic options analysis (which generates a single price or value for the options), or an optimization process (which uses a single objective function). EOA's range of metrics enables decision makers to compare alternative strategies from several perspectives simultaneously.

Table 11.3 illustrates the range of metrics EOA generates. In addition to the Expected Net Present Value, it reports:

- Measures of the dispersion of the results: the 10% Value at Risk, the 90% Value at Gain, and the Standard Deviation;
- The initial Capital Expenditure (Capex) of projects, to which many investors pay great attention when there is substantial risk.

As an aid to helping decisionmakers compare projects, Table 11.3 highlights in bold the best values of each metric. Looking at it, we can draw several conclusions:

Table 11.3 Multi-criteria decisionmaking table ($\alpha = 0.95$, LR = 10%, figures in $ Million) (Cardin et al. 2015)

Criterion	Fixed design	Timing strategy	Time + place strategy	Value of flexibility	Best strategy
Expected NPV	14.27	36.93	**43.17**	28.80	Time + place
Value at Risk, 10%	1.82	10.82	**11.06**	9.24	Time + place
Value at Gain, 90%	20.46	63.17	**80.09**	59.63	Time + place
Standard Deviation	**8.78**	18.91	25.31	NA	Fixed
Initial Capex	60.44	**27.50**	**27.50**	32.94	Either

- As often happens, different projects appear best according to different criteria.
- The fixed design has the lowest standard deviation, and thus might be labeled most "stable." This could be considered a good thing, but here this merely indicates that the fixed design performs uniformly poorly, as it cannot take advantage of upside opportunities.
- Overall, the flexible strategies appear to provide the most balanced overall solutions.

11.3.9 Guidance from Applying EOA to This Case

EOA offers practical, useful guidance for decisionmakers. In the case of the Liquid Natural Gas case in Australia this guidance is:

- Reject the fixed design.
- Its overall performance is inferior, as it has considerable downside risk with little upside opportunities (Figs. 11.4 and 11.5, Table 11.2).
- Its average performance is superior only in the extreme cases, in which there are the most favorable economies of scale and no learning (Fig. 11.7).
- Choose a flexible strategy using small modules. These dominate in terms of:

    - Overall performance, because they have less downside risk and enable considerable upside opportunities (Figs. 11.4 and 11.5, Tables 11.2 and 11.3); and
    - Average performance, which is generally far superior to the fixed strategy (Tables 11.2 and 11.3).

- Start the project with a small module, leaving open the question of whether to later adopt a strategy that accepts the possibility of locational flexibility in addition to timing flexibility.

Overall, commit to a policy inherent in flexible design: monitoring, incremental learning, and adaptive management as the future reveals itself.

11.4 Case Study 2: Water Management Infrastructure in the Netherlands: IJmuiden Pumping Station

This second case complements the first, providing insights about the use of EOA in the planning and design of water management infrastructure. The focus is less on economic profit and more on continued reliability of service.

This case explores the application of EOA to water management infrastructure, using the example of a pumping station on the North Sea Canal. The North Sea Canal is an important shipping and drainage channel in the Netherlands (Fig. 11.8). It connects inland waterways to the port of Amsterdam, and to international markets

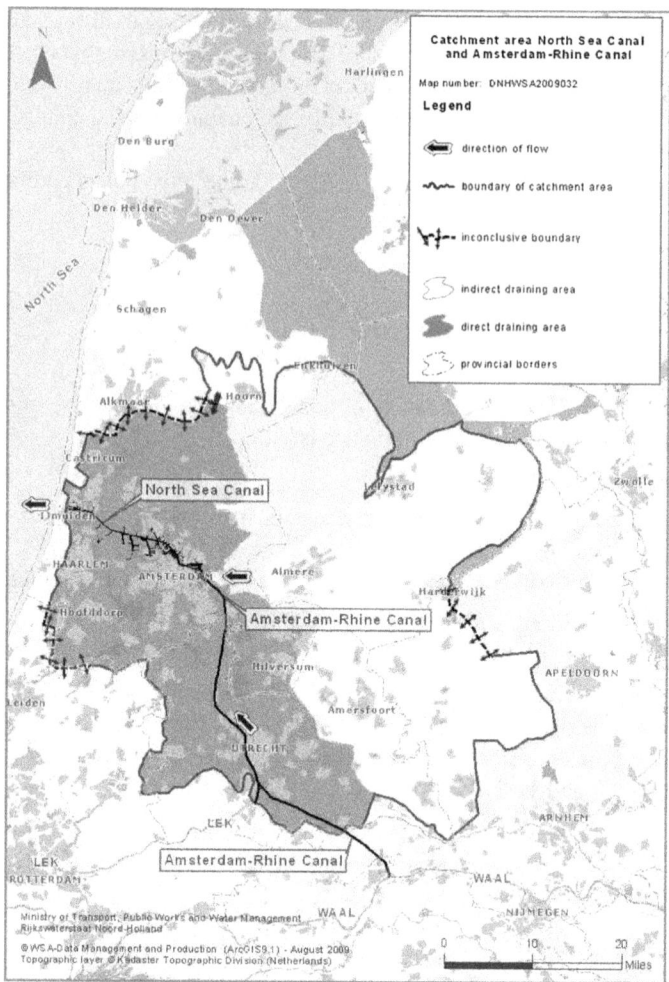

Fig. 11.8 Map showing IJmuiden and the North Sea Canal (The Netherlands Ministry of Transport, Public Works and Water Management 2009)

beyond. Construction of the canal necessitated piercing through a series of coastal sand dunes that historically offered flood protection to inland areas against high waters on the North Sea. At this interface between the canal and sea, known as IJmuiden, the Rijkswaterstaat (the agency responsible for the design, construction, management, and maintenance of main infrastructure facilities in the Netherlands) constructed a set of hydraulic structures, including shipping locks, discharge sluices, and a pumping station.

The IJmuiden pumping station is the largest in Europe, built in 1975 with an initial pumping capacity of 160 m^3/s (since expanded to 260 m^3/s in 2004). As it approaches the end of its design lifespan, the question of designing the next generation of structures is growing increasingly relevant. Accordingly, the Rijkswaterstaat-headed project, VONK, investigated possible ways of replacing these structures. This case study focuses specifically on using EOA to explore different possible designs for the replacement of the IJmuiden pumping station. The pumping station is multi-functional, fulfilling several different roles:

- *Flood defense*, serving as a barrier between the North Sea and inland areas, reducing the risk of flooding from high water on the North Sea.
- *Regulation of inland water levels*, discharging inland precipitation runoff to the North Sea.
- *Water quality management*, separating the saline water of the North Sea from the fresher water in the canal.
- *Ecological management*, facilitating the passage of fish.

This case study focuses only on IJmuiden's primary functions—the structure's role in *flood defense* against the North Sea and as *a regulator of inland water levels*. The ability of the structure to continue to fulfill these functions in the future depends on a variety of external factors, such as sea levels on the North Sea, and the impacts of precipitation and regional socio-economic development on canal runoff. Clearly, there is substantial uncertainty about future values of these key parameters, which complicates the question of planning and designing a replacement for the pumping station. The overall question is: given uncertainty about the future, how do different plans for structural replacement (each offering equivalent reliability of service) compare in terms of lifecycle costs? Which replacement strategy is a wise first step because it offers the ability to maintain service reliability as future uncertainty resolves itself?

11.5 Setup of the EOA Approach for the IJmuiden Pumping Station

11.5.1 Characterization of Sources of Uncertainty

This section characterizes the sources of uncertainty most relevant for long-term replacement planning for IJmuiden. Colleagues within the VONK project used expert consultation to determine the sources of uncertainty with the largest potential impact on the long-term functionality of the pumping station. As Table 11.4 indicates, the sources of uncertainty are station-function specific.

We can classify uncertainty into different levels, ranging from situations where there is uncertainty among well-defined alternatives with known likelihoods of occurrence, to those where neither the range of possible outcomes nor the associated probabilities are well characterized. For instance, while we can adequately describe day-to-day precipitation variability using stochastic variables informed by historic

Table 11.4 Salient sources of uncertainty for replacement planning for the IJmuiden pumping station

Station function	Source of uncertainty	Mechanisms by which uncertainty can have an impact
Flood defense	Sea level rise	Affects the adequacy of the installed flood defense height
Inland water level regulation	Sea level rise	Decreases the time that water discharges under gravity from the canal to the North Sea. Also increases the hydraulic head between the surface of the canal and the sea, thus reducing the pumps' discharge ability when pumping is required
	Precipitation intensity increase	Affects the volume of water entering the canal at a given time. Given limited storage in the canal, increased inflows may require expansion of existing discharge capacity

precipitation records, we can forecast long-term climate changes in extreme precipitation much less clearly.

Following the typology of uncertainty in Table 1.1 in Chapter 1, we divide the relevant sources of uncertainty in this case into two types:

- Level 2 uncertainties that we can model probabilistically with some confidence. The analysis models these using stochastic variables.
- Level 3 and Level 4 uncertainties[2] that we handle using a range of scenarios.

This case used two sea level rise scenarios and four precipitation scenarios (see Table 11.5) to inform its analyses over an 85-year project horizon to 2100. It did not assign probabilities to these discrete scenarios. By looking across all these scenarios, we can get a sense of how the performance of different courses of action vary across a wide spectrum of future scenarios, despite not having clear probabilistic information.

---

[2]Climate change is commonly mentioned as a source of deep uncertainty. The question of assigning probabilities to future scenarios of climate change is particularly controversial. While many argue that scientific uncertainty about emissions simply does not allow us to derive reliable probability distributions for future climate states, others counter by saying that the lack of assigned probabilities gives non-experts free rein to assign their own, less well-informed probability estimates. Grubler and Nakicenovic (2001), Pittock et al. (2001), Schneider (2001), Dessai and Hulme (2007), and Morgan and Keith (2008) present some of the arguments for and against assigning probabilities to climate scenarios.

Table 11.5 Specification of scenarios, indicating amount of change between 2015 and 2100[a, b]

Uncertain variable	Scenarios for 2100 relative to 2015			
Mean sea level	Low: +35 cm		High: +85 cm	
Mean winter precipitation	Low: +4.5%	High: +12%	Medium: +11%	Extreme: +30%

[a]These scenarios draw upon country-specific climate scenarios of the Royal Netherlands Meteorological Institute (KNMI). Details of the development of these can be found in van den Hurk et al. (2006), who describe the initial development of the first set of Dutch climate scenarios in 2006, and in KNMI (2012, 2014), which describe changes and improvements incorporated in the 2014 climate scenarios

[b]Note that there are only four joint scenarios. Although there are two scenarios for sea level rise and four for precipitation (a total of eight possible permutations), the physical correlation between these scenarios means that some combinations are unrealistic. For instance, the relatively modest temperature change resulting in the Low Sea Level scenario is highly unlikely to produce the large changes in atmospheric circulation necessary to produce the Extreme Precipitation scenario

Within each of these scenarios, the analysis included the following additional sources of uncertainty and natural variability as probabilistic variables:

- Uncertainty in the water heights associated with a particular flood return periods,
- Natural variability in precipitation, and
- Uncertainty in the precipitation-canal inflow relationship.

This approach—capturing deep uncertainty and "probabilistic uncertainty" by coupling scenarios with stochastic variables—is a pragmatic way of coping with different relevant types of uncertainty. Others have used it in the literature on water resources planning, for example Jeuland and Whittington (2013, 2014).

11.5.2 Design Alternatives

This case investigates several proposed replacement designs for the IJmuiden pumping station under the uncertain conditions described above. Table 1.6 schematically displays these design alternatives. Each of them maintains the same minimum function-specific level of service throughout the entire planning horizon. For flood defense, Dutch federal law (Netherlands Flood Defense Act 2009) mandates the protection level any structure should provide; IJmuiden should protect against North Sea levels exceeded on average once every 10,000 years. For the regulation of inland water level, this study used service levels consistent with the 2013 North Sea Canal Water Accord (Beuse 2013).

The differentiation among the design alternatives lies in the choice of initial structural design and how further capacity is added over time. The case study examined three design alternatives:

Table 11.6 Design alternatives considered in the IJmuiden case study

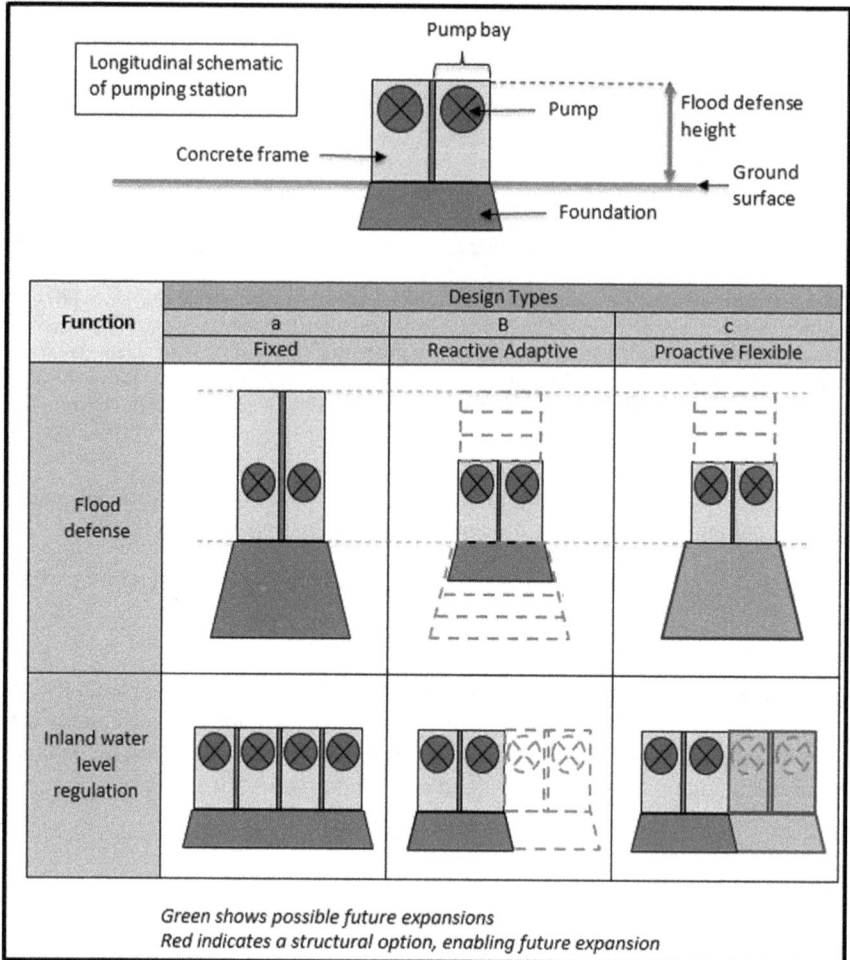

- *Fixed design*, consistent with the traditional predict-then-act approach to water resource planning. The structure provides at least the minimum level of service through to the end of its design life, with a safety margin added to buffer against any uncertainties that may not be captured in the analysis. It embodies the traditional engineering mindset, emphasizing over-dimensioning and taking advantage of any economies of scale (Table 11.6, Column a).
- *Reactive Adaptive design*, which acknowledges that a fixed structure may represent an over-investment and hence emphasizes designing for the best-available current information and making changes as needed as the future unfolds (Table 11.6, Column b). Designers size reactive adaptive designs for the short-term, but make no explicit preparations to facilitate possible future adaptations.

- *Proactive Flexible design*, which goes a step further than the reactive adaptive design in that it prepares for the future by choosing to include options within the initial structure (Table 11.6, Column c). Designers size flexible designs for the short term, but proactively incorporate options that enable easy adaptation in the future.

The Reactive Adaptive design features:

- A height able to withstand the best-estimate flood heights over a 25-year planning horizon. This relatively short planning horizon acknowledges the need to revisit this decision in the coming decades, but does not make any explicit preparations for possible later expansion. If a height expansion becomes necessary in the future, it will come at a considerable cost, because resizing of the structure's foundation will be necessary.
- A pumping capacity able to discharge the best-estimate canal inflow volumes over a 25-year planning horizon. If the addition of further pumping capacity becomes necessary over the structure's lifetime, a new "mini" pumping station will need to be installed adjacent to the current structure.

The Proactive Flexible design considers two function-specific options:

- The option to expand the flood defense function includes a larger-than-currently necessary foundation for the structure; this facilitates future height additions as needed.
- The option to expand the function to regulate the level of inland water includes additional pump bays in the concrete frame; these enable easy installation of additional pumps if/when necessary. Steel gates seal off these additional bays until the time managers install additional pumps.

Within this case, the factor that drives necessary modifications to the structure is the need to provide a continued level of service in spite of evolving external operating conditions. This contrasts with the LNG case presented earlier, in which the production and distribution system grew in response to changes in consumer demand. In both cases, appropriate decision rules signal when managers should initiate a change in the system. In the case of the pumping station, capacity is expanded whenever the required service level can no longer be maintained.

11.5.3 Details of the Analysis

The core of the quantitative analysis couples a physical and an economic module (Fig. 11.9). The:

- *Physical performance module* links changes in future operating conditions (such as higher sea level) to performance indicators of interest (such as water levels associated with specified return periods at a certain location). It generates many simulations of future environmental conditions, consistent with the different sea level

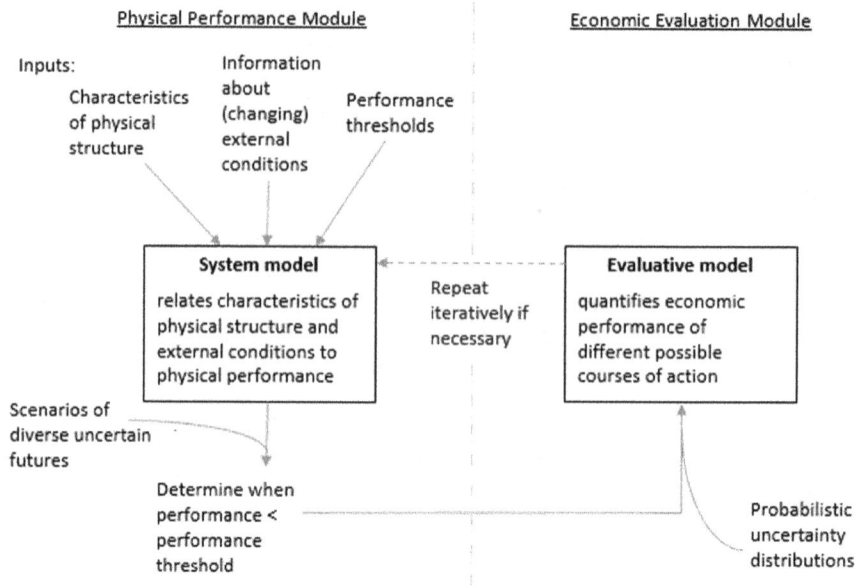

Fig. 11.9 Conceptual overview of the analysis

rise and precipitation change scenarios (Table 11.5). This module both indicates under what future conditions the current physical system becomes inadequate, and captures how different possible courses of action affect the future performance of the system. There is a different module for each of the two functions the case examined.

- *Economic evaluation module* uses the simulations of the physical system as input to compare different courses of action based on whichever performance indicator(s) the analyst considers most suitable. This case study analyzed lifecycle costs of the different structural designs (Table 11.6). It compared the alternatives based on total cost of ownership, including possible later expansion costs in addition to initial capital costs. All analyses applied a discount rate of 5.5%, consistent with a 2.5% risk-free rate and a 3% risk premium, which the Rijkswaterstaat uses for capital investment projects. However, the case study also generated results for a range of other discount rates for the purpose of sensitivity analysis. The lifecycle cost analysis used Monte Carlo simulation to evaluate 1000 different versions of the future for each of the different scenarios.

11.6 Results from Applying the EOA Approach to the IJmuiden Pumping Station

The analysis ultimately generates distributions of lifecycle costs, for each function in turn. These compare the performance of different designs, over many possible simulated futures. We present and discuss selected results below (to see a complete set of results, see Smet 2017).

11.6.1 Inland Water Level Regulation Function

To maintain the required water levels on the North Sea Canal throughout the project horizon, we explore three alternative strategies:

- *Fixed*, which establishes now the maximum pumping capacity that might eventually be needed;
- *Reactive Adaptive*, which builds what is needed now, and will upgrade the installed pumping capacity as dictated by emerging future conditions; and
- *Proactive Flexible*, which creates a pumping station with the maximum number of pump bays that might be needed, but defers purchasing and installing the pumps until actually necessitated by external developments, and thus saves on immediate costs.

Figure 11.10 shows the relative lifetime economic performance of these three different structural designs, for the two most extreme scenarios from Table 1.5, using discount rates of 0 and 5.5%.

In this case, both the Reactive Adaptive and Proactive Flexible designs always perform better than the Fixed design, across all scenarios and discount rates. The results for other scenarios and discount rates (not shown) are not appreciably different from those in Fig. 11.10. This demonstrates that, for the water level management function, Reactive Adaptive and Proactive Flexible designs can offer substantial gains compared to the traditional Fixed approach. Which of the two is better?

When there is no discounting of future costs (0% discount rate), the Proactive Flexible design generally outperforms the Reactive Adaptive design. In the riskier scenario (High Sea Level/Extreme Precipitation), the Proactive Flexible design dominates stochastically over the Reactive Adaptive design in delivering lower costs. In the less risky scenario (Low Sea Level/Low Precipitation), the Proactive Flexible design delivers lower costs on average and more reliably. In this case, the Proactive Flexible design leads to higher costs than the Reactive Adaptive design in about 15% of simulations, but still costs less about 25% of the time.

Comparing both of the above scenarios, the Proactive Flexible solution is more valuable in the riskier scenario. The intuition here is that the greater the number of expansions required over the course of the project horizon, the better the Proactive Flexible design, which enables relatively cheap expansions. In this it contrasts with

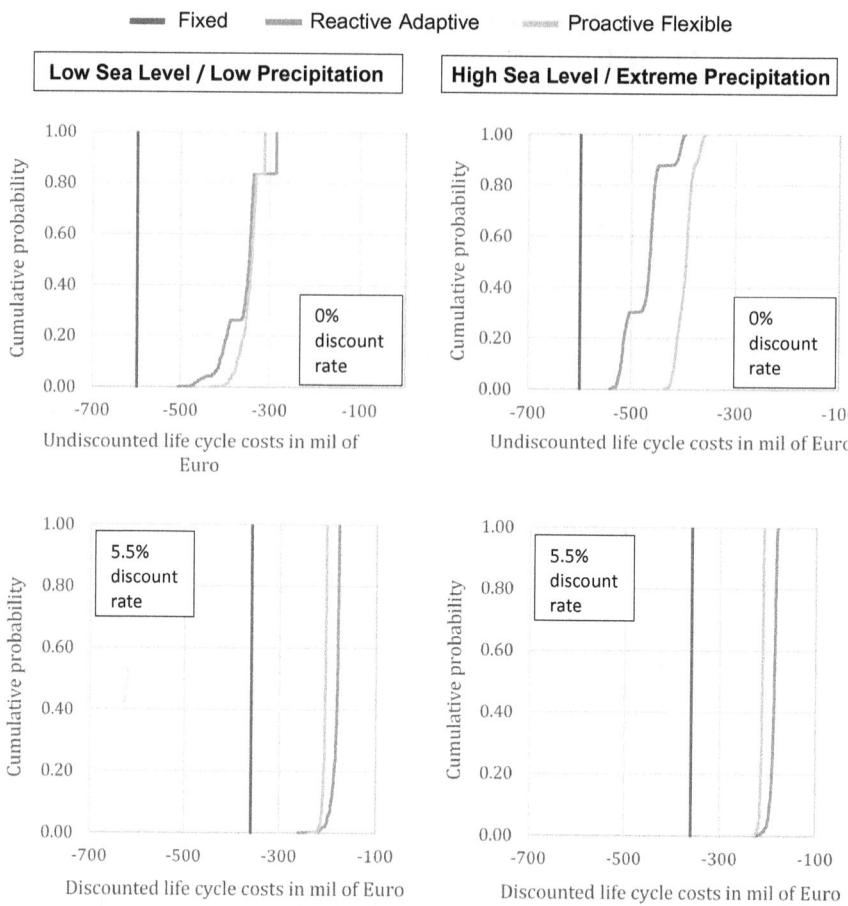

Fig. 11.10 Lifetime economic performance of design alternatives for regulating inland water level (across two extreme scenarios, for different discount rates)

the Reactive Adaptive design, which implicitly gambles on few, if any, expansions becoming necessary. This observation is consistent with our knowledge that the value of options grows as the degree of uncertainty about the future increases.

When future costs are discounted over the long-project horizon, this reduces the advantage of the Proactive Flexible design. In our case, the discount rate of 5.5% erases the disadvantage of expensive future adaptations, making the Reactive Adaptive design the dominant solution.

Finally, notice the consequence of using simulation in the EOA analysis: all results in Fig. 11.10 are in the form of distributions of outcomes. As in the LNG case, this is a major advantage of EOA over standard Real Options Analysis, which produces "value" as a singular measure of performance. By looking at entire distributions, we are able to gain insights about the specific conditions under which one design outperforms another. With results in this form, decisionmakers are able to balance different trade-offs, such as worst case versus average performance, as we did when discussing the relative merits of the Reactive Adaptive and Proactive Flexible designs in the undiscounted Low Sea Level/Low Precipitation Scenario.

When looking at the inland water regulation function, it is reasonable in this case to conclude that:

- Incremental Reactive Adaptive and Proactive Flexible designs outperform the Fixed design;
- The Proactive Flexible design dominates the Reactive Adaptive design in riskier futures, and when lower discount rates are applied, while the Reactive Adaptive design performs better in less risky futures;
- So the preferred choice between Reactive Adaptive and Proactive Flexible designs depends on the decisionmakers' belief about the future and their willingness to bear high-cost worst-case outcomes.

11.6.2 Flood Defense Function

To maintain the required 1:10,000 year flood protection throughout the project horizon, we again have three alternative designs:

- *Fixed*, which establishes now the maximum flood defense height that might eventually be needed;
- *Reactive Adaptive*, which builds what is needed now, and will upgrade the flood defense height as dictated by emerging future conditions; and
- *Proactive Flexible*, which creates the foundation on which to build the maximum flood defense height that might be needed, but defers raising the height until actually necessitated by external developments, and thus saves on immediate costs.

Figure 11.11 shows the relative lifetime economic performance of the three alternatives, using the standard 5.5% discount rate for the high sea level rise scenario. In this case, the dimensions of the foundation and other below ground structural components are central determinants of the capital cost. As the Fixed and Proactive flexible designs both provide for the same eventual height, they both require the same foundations. The Proactive Flexible solution is cheaper, because it defers or avoids an increment of height. The Reactive Adaptive design commits initially to lower height, and thus a smaller foundation. It is therefore markedly cheaper to build initially, but may lead to substantial eventual costs sometime in the future, when an upgrade in height, and thus reconfiguring of the foundation, may become necessary. Viewed in terms of present value, however, the discount rate greatly reduces these distant extra costs.

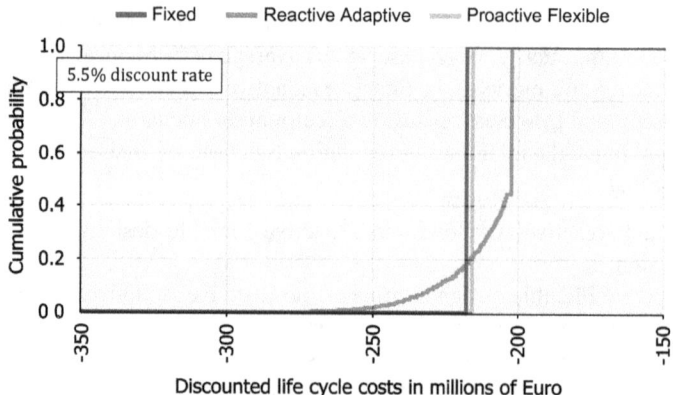

Fig. 11.11 Lifetime economic performance of design alternatives for flood defense (high sea level scenario)

While the Fixed and Proactive flexible designs have average lifecycle costs of approximately 216 million Euros, the Reactive Adaptive design has a lower average cost (210 million Euros), but a much larger range of possible outcomes (from 203 million to 290 million Euros in the worst case). With the results in this form of distributions, decisionmakers can explicitly decide whether they most value minimum average cost (i.e., choose the Reactive Adaptive design) or will accept a higher average cost in order to reduce the possible range of outcomes (i.e., choose the Fixed or Proactive Flexible design). When looking at the flood defense function, it might be reasonable to conclude that:

- The benefits of the Reactive Adaptive design do not justify its risks;
- There is little to choose from between Fixed and Proactive Flexible designs;
- So the preferred policy might be to adopt the Fixed design, and be done with it.

11.6.3 Guidance from Applying EOA to This Case

Again, the EOA produces practical, useful insights for decisionmakers. In the case of the IJmuiden pumping station in the Netherlands these include:

- For those design elements contributing to the pumping station's ability to regulate inland water levels:

  - Reject the Fixed design, which sees all the pumping capacity that might eventually be needed installed at the outset.
  - Choose one of the two incremental strategies, either the Reactive Adaptive design or the Proactive Flexible design.

- The Reactive Adaptive design is the preferred design for decisionmakers more willing to accept higher long-term costs in exchange for short-term savings by building a smaller structure.
- The Proactive Flexible design is the preferred design for decisionmakers who anticipate and want to be prepared for large degrees of environmental change in the future.
- For those design elements contributing to the pumping station's ability to *withstand future floods on the North Sea*:
  - Reject the Reactive Adaptive design because the short-term cost savings from choosing a smaller structure do not outweigh the future risks.
  - The Fixed and Proactive Flexible designs demonstrate comparable lifetime economic performance. Thus, all else being equal, the preferred policy may be to simply adopt the traditional Fixed design.

Overall, this case explored replacement of a complex structure that fulfills a number of different functions. For IJmuiden, the EOA analysis shows great added value from incorporating an option in the design of the pumping facilities, fulfilling the inland water regulation function. Conversely, in this case, the value of flexibility in the flood defense height of the structure does not substantially improve the lifetime cost performance, as compared to a traditional fixed design. As an approach, EOA offers the tools to structure an analysis of such complex, multi-functional systems, providing insights into which functionalities could benefit from a proactive and flexible approach and which conditions maximize these benefits.

11.7 Conclusions and Reflections for Practice and Theory

This section presents a few concluding reflections about the lessons learned for EOA practice and theory from these two case studies:

- EOA does not prescribe a single plan. The focus is on how to start on a path that leads to a line of possible desirable developments while avoiding threatening downside risks. EOA does this in the same way a chess master does not commit to the details of a strategy from the start, but begins with a single move, such as Pawn to King 4. Chess masters will choose the details of subsequent strategy as they learn about the intentions of their opponents, as their deep uncertainty gives way to more knowledge and understanding. Likewise, the project planners for the LNG plant can defer the details of their strategy until they learn more about the market for the product, the government intentions, energy prices, and the economies of scale and learning rates of the technology. So too, the project managers of IJmuiden can monitor changes in precipitation and sea level before taking appropriate next steps in their long-term investment plan.
- EOA can cope with diverse and deep uncertainties: The two cases presented here applied EOA to planning problems in which there were a diversity of sources of

deep uncertainty, ranging from future technological innovation, demand for LNG, sea level rise, and changes in precipitation. Taken together, these cases indicate that EOA can provide valuable insights for the decision process, when faced with immense uncertainty and in the absence of reliable probabilistic information.
- In contrast to traditional Options Analysis, EOA produces distributions of results, which provide valuable additional insights: These two cases produced results in the form of distributions, and demonstrate how the calculation of distributions of possible outcomes provides decisionmakers with useful information concerning worst-case outcomes, unavailable from average outcomes alone. Results in this form explicitly inform decisionmakers about the tradeoffs among objectives, helping them to identify preferred strategies.
- EOA is a versatile and rigorous analytical method: The core EOA method can be adapted for many different types of problems and purposes. The LNG case study emphasized the role of infrastructure size, economies of scale and learning rates on the development of a sound investment plan, while the IJmuiden case highlighted how a reactive versus proactive approach to investment can result in substantially different outcomes. The basic method lends itself to diverse modifications and additions.

References

Beuse, P. H. (2013). *Waterakkoord voor het Noordzeekanaal en Amsterdam-Rijnkanaal 2013*. Rijkswaterstaat West-Nederland Noord. [online] Available at http://www.infomil.nl/publish/pages/72267/waterakkoordnoordzeekanaal-amsterdam_rijnkanaal_2013.pdf. Accessed September 15, 2017.

Cardin, M.-A, Ranjbar-Bourani, M., & de Neufville, R. (2015) Improving the lifecycle performance of engineering projects with flexible strategies: Example of on-shore lng production design. *Systems Engineering, 18*(3), 253–268. http://doi.org/10.1002/sys.21301.

de Neufville, R., & Scholtes, S. (2011). *Flexibility in engineering design*. Cambridge, MA: MIT Press.

Dessai, S., & Hulme, M. (2007). Assessing the fixedness of adaptation decisions to climate change uncertainties: A case study on water resources management in the East of England. *Global Environmental Change, 17*(1), 59–72.

Grubler, A. N., & Nakicenovic, N. (2001). Identifying dangers in an uncertain climate. *Nature, 412*, 15.

Jeuland, M., & Whittington, D. (2014). Water resources planning under climate change: Assessing the fixedness of real options for the Blue Nile. *Water Resources Research, 50*, 2086–2107.

Jeuland, M., & Whittington, D. (2013). Water resources planning under climate change: A "real options" application to investment planning in the Blue Nile. *Environment for Development*. Discussion Paper Series 13–05.

Morgan, M. G., & Keith, D. W. (2008). Improving the way we think about projecting future energy use and emissions of carbon dioxide. *Climatic Change, 90*, 189–215.

Netherlands, Flood Defense Act. (2009). *Wet van 29 januari 2009, houdende regels met betrekking tot het beheer en gebruik van watersystemen (Waterwet)*. The Hague, The Netherlands: SDU Tweede Kamer der Staten-Generaal.

Netherlands, Ministry of Transport, Public Works and Water Management. (2009). *Catchment area North Sea Canal and Amsterdam-Rhine Canal*.

Pittock, B. A., Jones, R. N., & Mitchell, C. D. (2001). Probabilities will help us plan for climate change. *Nature, 413,* 249.

Ranjbar-Bourani, M. (2015). *An integrated multi-criteria screening framework to analyze flexibility in engineering systems design: Applications in LNG infrastructures.* Ph.D. dissertation, Department of Industrial and Systems Engineering, National University of Singapore. http://pqdtopen.proquest.com/doc/1761163141.html?FMT=AI.

Royal Netherlands Meteorological Institute (KNMI) (2012*). Advisory board report: Towards the KNMI'13 scenarios.* Climate change in the Netherlands.

Royal Netherlands Meteorological Institute (KNMI). (2014). *KNMI'14: climate scenarios for the 21st century—A Netherlands perspective* (B. van den Hurk, P. Siegmund, & A. Klein Tank (Eds.)). Retrieved September 10, 2017, from http://bibliotheek.knmi.nl/knmipubWR/WR2014-01.pdf.

Schneider, S. H. (2001). What is 'dangerous' climate change? *Nature, 411,* 17–19.

Smet, K. S. (2017). *Engineering options—Proactive planning of water resource infrastructure under climate change uncertainty.* Ph.D. dissertation, School of Engineering and Applied Sciences, Harvard University.

van den Hurk, B., Tank, A. K., Lenderink, G., van Ulden, A., van Oldenborgh, G. J., Katsman, C., van den Brink, H., Keller, F., Bessembinder, J., Burgers, G., Komen, G., Hazeleger, W., & Drijfhout, S. (2006). *KNMI climate change scenarios 2006 for The Netherlands.* KNMI Scientific Report WR 2006-01.

Prof. Richard de Neufville (Massachusetts Institute of Technology, Institute for Data, Systems, and Society) has worked on decisionmaking under uncertainty since the 1970s. He focuses now on practical application of Engineering Options Analysis. He co-authored: 'Flexibility in Engineering Design' (MIT Press, 2011) and 'Flexibility and Real Estate Evaluation under Uncertainty: A practical Guide for Developers' (Wiley, 2018). He applies these approaches internationally to the design of infrastructure systems and 'Airport Systems Planning, Design, and Management' (McGraw-Hill, 2013).

Dr. Kim Smet (University of Ottawa, Smart Prosperity Institute) is a Postdoctoral Researcher at the Smart Prosperity Institute, an environmental policy think tank and research network based at the University of Ottawa. She completed her Ph.D. in Environmental Engineering at Harvard University in 2017, where she explored how Engineering Options can be incorporated into the redesign of aging water resources infrastructure under uncertainty. Through her research, she has worked with the US Army Corps of Engineers and the Rijkswaterstaat in the Netherlands.

Dr. Michel-Alexandre Cardin (Imperial College London, Dyson School of Design Engineering) is a senior lecturer in Computational Aided Design. His work focuses on the development and application of analytical procedures for Flexibility and Real Options Analysis in Engineering Design, in particular in the areas of infrastructure and financial systems. His work has been published in scientific journals such as Energy Economics, IISE Transactions, IEEE Transactions on Systems, Man, and Cybernetics; Journal of Mechanical Design, Research in Engineering Design, Systems Engineering, Transportation Research, and Water Research.

Dr. Mehdi Ranjbar-Bourani (University of Science and Technology of Mazandaran, Department of Industrial Engineering) is an Assistant Professor at the University of Science and Technology of Mazandaran, Behshahr, Mazandaran, Iran. He received his Ph.D. degree in Industrial and Systems Engineering from the National University of Singapore (NUS) in 2015. He has professionally collaborated with the National Elite Foundation and the University of Tehran in Iran, A*STAR, and Keppel Offshore and Marine technology center (KOMtech) in Singapore, and the Engineering Systems Division at MIT in the USA.

Open Access This chapter is licensed under the terms of the Creative Commons Attribution 4.0 International License (http://creativecommons.org/licenses/by/4.0/), which permits use, sharing, adaptation, distribution and reproduction in any medium or format, as long as you give appropriate credit to the original author(s) and the source, provide a link to the Creative Commons licence and indicate if changes were made.

The images or other third party material in this chapter are included in the chapter's Creative Commons licence, unless indicated otherwise in a credit line to the material. If material is not included in the chapter's Creative Commons licence and your intended use is not permitted by statutory regulation or exceeds the permitted use, you will need to obtain permission directly from the copyright holder.

Part III
DMDU-Implementation Processes

Chapter 12
Decision Scaling (DS): Decision Support for Climate Change

Casey Brown, Scott Steinschneider, Patrick Ray, Sungwook Wi, Leon Basdekas and David Yates

Abstract

- Adaptation planning and climate risk management are examples of decision processes made under climate uncertainty.
- A variety of approaches exist for helping an analyst to evaluate alternatives over future unknown states of the world. However, climate uncertainty requires additional considerations, including how to use available climate information, such as climate change projections, to inform the decision process without overwhelming it.
- Decision Scaling is specifically designed to support decisionmaking under climate uncertainty while it is general enough to address other uncertainties. The process is designed to make the best and most efficient use of uncertain but potentially useful climate change projections.
- DS consists of three steps: Decision Framing, Climate Stress Test, and Estimating Climate-Informed Risks.
- This is accomplished by using weather generator tools and systematic sampling algorithms to create an unbiased description of system response to plausible climate changes.
- Climate information is incorporated as a sensitivity factor in the last stage of analysis for aiding the process of prioritizing risks or choosing among adaptation

---

C. Brown (✉) · S. Wi
University of Massachusetts, Amherst, MA, USA
e-mail: casey@engin.umass.edu

S. Steinschneider
Cornell University, Ithaca, NY, USA

P. Ray
University of Cincinnati, Cincinnati, OH, USA

L. Basdekas
Black and Veatch, Colorado Springs, CO, USA

D. Yates
National Center for Atmospheric Research, Boulder, CO, USA

options through evaluation of probabilities of underperforming strategies and the need for adaptation.
- An important benefit of using DS is the establishment of open lines of communication and trust among the analysts, decisionmakers, and representative stakeholders. This is achieved through frequent consultation and validation of models that will be used for the analysis.

12.1 Introduction

Decision Scaling (DS) originated in response to questions regarding the best approaches to process and use climate change projections for adaptation planning. At the time, and as continues to be the case, planners face an overwhelming number of choices of climate change projections, which vary in terms of climate model, downscaling approaches, emission scenarios, etc., and no clear guidance to navigate among them. DS uses a decision analytic framework and structured, physically based, multidimensional sensitivity analysis to first identify the priority climate-related concerns, reserving climate projections for use in latter stages of the analysis, to inform the level of concern for climate vulnerabilities that are identified. A formal probabilistic framework is used to characterize the information from climate projections in the spatial and temporal scales that maximize credibility.

The application of decision analysis techniques has evolved from an intent to identify an optimal decision to an intent to fully explore the consequences and tradeoffs of alternative decisions, often to select a reduced set of best-performing or "noninferior" decisions. Public resource problems consist of multiple objectives, often the goals of different groups or people or constituencies, and so are not reducible to a single numeraire. For public resource problems, there are important considerations:

1. Multiple irresolvable preferences for the choice of a solution;
2. Full exploration of the performance space over various uncertainties and in terms of multiple objectives;
3. Multiple beliefs about the future, possibly conflicting information regarding relative likelihoods or simply indications of plausibility of different futures.

A planning approach for climate change must address each of these considerations. Methods for decisionmaking under uncertainty have long existed for exploring multiple objectives and irresolvable preferences among those objectives, including concepts such as preference dominance and Pareto optimality.

However, traditional methods do not address the issue of multiple beliefs about future states of the world. Typically, the scenarios used to explore the outcomes of different decisions include embedded beliefs that are not explicitly recognized. For example, the choice of a particular set of climate simulations, or downscaling approach, or emissions scenario, embeds the assumptions that those projections represent, and precludes other possible climate futures. Thus, the performance space of alternative decisions is affected and potentially biased by the choice of the climate futures considered, including those derived from climate models. Climate may change in a number of ways. There are two approaches in common practice: (1) the use of climate change projections from climate models (by far the most common) and (2) the use of climate narratives, such as used in a scenario planning approach.

Scenario planning typically uses a small number of narratives, representing internally consistent and mutually exclusive possible futures that are developed independently of climate projections or else informed by them (for example, considering the range of projections). For many planning exercises, this is probably adequate. But for natural resource systems or complex-coupled human-natural systems, such as a water resource system, a small set of narratives provides a limited view of potential climate change effects. Consequently, the resultant plans may be vulnerable to climate changes that could have been identified with a more comprehensive approach. (So, at best, scenario planning is able to deal with Level 3 uncertainties, but not Level 4 uncertainties.)

The most common approach to exploring climate futures is the use of climate change projections from general circulation model (GCM) simulations. Climate projections have the imprimatur of authority because they are used extensively by the climate science community to better understand the earth's climate, the effect of anthropogenic emission of greenhouse gases and to inform key decisions regarding the regulation of these emissions. However, they may not be the best source of climate futures for use in adaptation planning. Most significantly, GCM projections are inefficient and biased samplers of possible future climate changes. They do not explore the full range of climate changes, but rather the "minimum range of the maximum uncertainty." Projections typically have biases in terms of climate variability and extremes, which are sometimes the most important climate statistics in terms of impacts. Using climate projections as scenario generators requires processing steps that require many choices that can be controversial themselves. Many a climate change study becomes bogged down in the evaluation of alternative downscaling approaches and choices of which climate models to use. In the end, the results are dependent on these choices, and would be different if other choices were made, further confusing the results.

In practice, adaptation planners are often overwhelmed by the many choices involved in using climate projections for scenario analysis, including emissions scenarios, downscaling methods, model selection, and bias correction. In addition, with new sets of climate models, or new downscaling methods introduced every few years, practitioners feel compelled to redo the entire analysis to see if results have changed. Consequently, when using climate projections as the starting point, the analysis is never complete, and the planner will (and should) always wonder if the results would be different if a different set of projections were used.

DS reveals vulnerabilities to climate changes independent of climate projections, thus negating the time-consuming and expensive debates on the choices related to climate projection use. It does this through the application of a "climate stress test" algorithm, which generates physically realistic climate changes over the widest plausible range. It generates a comprehensive, unbiased estimation of climate effects on the system or decision of interest, without the assumptions and biases embedded in climate change projections from climate models. Climate information from projections or other information sources are used only in the final stages of analysis, to inform the vulnerabilities or differential performances that are identified in the climate stress test. In this way, climate information is used as a sensitivity factor. The understanding of climate change effects on a system only change if the system changes—the understanding is not dependent on the climate projections that happen to be used.

DS shares the general problem formulation and structure of decision analysis frameworks (c.f., Schlaifer and Raiffa 1961), with the formal structuring of decisions in terms of objectives, performance measures or "rewards," unknown future states of the world, and alternative choices. It is designed to be incorporated into public decisionmaking processes, including with diverse sets of stakeholders holding different objectives and preferences for the matter at hand. It accommodates multi-objective analysis and multiple "beliefs" or viewpoints on the more likely future states of the world. This chapter explains the theoretical derivation of Decision Scaling, explains the process in detail, and summarizes a recent application to the challenge of assessing climate risks to the water supply system for Colorado Springs.

12.2 Technical Approach

12.2.1 Overview

DS was designed as a "fit-to-purpose" decision framework for the use of climate change information in climate risk management and adaptation planning. "Decision Scaling" (DS) is positioned as a method to use climate information to improve decisions made under climate uncertainty. DS inverts the usual order of forecast information used in decision analysis, focusing on understanding how decisions are sensitive to changing climate, and using that insight to tailor the climate information provided by GCMs to provide the most credible information to inform the decision. The approach is based on an implicit acceptance of the inherent uncertainty of future climate, and the difficulty of attempting to reduce that future uncertainty. Instead, the goal is to characterize the uncertainty in terms of its implications for decisions and identify the best decisions in view of this uncertainty.

DS consists of three steps (see also Sect. 1.5):

1. Decision Framing
2. Climate Stress Test
3. Estimating Climate-Informed Risks.

The decision framing step is used to identify the mission objectives, and metrics for quantifying them, the uncertain factors that affect the decision, such as future climate, the models or functional relationships needed to represent the system being investigated, and if adaptations are being considered, any choices among adaptation alternatives. The climate stress test is a pragmatically designed multidimensional sensitivity analysis that reveals the fundamental sensitivity of the sector to climate changes, and/or, other uncertain factors. In doing so, it exposes the climate conditions that are problematic for the sector. The final step is to prepare climate information, such as downscaled climate projections, to assess the level of concern that one might assign to the problematic conditions. This can be accomplished using both informal approaches, such as "weight of evidence", and formal approaches, such as probabilistic methods.

DS is expected to yield benefits in at least three ways. First, the process will provide a clear delineation of the climate risks that are problematic for a specific sector, and the climate changes by which a sector is not threatened. These results are independent of climate projections, and thus are not subject to the various choices and uncertainties associated with processing steps such as downscaling. They also do not require updating with every new generation of climate change projections. Second, the approach allows the characterization of climate change projections in terms relative to decisions and the revealed vulnerabilities. The effects of alternative methods for processing of climate change projections, including downscaling approaches (e.g., statistical vs. dynamic) and methods for estimating probability distributions of changes in climate variables, can be presented in terms of their implications regarding the response of the system or sector to climate changes. The information provided by the projections can inform judgments made relative to the level of concern associated with any revealed problematic climate conditions or vulnerabilities of a particular sector. Similarly, the information can be used to inform judgments made relative to adaptation. In this way, the DS approach does not reject the use of climate change information or projections, but rather is designed to use them in the most decision-relevant and helpful way. Third, the approach is designed to facilitate robust decisionmaking approaches, such as in the approaches described and applied in Parts I and II of the book. DS includes quantification of the robustness of alternative system configurations or other policies, and a clear indication of the expected risk reduction through alternative adaptations. This final benefit is not demonstrated explicitly here, although the results should provide a conceptual understanding of how this would be possible.

DS is based on traditional decision analytic approaches, in particular, the analysis technique known as pre-posterior analysis (Schlaifer and Raiffa 1961). Pre-posterior analysis involves identifying optimal decisions that you would make given new information (posterior to the new information) before you actually know what the new information will tell you (thus, "pre"–posterior). For example, a user may be interested in knowing the expected value of a forecast (say a weather forecast) before paying for it. One would have to account for the value of the improved decision that results from receiving the forecast. But since the forecast result is unknown prior to receiving it, one needs to evaluate the optimal decision for all possible future forecast results, and then account for the probability of receiving each of these forecasts. Application of this framework yields the expected value of perfect information

(EVPI) and the expected value of including uncertainty (EVIU), which are two useful quantities for evaluating whether pursuing forecast information and considering uncertainties is worth the effort. EVPI is the difference in value when a decision is made with the uncertain outcome known and when a decision is made without considering a forecast. This serves as a check for whether to consider forecasts, since the value of decisions with actual operational forecasts cannot exceed this upper bound. EVIU is the value of a decision made taking account of the uncertain outcomes compared to the value of that decision when uncertainty is ignored. Again, it provides a check as to whether these more analytically demanding approaches are likely to be worth the effort.

While the classic decision analytic framework is typically used to evaluate information for decisionmaking prior to receiving it, it reveals insights about the decision that are more broadly useful. In particular, through a systematic exploration of all future states of the world, it produces a mapping of optimal decisions for each of these states. More critically, since the analysis assumes the forecast is not yet received, the mapping is created independent of the forecast or any expectations of future likelihoods. This creates a clear separation between our understanding of the decision at hand, and the estimates of probabilities of future states of the world.

DS exploits the advantages of this theoretical framework, using Monte Carlo stochastic sampling tools and currently available high-powered computers to create the mapping of the computed optimal decisions over the wide range of possible futures. The mapping is then used to identify the scenarios where one decision is favored over another and, in doing so, revealing what information would be most helpful in selecting one decision over another. It also can be used to identify and evaluate risks to an existing system or a planned system design.

DS provides several advantages for addressing uncertainties related to climate change. In typical climate change analyses, there are multiple sources of climate information available. These sources may be projections from general circulation models (GCMs) or regional circulation models (RCMs), stochastic models based on historical data, information drawn from paleoclimatological records, the historic observed record, or some combination of these sources. Often the views they offer for future climate vary widely and there is no clear guidance for choosing among them, because estimating their skill in future projections is difficult or impossible (Gleckler et al. 2008). Yet, because they vary, the results of any analysis are likely to be highly influenced by the choice of futures to consider. In addition, in a multi-stakeholder and multi-objective decisionmaking process, the stakeholders may hold strongly divergent beliefs about future climate. Consequently, an analysis that is driven by belief about future climate may not be responsive to some stakeholders, and the process for selecting the future scenarios to consider can become contentious and stall the process.

DS uses the mapping generated from a pre-posterior analysis framework to generate an understanding of optimal decisions prior to receiving "the forecast," in this case, a particular set of climate futures. Since the mapping is independent of any set of futures, the selection of a set of futures to evaluate is no longer required. Instead of being a potential sticking point, alternative beliefs about the future can be used as a sensitivity factor. Indeed, the concept of "belief dominance" can be used to select the decisions that are best performing across the unknown probabilities of future climate. This results in a non-inferior set of decisions across beliefs, analogous to the more familiar preference dominance, which reveals the Pareto set of non-inferior solutions across preference weightings.

The key methodological challenge for applying this framework is the development of an approach for sampling the plausible range of climate changes. Known issues with climate projections, including their limited sampling of changes in extremes and variability, the debates regarding choices for processing the projections (e.g., "downscaling"), and the fact, as Stainforth et al. (2007) state that they sample "the minimum range of the maximum uncertainty," preclude their utility for this purpose. Instead, a fit for purpose climate and weather sampler algorithm was created, allowing efficient and systematic sampling of plausible climate changes (Steinschneider and Brown 2013).[1]

12.2.2 Step 1. Decision Framing

A premise of DS is that attributes of decisions have a significant and possibly critical effect on the utility of information produced to improve a given decision. In the context of climate change, this implies that the value of climate information and the best means of providing that information (in terms of both its attributes and the effort and expense to provide it) is best discovered by investigating vulnerabilities and decisions, rather than solely investigating the various ways of producing climate information. Therefore, the first step of the analysis is to frame decisions in terms of specific decision attributes and their context relative to climate change.

Following a traditional decision analytic framework, the process begins with gathering information on the decision at hand in a structured fashion, using four categories:

- Choices (e.g., to adapt or not; plan A vs. plan B)
- Uncertainties (e.g., future climate; future population)
- Consequences (e.g., net benefits, damages)
- Connections (e.g., system diagram, system model)

This is a general problem formulation framework common to many decision analytic frameworks; classic texts often use the alternative terminology "actions," "states of the world," "rewards," and "models" (c.f., Winston and Goldberg 2004). In this book, the same framework is described as XPROW (X = exogenous uncertainties,

---

[1] For some applications, simpler methods could be used—for example, cases in which temporal and spatial patterns of weather and climate are not critical to preserve. In addition, tools such as Latin Hypercube Sampling and application of PRIM for scenario identification (e.g., Groves and Lempert 2007) have been productively applied.

P = policies, R = relationships, O = outcomes, W = weights) (see Fig. 1.2)[2]. The framework described here is intended to be intuitive for a general participant. The articulation of **choices** should include not only the specific alternatives, but also characteristics of the decisions, such as the decision hierarchy (at what hierarchical levels are decisions made?) and the temporal nature of the decision (one shot, reversible, sequential, etc.). **Uncertainties** should include not only climate changes, but also other relevant external factors, which will vary based on the specific decision problem. **Consequences** should be quantifiable outcomes (also called performance measures) that are meaningful to the decisionmaker and other stakeholders. Ideally, they would represent the outcomes currently used to assess the performance of the system or activity. Finally, **connections** are the way in which decisions yield consequences, and the way by which uncertainties affect them. The connections between climate and consequences often exist in functional relationships between weather or climate and existing activities. In some cases, existing models quantify those functional relationships. In other cases, historical data may be used to characterize the relationships. In still other cases, such as for water supply, new models or model combinations that relate climate to the reliability of water serving a specific installation may be required.

The decision framing serves to organize what can often seem a nebulous discussion of information and desires into categories that fit directly into the analytical framework. That is, the decisionmaker faces choices that are to be evaluated in terms of their consequences that result from different realizations of the uncertainties. The consequences of choices are estimated based on our understanding of the connections between them, often represented by models or functions.

In many applications, the decision process will involve multiple stakeholders convening to incorporate multiple viewpoints on the objectives and important considerations for analysis. In public decisionmaking processes with many potentially affected people, it is essential to involve representatives of these groups, and understand their preferences and concerns. Otherwise, a decision may be optimal for a narrow set of planners but judged to be far from optimal by those affected by the decision. Trade-offs among objectives can be described and assessed in Step 3.

The results of the first stage of the analysis provide the inputs and information needed for the following stages. These include an articulation of alternative decisions, a list of key uncertainties to incorporate into the analysis, a list of the outcomes used to evaluate the consequences of different decisions, and the models or relationships used to represent the decision problem. The final key product of Step 1 is the establishment of open lines of communication and trust among the analysts, decisionmakers, and representative stakeholders. This is achieved through frequent consultation and validation of models that will be used for the analysis.

---

[2]In RDM the same framework is described as XLRM (X = exogenous uncertainties, L = levers, R = relationships, and M = measures of performance (see Chap. 2)

12.2.3 Step 2. Climate Stress Test

The climate stress test is the term given to the multidimensional sensitivity analysis that is used to reveal the effects of possible climate changes, and other uncertain factors, on the activity or system of interest. This step can be a more general stress test, since the sensitivity factors need not be climate-related. However, since climate requires special handling due to its distributed spatial and correlated temporal characteristics, this chapter focuses on the climate stress test. The approach is to parametrically vary climate variables in a representation of the system or activity, and infer from the results the response of the system to a wide range of climate changes. The representation of the system can range from a simple empirical relationship to sophisticated models or sequences of models. In most cases, a formal model of the natural, engineered, or socio-economic system is created that relates climate conditions to the outcomes identified in Step 1. The models are mathematical representations of physical, social, or economic processes that allow the analysts to systematically explore the potential effects of changes in climate on the system. The models might be complex or simple, depending on the attributes of the decision and the available resources for the analysis. The only requirement is that the representation includes some climate or weather inputs. Models are validated with available data to ensure that they appropriately represent the system of interest and the outcomes of interest in the terms the stakeholders utilize for decisionmaking. The climate stress test approach is quite general. In principle, any model that could be used for assessing climate change impacts via climate projections can be used for the climate stress test, and likely more.

The climate stress test represents advancement over traditional sensitivity analysis. In the past, single variable sensitivity analysis has been criticized because varying a single factor individually will fail to reveal sensitivities that are caused by the correlated behavior of multiple variables. For example, separately varying temperature and precipitation would not reveal problems that occur when both change together. However, varying multiple variables simultaneously requires preserving the physical relationships among these variables, if they are not independent, for the results to be physically meaningful. In the case of weather and climate variables, the challenge is maintaining spatial and temporal relationships in aspects such as precipitation, temperature, and wind. The answer to this challenge resides in stochastic weather generators, which are statistical models that are designed to produce stochastically generated weather time series and have been used in the past in crop modeling and hydrologic modeling to create alternative historical time series. They are used to investigate the effects of possible alternative realizations of weather and climate variability. These models can serve as the basis of climate stress testing.

The required attributes for a multipurpose climate stress testing algorithm include:

1. Physical fidelity—the algorithm must preserve known physical relationships among climate variables.
2. Spatial and temporal cogency—the algorithm must produce results that have a physically meaningful spatial and temporal scale (e.g., average annual tempera-

ture for a 30-year period over the spatial area of study). This allows the results to be linked to climate information, such as climate projections, which are always positioned in time and space.
3. Modifiable—the climate conditions must be able to be changed in a controlled and predictable way, to allow exploration of climate changes.
4. Representative variability—the representation of internal or "natural" variability, the unpredictable chaotic nature of weather time series, should be accurate to the degree possible. This is, especially, important because internal variability is typically dominant at the temporal and spatial scales of adaptation decisionmaking and risk assessment. For example, at the spatial scale of a water resources system and over a 30-year planning period, natural variability is likely to have a larger explanatory role in the conditions experienced than any trends in climate change.

A climate stress testing algorithm was designed with these desired qualities. It is fully described in Steinschneider and Brown (2013). As illustrated in that paper, the climate stress testing algorithm can create a wider range of climate changes than could have been derived from a typical downscaled, multi-model ensemble of projections. More important, in some cases, the algorithm can more efficiently (in terms of computational effort) sample a wide range of climate changes, due to its systematic approach. This is important when the model or models used to represent a specific system are expensive in terms of computation time (e.g., a model representing the California water supply system). Not all applications require such a sophisticated approach. Water supply systems require careful simulation because they are expected to provide water at very high reliability, and will only show sensitivity during critical periods, or rare events consisting of very dry conditions. In addition, they typically collect water from wide spatial areas, such as river basins, and the transport and storage of water defies a simple linear representation. As a result, the detailed representation of temporal and spatial variability is required to accurately assess these systems. Other systems (i.e., systems that have no spatial distribution) can be assessed using average conditions at a point location and do not require the full features of a climate stress testing algorithm.

The stress test provides the analyst and decisionmaker with a response function that relates a change in decision-relevant outcomes measuring performance to changes in the climate and socio-economic system, enabling the analyst to parse the space of future conditions into regions of "acceptable" and "unacceptable" performance (throughout this book, this process is termed Scenario Discovery (SD)). This functional relationship is extremely powerful in the decisionmaking process. For instance, it may indicate that the system is relatively insensitive to changes in climate or other stressors and further analysis is unnecessary. Alternatively, the stress test may reveal that the system is extremely sensitive to even a modest change in one component of the climate system or small amounts of population growth, suggesting that proactive measures may be needed sooner rather than later.

Although climate change is typically the focus of adaptation planning and climate risk assessment, other changing factors or uncertainties may be as important or even more so. The stress test incorporates these non-climate factors as well, in which case it is a more general type of stress test. The stress test uses stochastic sampling techniques to determine the system response and the performance of alternatives over the full range of uncertainties considered. The framework allows both simulation modeling and optimization modeling, and thus the response can be in terms of the system performance (simulation) or optimal decision (optimization) for each combination of factors considered.

Designing the climate stress test requires two additional considerations: the range of the factors to be sampled and the structure of the sampling algorithm. The goal of range setting on the uncertain factors is to sample all plausible values and not preclude unlikely but plausible outcomes. The range should be broad enough that there is no question that all plausible values are sampled. There should be little concern that implausible values might be sampled, since the goal of this step is to understand the response of the system, not to assess risks or vulnerabilities. If vulnerabilities are identified near the edge of the sampling range, they can be disregarded during Step 3. That is the appropriate time to make judgments on the plausibility of identified vulnerabilities.

The second consideration is the structure of the sampling algorithm and the preservation of correlations among uncertain factors, if necessary. The climate stress testing approach described earlier is specifically designed to preserve spatial and temporal patterns in weather and climate statistics. This is necessary in order to sample physically realistic values. Likewise, there may be a need to preserve correlations among other uncertain values. For example, the future demand for water may be an important uncertain factor in water supply planning that is to be sampled over a wide range of values. Depending on local conditions, water demand might be influenced by temperature and precipitation. Therefore, there may be a need to ensure that this relationship is preserved in the sampling technique—for example, by enforcing a positive correlation between water demand and temperature. While the procedures for doing so go beyond the scope of this chapter, there are several possible approaches, including parametric correlation functions and nonparametric (data-based) sampling techniques.

A water supply climate stress test provides an illustration of the approach. In the case of assessing vulnerabilities of a water system, a series of models is developed that can be used to identify the vulnerabilities of all components of the water resources system that serve the water demands of the installation. These models typically include the following three steps: (1) climate/weather generation, (2) hydrologic modeling, and (3) water resources system modeling (Fig. 12.1).

Stochastic weather generators can create new sequences of weather consistent with current or a changed climate that simultaneously exhibit different long-term mean conditions and alternative expressions of natural climate variability. The scenarios created by the weather generator are created independent of climate projections, allowing for a systematic exploration of future climates. The scenarios are designed to maintain physical attributes of weather and climate, such as spatial and temporal patterns and correlations. Furthermore, climate scenarios exhibiting the same mean climate changes can be stochastically generated many times to explore the effects of internal climate variability. In this way, the climate-weather generator (i.e., a

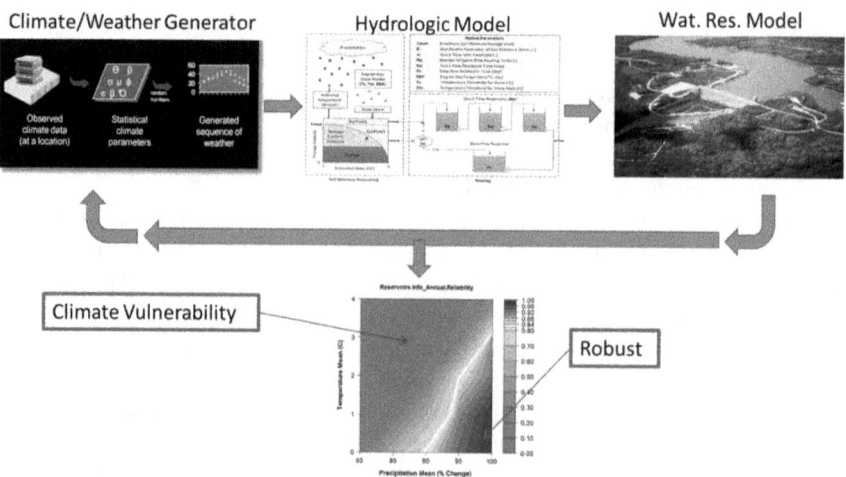

Fig. 12.1 Three sequential elements of the vulnerability assessment modeling environment

weather generator linked to specific mean climate conditions) can be used to fully describe the response of the system to possible climate changes without making strong assumptions about how likely the particular climate changes are.

The climate information from the weather generator is fed into a hydrologic model. The hydrologic model translates changes in weather variables to hydrologic variables of interest (e.g., streamflow at inflow points to water infrastructure). The water resources model accepts hydrologic variables as inputs, as well as other driving variables (e.g., water demanded from domestic or agricultural users; environmental release requirements), simulates the infrastructure operation and conveyance plans that determine the flow of water through the engineered system, and calculates the variables that are of interest for policy and management (e.g., reservoir storage, water delivered to users, water released to the environment). This process is repeated a large number of times (total scenario runs may number in the thousands to fully sample climate and other changes), the performance of proposed plans is calculated, and the results are presented on a "climate response surface" that displays the climate changes that are problematic (Fig. 12.2). [In the rest of the book, this process is called Exploratory Modeling (EM).]

Figure 12.2 is an illustration of the results for a typical climate stress test. It shows a map of changes in climate (both precipitation and temperature) and the resulting impact. In this case, the area in red indicates that the impacts have exceeded an impact level, adaptation tipping point, or threshold beyond which adaptation would be required. Thus, the climate changes represented by the red area indicate the climate changes that would cause adaptation to be necessary. The way of determining threshold levels is not necessarily dependent on any specified threshold; thresholds are completely malleable and can be specified by the individual analyst for the question of interest.

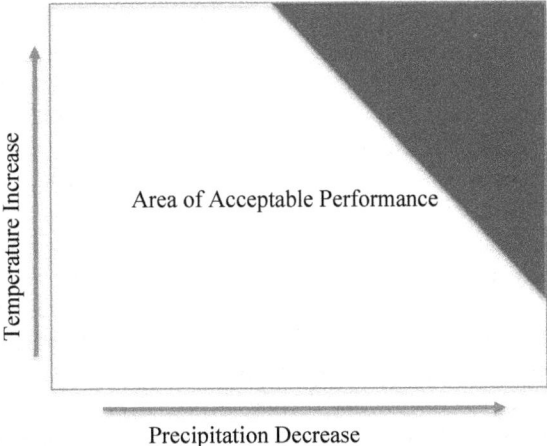

Fig. 12.2 Illustration of climate stress test results indicating the climate conditions that cause unacceptable performance (red area) (Color figure online)

The climate response surface is a two-dimensional visualization of the response of the system to climate change. In this case, the response variable is the reliability of water supply calculated for a 50-year simulation. The climate response surface summarizes the effects of climate change on the system, including changes in mean climate and the range of variability. In this case, the effects of mean climate changes were estimated with 13 stochastically generated weather time series of 50 years in length. The carefully designed sampling of mean climate changes and variability produces the best unbiased estimate of future system performance, and accounts for the internal variability of the climate system.

In a multidimensional uncertainty analysis, more than two dimensions may be needed to display all the effects of uncertain factors. There are a number of ways to display multidimensional results, including "parallel coordinate plots" and "tornado plots". However, experience has shown that stakeholders are most receptive to two-dimensional visualizations, and these 2D figures can be created as slices of the multidimensional response surface for any two variables. The representation of the climate response surface is best determined through iteration with the stakeholder partners.

12.2.4 Step 3. Estimation of Climate-Informed Risks

The final step of DS is the characterization of risk associated with the problematic conditions for a given system. Here the term risk is used to denote the probability of occurrence of the problematic conditions. Up to this point, the analysis is free of the use of probabilities for particular scenarios. This allows exploration of the decision space that is not biased by prior probability assumptions that are weakly supported or not agreed, and that are thus deeply uncertain. At this step, the available sources of information regarding possible future outcomes is described quantitatively using a probabilistic framework. The information is then used as a sensitivity factor to inform final decisions. The use of a probabilistic framework facilitates accounting for the sampling characteristics of the source information, such as the sample size and the dependence among different sources. Ignoring sampling characteristics has been a common mistake in decisionmaking under uncertainty approaches to climate change uncertainty (e.g., the assumption that all projections are equally likely).

Because climate change projections are often commonly used and misused for long-term scenarios, the tailoring of climate information is described in detail here. The climate stress test is designed to create physically representative time series of weather variables at specific spatial scales that can match the source of climate information that is most credible. This enables the mapping of climate information from climate projections or other sources directly to the climate response surface, and thus to the decision. In the case of climate projections, the weather generator design can allow the projections to be used at the spatial and temporal scales where they have most skill (generally coarse scales, large areas (100's of kilometers), and long averaging periods [e.g., 30 years]). Then inference on the relative likelihood of climate changes of interest can be made using the most credible scales of the climate projections.

Another way to increase the credibility of estimations of future climate changes is to create categories of interest and to estimate the likelihood of these categories, rather than the full probability distribution of outcomes. This approach has long been used in seasonal climate forecasts. Seasonal climate forecasts are forecasts made of the mean climate expected for an upcoming season, such as mean temperature and mean precipitation. The basis of these forecasts is typically deterministic or near-deterministic components of the climate system, which typically includes persistence of ocean temperatures and the resulting effects on atmospheric circulation. The El Niño/Southern Oscillation is the most well-known example. Given the amount of uncertainty associated with these forecasts, they are typically made as probabilities assigned to terciles of outcomes. For example, a typical forecast assigns probabilities to precipitation being Above Average, Near Average, or Below Average. This reflects the skill of the forecasts (no more precise forecasts can be made credibly) and the potential utility of the categories for stakeholders (it is useful to know whether precipitation may be above average).

The same approach can be used to improve the credibility of information about future climate change. Indeed, the DS approach enables the creation of categories that are directly relevant to the decision at hand, and also directly coordinated with the spatial and temporal scales at which climate information is most credible. Groves and Lempert (2007) describe the use of a cluster analysis approach to derive *ex post* socio-economic scenarios from the results of a sensitivity analysis. This approach can be employed with the results of a climate stress test, as described in Brown et al. (2012). This creates *ex post* categories of climate change, to which probabilities of occurrence can be assigned. The probabilities are conditional on the source (e.g.,

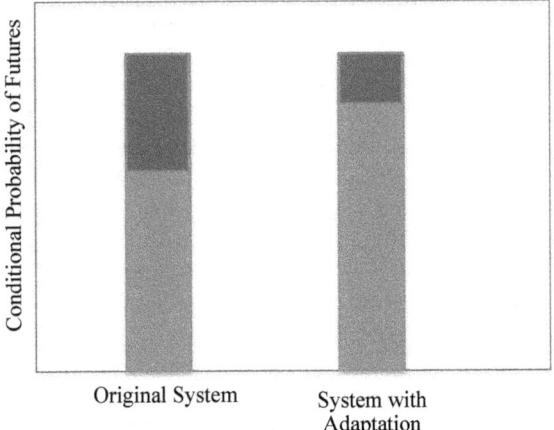

Fig. 12.3 Fraction of probable futures for which the system (left) and the adapted system (right) provide acceptable performance (blue) and unacceptable performance (red) based on stakeholder-derived performance thresholds (Color figure online)

climate projections under a specific emissions scenario). The use of categories allows the information to be conveyed in terms that are meaningful for decisionmakers, and is a more credible prescription of a full probability distribution.

When the system is found to be sensitive to certain future scenarios, the process can be repeated and applied to evaluate possible adaptation strategies as well, by repeating the analysis for the new alternatives considered. The results of this analysis reveal how well each adaptation strategy can preserve system performance over a wide range of futures. Specifically, this analysis determines which adaptations can preserve outcomes above their thresholds across the range of future scenarios. The decision to accept a specific adaptation strategy then relies on an appraisal of the resilience of the original system and the alternatives considered. The climate risk analysis can be used to compare alternative plans or to evaluate the additional robustness that a specific planned adaptation might provide. Figure 12.3 is an illustration of how comparative analysis could be conducted using the results of the climate risk screening. The figure shows the probable fraction of futures for which the current system and the adapted system provide acceptable performance. In this case, the adapted system is more robust, as it performs acceptably over more of the future space. Note that these results are conditional on the assumed probability distribution, and in some cases it will be beneficial to use multiple distributions and compare across them. For example, see Moody and Brown (2013), where alternative adaptations are evaluated conditional on alternative future probability distributions of climate change, including those based on historical conditions and climate change projections from GCM.

This discussion has focused primarily on the particular treatment of climate information for informing decisions related to long-term plans that may be affected by climate change. In cases where climate change is not a key consideration, the analysis methods presented here are likely unnecessary. Other important aspects of formal decisionmaking processes are also not discussed here. These include the process of problem formulation, facilitating stakeholder discussions, and the enumeration of possible adaptation options. In addition, the selection of outcomes is an important component of decisionmaking. While the discussion here used robustness for illustrative purposes, the best or optimal performance for each alternative, the expected value of performance, and regrets are other useful means for evaluating performance. A rigorous analysis would use multiple outcomes to compare and evaluate alternatives.

12.3 Case Study: Assessing Climate Risks to the Water Supply for Colorado Springs, Colorado, USA

The Colorado Springs Utility (CSU) embarked on a long-term strategic planning process to create an Integrated Water Resource Plan (IWRP). The ultimate goal of this plan is to provide "reliable and sustainable water supply to customers in a cost-effective manner." The overall planning process incorporates consideration of water supply and demand, water quality, infrastructure, and regulatory and financial issues. It includes activities such as stakeholder outreach, internal planning with Board leadership, and technical studies involving modeling and analysis. The planning time frame is defined as a future scenario described as "Build-Out," meaning 50 or more years in the future when the city reaches a sustained equilibrium population. The planning process sought to answer these specific questions:

- What is an acceptable level of risk in addressing future water demands?
- What is an appropriate approach for CSU to follow in meeting regional water demands within the Pikes Peak region?
- What role do different supply options contribute to achieving a balanced water supply portfolio?
- How do we ensure a proper level of investment in CSU's existing and future water system to maintain an acceptable level of risk?

As part of the planning process, CSU in collaboration with the University of Massachusetts and the National Center for Atmospheric Research, applied DS to investigate the effects of climate change on the reliability of their water supply deliveries. The analysis leveraged the planning processes that identified the key questions described above, as well as a cataloging of performance measures and possible investments. Much of this information was compiled through an earlier climate change study that incorporated aspects of Robust Decision Making. CSU maintain hydrologic models and water system models that were available for this analysis. This case study describes the application of DS to characterize the risk that climate change poses to the water supply reliability for Colorado Springs. Although the framework could be used to also evaluate alternative investments to address vulnerabilities, this was not part of the analysis.

In this case study, focus is given to a novel component of the analysis that relates to how we assess climate change uncertainty and integrate that evaluation with the vulnerability assessment to characterize climate-related risks to water supply. Specifically, we explore the processing of climate change projections, with particular attention to climate model similarity, and demonstrate how this issue can affect decisions related to adaptation measures taken by the water utilities.

12.3.1 Step 1: Decision Framing

A key principle of the DS framework is tailoring the analysis to address the primary concerns of planners and decisionmakers. In this case, in order to provide useful information to inform municipality-level adaptations for water supply, it is critical to understand the water supply objectives of the municipality, identify quantitative outcome indicators that can measure those objectives, determine thresholds for those objectives that would indicate unacceptable water supply performance, determine the context and scale of a municipality's water supply system, and identify the adaptation options available to the municipality. All of these components are necessary to adequately frame the future water supply risks facing a municipality, and the possible actions that can be taken to manage those risks.

In this case study, much of the decision framing had been established as part of the IWRP process. CSU established desired attributes for their strategy, including robustness to a wide variety of future conditions, economic sustainability (meaning that the strategy could be supported by available resources), reliability of water delivery services, and that the strategy was ultimately explainable and acceptable to customers and stakeholders. The analysis would look forward roughly 50 years (long term), and considered a "Build-Out" scenario as described above. It would also consider a wide range of possible investment options to meet its goals of reliable and sustainable water service. These included four different levels of demand management: multiple water reuse and non-potable water supply options, new agricultural transfers, construction of new reservoirs, and enlargement of existing reservoirs. An extended list of performance measures was also created for evaluation of current and future performance and for evaluation of the various options. For the purpose of this case study, reliability of water supply was used as an illustrative performance measure.

The spatial extent of the study included the entire water collection system (which extends to the continental divide in western Colorado) and the entire service area. The CSU water collection system serves an estimated 458,000 people, including the residents of Colorado Springs, the Ute Pass communities west of the city, and several military installations, including the United States Air Force Academy. Currently, the firm yield of potable and non-potable water for the system is about 187 million cubic meters per year, with potable deliveries at approximately 22 billion gallons per year. CSU estimates that they currently have enough water to meet the demand of their customers until approximately 2040, assuming average projections of population

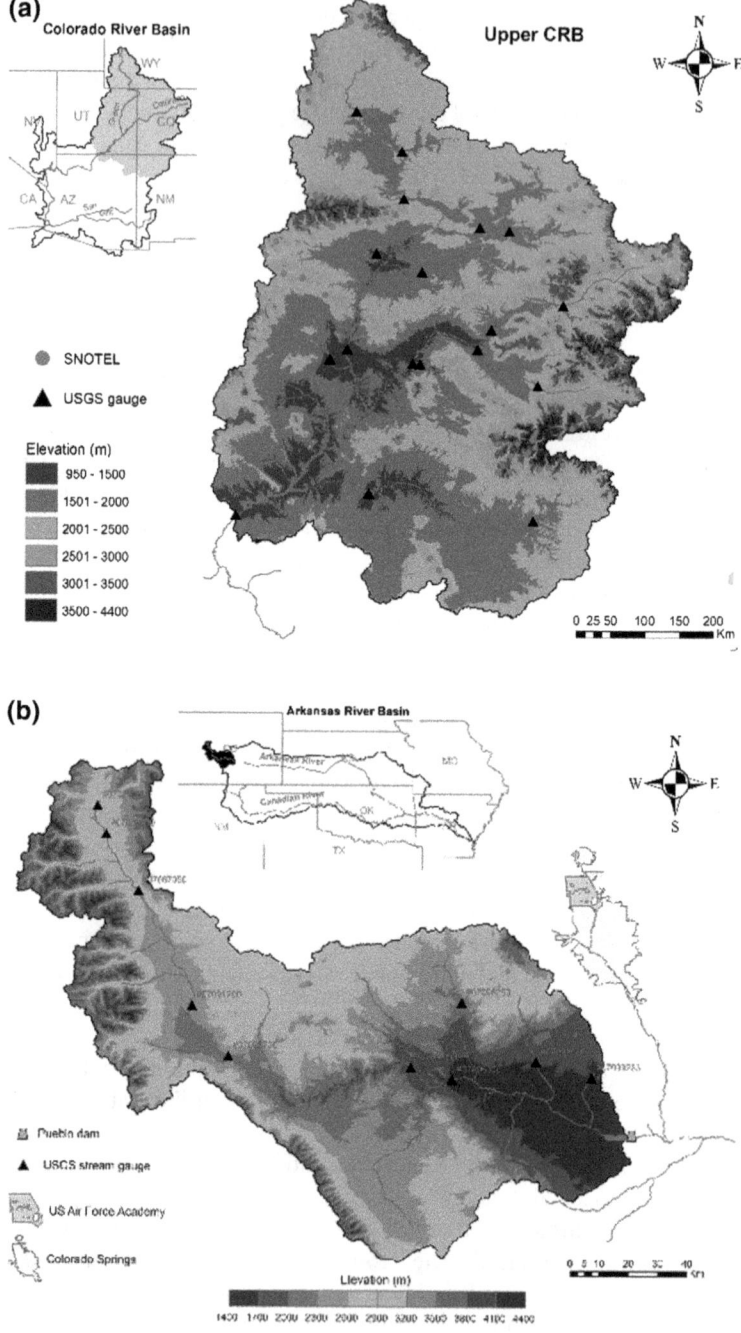

Fig. 12.4 Source water for the Colorado Springs Utilities water collection system, **a** map of the Upper Colorado River Basin, **b** map of the Upper Arkansas River Basin

growth, per capita water demand changes, and the completion of planned infrastructure projects. These assumptions do not account for any changes in climate, however, and beyond 2040 additional water demands also become a substantial concern.

The CSU water collection system acquires its water from two primary sources—the Upper Colorado River Basin (Fig. 12.4a) and the Arkansas River Basin (Fig. 12.4b). The Upper Colorado River Basin is one of the most developed and complicated water systems in the world. Waters from the Colorado serve people in seven states and Mexico, and are allocated according to a complex set of compacts and water rights provisions. CSU holds one such water right, albeit a junior right, and acquires approximately 70% of its annual water supply through four transmountain diversions that divert water across the continental divide into storage reservoirs operated by CSU. In this way, the water supply security of the CSU is directly linked to the broader water supply risks facing the entire Colorado River Basin. The remaining 30% of the CSU water supply is derived from local runoff in the Arkansas River Basin that must also be shared with users downstream of Colorado Springs. In order to assess the climate-related risks to the CSU water supply, both the Upper Colorado and Arkansas systems need to be accounted for in the analysis. This presents a substantial challenge, which required significant modeling efforts and persistent collaboration and communication with the CSU engineering team, and composed a substantial portion of the ongoing CSU Integrated Water Resources Planning process.

12.3.2 Step 2: Climate Stress Test

The goal of the climate stress test is to identify the vulnerabilities of the CSU water delivery system to climate change. Step 2 starts with a vulnerability assessment. In this assessment, CSU system performance is systematically tested over a wide range of annual mean climate changes to determine under what conditions the system no longer performs adequately. The approach is illustrated in Fig. 12.5. This stress test is driven using a daily stochastic weather generator that creates new sequences of climate that simultaneously exhibit different long-term mean conditions and alternative expressions of natural climate variability. These weather sequences are passed through a series of hydrosystem models of the relevant river basins and infrastructure network to estimate how these changes in climate will translate into altered water availability for the customers of CSU, including the Air Force Academy. The results of the stress test are summarized in a climate response surface, which provides a visual depiction of changes in critical system outcomes due to changes in the climate parameters altered in the sensitivity analysis. The different components of the vulnerability assessment are detailed below.

Stochastic Climate Generator

This work utilizes a stochastic weather generator (Steinschneider and Brown 2013) to produce the climate time series over which to conduct the vulnerability analysis. The weather generator couples a Markov Chain and K-nearest neighbor (KNN)

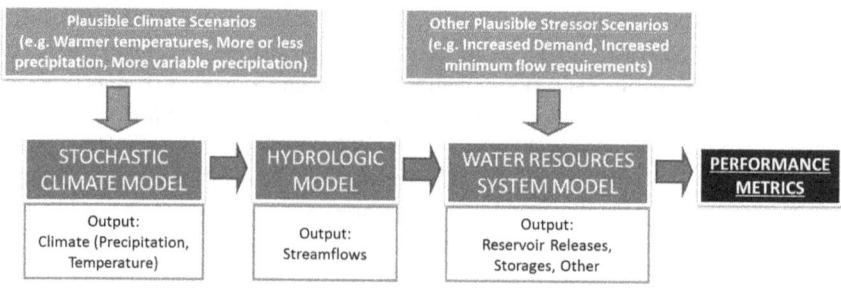

Fig. 12.5 Vulnerability assessment flow chart

resampling scheme to generate appropriately correlated multisite daily weather variables (Apipattanavis et al. 2007), with a wavelet autoregressive modeling (WARM) framework to preserve low-frequency variability at the annual time scale (Kwon et al. 2007). A quantile mapping technique is used to post-process simulations of precipitation and impose various distributional shifts under possible climate changes; temperature is changed using simple additive factors. The parameters of the model can be systematically changed to produce new sequences of weather variables that exhibit a wide range of characteristics, enabling detailed climate sensitivity analyses. The scenarios created by the weather generator are independent of any climate projections, allowing for a wide range of possible future climates to be generated. Furthermore, climate scenarios exhibiting the same mean climate changes can be stochastically generated many times to explore the effects of internal climate variability. The preservation of internal climate variability is particularly important for the CSU system, because precipitation in the region exhibits substantial decadal fluctuations that can significantly influence system performance (Nowak et al. 2012; Wise et al. 2015). The stochastic model is designed to reproduce this low-frequency quasi-oscillatory behavior; many downscaled climate projections often fail in this regard (Johnson et al. 2011; Rocheta et al. 2014; Tallaksen and Stahl 2014).

The WARM component of the weather generator was fit to annual precipitation data over the Upper Colorado River Basin. These data were provided by CSU, and are derived from the DAYMET database (Thornton et al. 2014). The WARM model is used to simulate time series of annual precipitation averaged over the Upper Colorado Basin, with appropriate inter-annual and decadal variability. A Markov Chain and KNN approach is then used to resample the historic daily data to synthesize new daily time series, with the resampling conditioned on the annual WARM simulation. The data are resampled for both the Upper Colorado and Arkansas River Basins to ensure consistency between the synthesized climate data across both regions. In this way, the major modes of inter-annual and decadal variability are preserved in the simulations, as is the daily spatio-temporal structure of climate data across both the Upper Colorado and Arkansas River Basins.

CSU HydroSystems Models

The climate scenarios from the weather generator are used to drive hydrosystem models that simulate hydrologic response, water availability and demand, and infrastructure operations in the river basins that provide water to CSU. The output of the hydrosystem model simulations under each climate time series is used to create a functional link between water supply risk and a set of mean climate conditions. The CSU system requires two separate hydrosystem models, because water is sourced from both the Upper Colorado River Basin on the western side of the continental divide and the headwaters of the Arkansas River on the eastern side of the divide. These models are described in more detail below.

Upper Colorado River Basin Hydrosystems Model

The Water Evaluation and Planning System (WEAP) model (Yates et al. 2005) was used to simulate the hydrologic response, reservoir operations, withdrawals, and transmountain diversions of the Upper Colorado River Basin system. By necessity, the WEAP model simplifies the extreme complexity of the Upper Colorado system, yet still requires nearly 40 minutes per 59-year (period-of-record) run on a standard desktop computer (HP Z210 Workstation with a 3.40 GHz processor and 18.0 GB of RAM). The WEAP model approximates how the climate scenarios developed above translate into changes in water availability for users throughout the Colorado River system. Importantly, the WEAP model of the Upper Colorado simulates the availability of water for transfer across four transmountain diversion points that feed into the CSU system. Changes in these diversions substantially alter the water available for CSU and its customers.

Upper Arkansas River Basin Hydrosystems Model

The transmountain diversions estimated by the WEAP model are used to force a MODSIM-DSS model (Labadie et al. 2000) that represents the Eastern slope waterworks system operated by CSU. In addition to these diversions, the MODSIM model requires additional inflow data to a variety of nodes. But these inflows cannot be modeled as natural hydrologic response to meteorological forcings, because there are legal constraints on the inflows not accounted for by MODSIM. Therefore, historical years of inflow data, which implicitly account for legal constraints, are resampled from the historic record in all future simulations. To ensure that these flows are correctly correlated with the Western slope simulations from WEAP, the stochastic weather generator is used to produce synthetic weather simultaneously across both Western slope and Eastern slope regions. Natural streamflow response from the Eastern slope system under synthetic climate is estimated using the hydrologic model in WEAP calibrated to naturalized flows in the Arkansas River. A nearest-neighbor resampling scheme is then used to resample historic years based on a comparison between historical, naturalized Arkansas River streamflow, and modeled hydrology of the Arkansas River under synthetic climate. Inflow to all MODSIM nodes besides those associated with Western slope diversions are then bootstrapped for use in future simulations based on the resampled years. One major drawback of this approach is the simplicity of the WEAP hydrologic model used to simulate natural flow in the Arkansas River.

Climate and Demand Alterations Considered

The weather generator described above is used to generate day-by-day, 59-year (period-of-record length) climate sequences with different mean temperature and precipitation conditions that maintain the historic decadal variability in the observed data (McCabe et al. 2004, 2007). To impose various climate changes in simulated weather time series, multiplicative (additive) factors are used to adjust all daily precipitation (temperature) values over the simulation period, thus altering their mean annual values. Annual changes are most important for the long-term planning purposes of the CSU system, because the significant reservoir storage on both sides of the continental divide largely mitigates the impact of seasonal changes to runoff and snowmelt timing, consistent with classical reservoir operations theory (Hazen 1914; Barnett et al. 2005; Connell-Buck et al. 2011). This does not preclude the importance of other hydrologic characteristics for long-term planning, such as the effects of climate extremes on flood reduction capacity or water quality, but these issues are not addressed in this study. Annual changes to the precipitation mean were varied from -10% to $+10\%$ of the historic mean using increments of 5% (5 scenarios altogether), while temperature shifts were varied from approximately -1 to $+4$ °C using increments of 0.5 °C (10 scenarios). These changes were chosen to ensure the identification of climate changes that cause system failure. Each one of the $50 = 5 \times 10$ scenarios of climate change is simulated with the weather generator seven times to partially account for the effects of internal climate variability while balancing the computational burden of the modeling chain, leading to a total of $350 = 5 \times 10 \times 7$ weather sequences.

The seven realizations of internal climate variability were selected in a collaborative process with the CSU engineering team as part of their Integrated Water Resources Planning process. The seven series were chosen among 10,000 original weather generator simulations to span the range of natural climate fluctuations that could influence the system. This selection proceeded in two steps. First, 40 simulations were selected from the original 10,000 to symmetrically span the empirical distribution of a precipitation-based drought outcome indicator preferred by the utility. Second, the subset of 40 simulations was run through the hydrosystem models (described above), and seven final simulations were chosen that spanned the empirical distribution of minimum total system reservoir storage across the 40 runs. Long-term climate changes were then imposed on these final seven climate simulations, and were used to force the hydrosystem models, producing a comprehensive vulnerability assessment that maps CSU system performance to long-term climate changes while also accounting for the effects of internal climate variability.

CSU considers two scenarios for water demands on their system. The first scenario is called the Status Quo. It reflects current water demands after accounting for the connection of several communities to the system's supply that was to be completed by 2016. The second demand scenario, referred to as "Build-Out Conditions," reflects a substantial increase in system demand. No date is associated with the water demands of the "Build-Out" scenario, but it is assumed that under current growth projections this level of demand will be reached around the year 2050. The stress test is repeated for both of these water demand scenarios to enable an assessment of the relative importance of climate and water demand changes on water resources vulnerability and risk.

12.3.3 Step 3. Estimation of Climate-Informed Risks

Over the past decade, the climate science community has proposed different techniques to develop probabilistic projections of climate change from ensemble climate model output. The most recent efforts (Groves et al. 2008; Manning et al. 2009; Hall et al. 2012; Christierson et al. 2012) for risk-based long-term planning have relied on climate probability density functions pdfs from perturbed physics ensembles (PPEs) (Murphy et al. 2004), a Bayesian treatment of multi-model ensembles (MMEs) (Tebaldi et al. 2005; Lopez et al. 2006; Smith et al. 2009; Tebaldi and Sanso 2009), or a combination thereof (Sexton et al. 2012). Of interest here, probabilities of change based on MMEs often assume each individual climate model in an MME serves as an independent representation of the Earth system. This assumption ignores the fact that many GCMs follow a common genealogy and supposes a greater effective number of data points than are actually available (Pennell and Reichler 2011). Following Masson and Knutti (2011) and Knutti et al. (2013), we consider models to share a common genealogy (or to be within the same family) if those models were developed at the same institution or if one is known to have borrowed a substantial amount of code from the other (e.g., the entire atmospheric model).

To address this issue, recent work has explored methods to optimally choose a subset of models to capture the information content of an ensemble (Evans et al. 2013) or to weight models based on the correlations in their error structure over a hindcast period (Bishop and Abramowitz 2013). This latter approach of independence-based weighting has recently shown promise in ensuring that observations and the ensemble of projections are more likely to be drawn from the same distribution, and consequently improve estimates of the ensemble mean and variance for climate variables of interest (Haughton et al. 2015). As noted in Haughton et al. (2015), improvements from independence-based weighting could provide substantial gains in projection accuracy and uncertainty quantification that may be relevant for informing adaptations to large climate changes.

For this case, we used an ensemble of projections from the Coupled Model Intercomparison Project Phase 5 (CMIP5) to develop probabilistic climate information, with and without an accounting of inter-model correlations for the river basins serving the CSU system, and use the pdfs to estimate mid-century climate-related risks to the water supply security of CSU. Climate change pdfs from both methods are coupled with the previously described vulnerability assessment of the CSU water resources system to estimate climate-related risks to water supply. The probability that climate change will lead to inadequate future performance is estimated by sampling 10,000

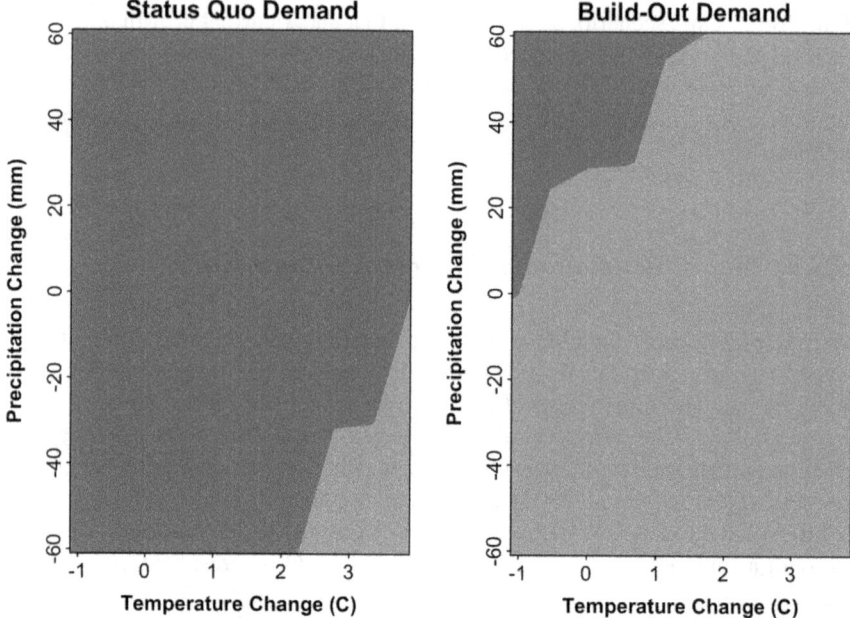

Fig. 12.6 Climate response surfaces for the CSU system based on regions of climate change space that have 100% indoor water supply reliability. Acceptable (blue) and unacceptable (red) regions of performance are highlighted for both the Status Quo and Build-Out demand scenarios (Color figure online)

samples of ΔT and ΔP using the pdfs from above and counting the fraction of samples that coincide with climate changes in the vulnerability assessment with indoor water demand shortfalls (Moody and Brown 2013).

Vulnerability Assessment Results

There are many outcome indicators that can be used to assess the performance of the CSU system, but for the screening purposed in the Integrated Water Resources Planning process, CSU initially wanted to focus on two measures: (1) storage reliability and (2) supply reliability. Storage reliability reflects the frequency that total system storage drops below a critical threshold set by the utility, while supply reliability represents the percentage of time that indoor water demands are met in a simulation. At the most basic level, system performance is considered adequate for a particular climate sequence if indoor water demands are met for the entire simulation (i.e., 100% supply reliability), since any drop in supply reliability suggests that the tap runs dry for some customers, which is considered unacceptable. We note that indoor water demand is a representative rather than encompassing performance measure, but will be the outcome we focus on in this assessment.

We first present the results of the vulnerability assessment without any consideration of climate model output. Figure 12.6 shows the climate response surface of the

CSU system to changes in mean precipitation and temperature under the Status Quo and Build-Out demand scenarios. The response surfaces, developed without the use of any projection-based data, show the mean precipitation and temperature conditions under which the utility can provide adequate water services, and those climate conditions under which the reliability of their service falls below an acceptable level.

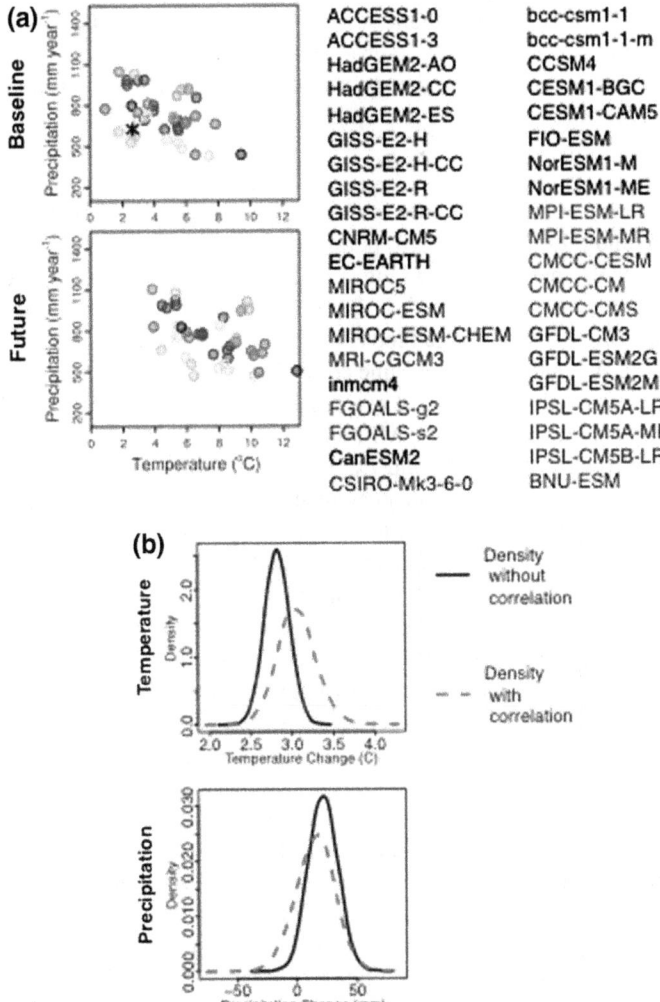

Fig. 12.7 a Scatterplot of mean temperature and precipitation over the Upper Colorado and Arkansas River Basin area from GCMs for a baseline (1975–2004) and future (2040–2070) period. The different models are colored according to their associated "families", b Marginal probability density functions for mid-century temperature and precipitation change across the Upper Colorado and Arkansas regions developed with (red-dashed) and without (black solid) accounting for intra-family model correlations (Color figure online)

Here, we define unacceptable performance as an inability to meet indoor municipal water demands. For the Status Quo system, the response surface suggests that the system can effectively manage moderately increasing temperatures and declining precipitation, but large changes beyond +2.2 °C, coupled with declining precipitation, will cause the system to fail. For the Build-Out demand scenario, the current system cannot adequately deliver water even under baseline climate conditions (no changes in temperature and precipitation), let alone reduced precipitation or increased temperatures. These results highlight that the CSU system is at risk of water supply shortages simply due to the expected growth of water demands over the next several decades. These risks grow when the specter of climate change is considered, which is considered next.

Likelihood of Future Climate Changes

Regional mean annual temperature and accumulated precipitation for a baseline (1975–2004) and future (2040–2070) period averaged over the Upper Colorado and Arkansas River Basins are shown in Fig. 12.7a for the Representative Concentration Pathway (RCP) (Meinshausen et al. 2011) 8.5 scenario. Models that originate from the same institution or share large blocks of code are grouped into the families used in this analysis and denoted by the same color (Knutti et al. 2013). For example, the models within the NCAR family all use key elements of the CCSM/CESM model developed at NCAR. Likewise, the MPI and CMCC models are combined because they are based on the ECHAM6 and ECHAM5 atmospheric models, respectively. Being in the same family does not guarantee that regional climate characteristics of related models will cluster, but global clustering analyses (Masson and Knutti 2011; Knutti et al. 2013) suggest an increased likelihood of clustering even on small regional scales, a hypothesis which we test here.

By visual inspection of Fig. 12.7a, there is nontrivial clustering in both temperature and precipitation among models belonging to the same family. A formal hierarchical clustering of the baseline and future climatology (not shown) confirms the tendency of models within the same family to cluster with respect to simulated regional precipitation and temperature averages. The degree to which clustering occurs within each family depends on the model family being considered (GISS and HadGEM cluster well, IPSL/CMCC less so). There is a tendency for models with very similar atmospheric structures to cluster, even if other components are different.

Probability models are fit to the climate model data with and without an accounting of the correlation among individual models within a family. The estimate of intra-family model correlation for both annual mean temperature and precipitation are statistically different from zero at the 0.05 significance level. Figure 12.7b shows pdfs of annual mean temperature and precipitation change with and without an accounting of within-family correlation. When inter-model correlations are included, an increase in variance is clear for both variables due to increased sampling uncertainty associated with a reduction in the effective number of data points. Beyond the increased variance of projected climate changes, high inter-model correlations also shift the mean climate change estimate, since entire families of models are no longer regarded as independent data points, allowing, for example, centers with only a single model to assume more weight in the calculation.

The pdfs of regional climate change can be used to determine the risk posed to the CSU water system. Figure 12.8 shows the climate response surface for the Colorado water utility under 2016 demand conditions presented previously, but with the bivariate pdfs of annual temperature and precipitation change with and without intra-family correlations superimposed. We do not show the Build-Out demand conditions because system performance is unacceptable under that scenario even without climate change. The degree to which the pdfs extend into the region of unacceptable system performance in Fig. 12.8 describes the risk that the water utility may face from climate change. Visually, it is clear that the tails of the pdf developed with an accounting of intra-family model correlations extend into the region of unacceptable system performance, while those of the pdf with an independence assumption do not. A climate robustness metric is used to summarize the risk by numerically integrating the pdf mass in the region of unacceptable performance. For the pdf that does not account for model correlation, essentially 0% of its mass falls into the region of unacceptable performance. When correlations are accounted for, the metric increases to 0.7%. While still small, this non-negligible probability (similar in magnitude to a 100-year event) is important because of the intolerable impact that such shortfalls would have on the local community. Any nontrivial probability that indoor water use will have to be forcibly curtailed would motivate the water utility to

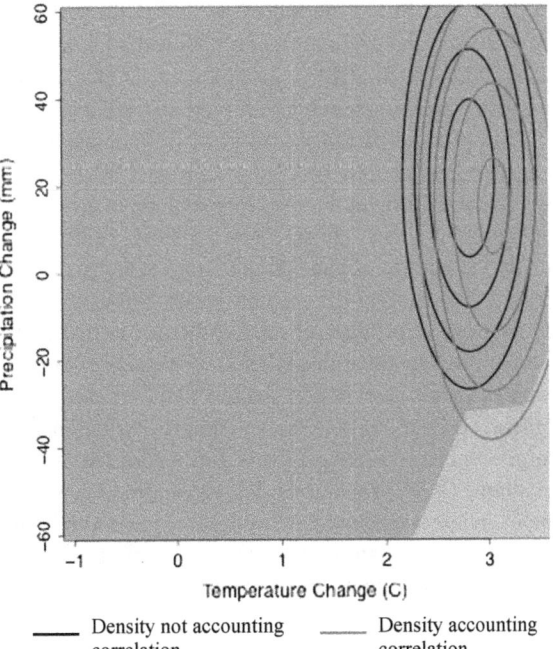

Fig. 12.8 Climate response surface displaying the conditions of mean precipitation and temperature under which CSU can (blue) and cannot (red) provide reliable indoor drinking water. Bivariate pdfs of mean temperature and precipitation are superimposed on the response surface. The red pdf was developed with an accounting of intragroup model correlation, while the black pdf was not. Both pdfs are contoured at the same levels, with the final level equal to 1.5×10^3 (Color figure online)

invest in measures to prevent such an outcome. Thus, there is an important increase in decision-relevant, climate-change-related risk facing the water utility when we alter our interpretation of the information content present in the model ensemble.

12.4 Conclusions

DS leverages traditional decision analytic methods to create a decision analysis approach specifically designed for the treatment of climate change uncertainty, while also incorporating other uncertain and deeply uncertain factors. A distinctive attribute of the approach is the use of a climate stress-testing algorithm, or stochastic weather sampler, which produces an unbiased estimation of the response of the system of interest to climate change. This avoids the numerous difficulties that are introduced when climate projections from climate models are used to drive an analysis. Instead, the information from climate projections can be introduced after the climate response is understood, in order to provide an indication of whether problematic climate changes are more or less likely than non-problematic climate changes. The approach is effective for estimating climate risks and evaluating alternative adaptation strategies. It can be incorporated into typical collaborative stakeholder processes, such as described in Poff et al. (2015). DS has been applied in a number of cases around the world, including adaptation planning for the Great Lakes of North America (Moody and Brown 2012), evaluation of climate risks to US military installations (Steinschneider et al. 2015a, b), evaluation of water supply systems (Brown et al. 2012; Whateley et al. 2014), and evaluation of long-lived infrastructure investments (e.g., Ghile et al. 2014; Yang et al. 2014), among others. It also serves as the basis for a climate risk assessment process widely adopted at the World Bank (Ray and Brown 2015).

The process of DS is designed to generate insightful guidance from the often confusing and conflicting set of climate information available to decisionmakers. It generates information that is relevant and tailored to the key concerns and objectives at hand. The process bridges the gap in methodology between top-down and bottom-up approaches to climate change impact assessment. It uses the insights that emerge from a stakeholder-driven bottom-up analysis to improve the processing of GCM projections to produce climate information that improves decisions. The process is best applied to situations where the impacts of climate change can be quantified and where models exist or can be created to represent the impacted systems or decisions. Historical data are necessary. The process can be applied both in conjunction with large climate modeling efforts and where the analysis depends simply on globally available GCM projections. It is most effective when conducted with strong engagement and interaction with the decisionmakers and stakeholders of the planning effort. The transparent nature of the process attempts to make the analysis accessible to non-technical participants, but having some participants with technical backgrounds is beneficial.

DS was one activity in the large set of activities that comprised the CSU IWRP development process. Through the approach, the climatic conditions that caused the system to not be able to meet their performance objectives were identified. Indeed, these conditions were identified in a way that preserves the findings from the usual uncertainty associated with projecting future climate conditions. In this case, it was found that the current system is very robust to climate changes when considering the current water demand. The problematic climate conditions were a precipitation reduction of greater than 20% from the long-term average when accompanied by a mean temperature increase of 3 °C or more. Such extreme changes were considered to have a low level of concern, as is discussed below. However, for the "Build-Out" scenario, which includes an increase in water demand, the climate conditions that cause vulnerability were much wider. In this case, the system is not able to meet their performance objectives unless precipitation increases substantially and mean temperatures do not increase more than 2 °C. This scenario was considered to have a much higher level of concern.

The level of concern associated with these vulnerability scenarios could be estimated from discussion with experts on Colorado climate change, derived from an ensemble of skillful climate projections, or (preferably) discussion with experts that are informed by skillful climate projections. In this case, careful attention was given to producing the most meaningful climate information available from an ensemble of climate projections. For example, the skill of the GCMs as indicated by past performance, and the fact that many GCMs are not truly independent, were accounted for in our estimation of probabilities from the ensemble. These probabilities are not intended to represent the known probabilities of future climate change for the region. Rather, they are a quantitative representation of the information contained in the ensemble of projections. In this case, the results show that the projections indicate a very low probability of the problematic scenario identified for current demand. On the other hand, they also indicate substantial probability for the problematic conditions identified for the future Build-Out demand.

What may be required are new tools that can effectively approximate and propagate the uncertainty at each stage of the modeling chain with greater efficiency to reduce the computational burden. This is an important avenue of future work to emerge from this research.

References

Apipattanavis, S., Podesta, G., Rajagopalan, B., & Katz, R. W. (2007). A semiparametric multivariate and multisite weather generator. *Water Resources Research, 43*, W11401. https://doi.org/10.1029/2006WR005714.

Barnett, T. P., Adam, J.C., & Lettenmaier, D. P. (2005). Potential impacts of a warming climate on water availability in snow-dominated regions. *Nature, 438*, 303–309. https://doi.org/10.1038/nature04141.

Bishop, C., & Abramowitz, G. (2013) Climate model dependence and the replicate Earth paradigm. *Climate Dynamics, 41*(3–4), 885–900.

Brown, C., Ghile, Y., Laverty, M., & Li, K. (2012). Decision scaling: linking bottom-up vulnerability analysis with climate projections in the water sector. *Water Resources Research, 48,* W09537. https://doi.org/10.1029/2011wr011212.

Christierson, B. V., Vidal, J., & Wade, S. D. (2012). Using UKCP09 probabilistic climate information for UK water resource planning. *Journal of Hydrology, 424,* 48–67.

Connell-Buck, C. R., Medellin-Azuara, J., Lund, J. R., & Madani K. (2011). Adapting California's water system to warm vs. dry climates. *Climate Change, 109,* 133–149.

Evans, J. P., Ji, F., Abramowitz, G., & Ekstrom, M. (2013). Optimally choosing small ensemble members to produce robust climate simulations. *Environmental Research Letters, 8*(4). https://doi.org/10.1088/1748-9326/8/4/044050.

Ghile, Y. B., Taner, M. U., Brown C. M., & Grijsen, J. G. (2014). Bottom-up climate risk assessment of infrastructure investment in the Niger River Basin. *Climatic Change, 122*(1–2), 97–111.

Gleckler, P. J., Taylor, K. E., & Doutriaux, C. (2008). Performance metrics for climate models. *Journal Geophysical Research, 113,* D06114.

Goddard, L., Aitchellouche, Y., Baethgen, W., Dettinger, M., Graham, R., Hayman, P., et al. (2010). Providing seasonal-to-interannual climate information for risk management and decision-making. *Procedia Environmental Sciences, 1,* 81–101.

Groves, D. G., & Lempert, R. J. (2007). A new analytic method for finding policy-relevant scenarios. *Global Environmental Change, 17,* 73–85.

Groves, D. G., Yates, D., & Tebaldi, C. (2008). Developing and applying uncertain global climate change projections for regional water management planning. *Water Resources Research, 44.*

Hall, J., et al. (2012). Towards risk-based water resources planning in England and Wales under a changing climate. *Water and Environment Journal, 26,* 118–129.

Haughton, N., Abramowitz, G., Pitman, A., & Phipps, S. J. (2015). Weighting climate model ensembles for mean and variance estimates, Climate Dynamics. https://doi.org/10.1007/s00382-015-2531-3.

Hazen, A. (1914). Storage to be provided in impounding reservoirs for municipal water supply. *Transactions of the American Association of Civil Engineers, 77,* 1539–1669.

Johnson, F., Westra, S., Sharma, A., Pitman, A. J. (2011). An assessment of GCM skills in simulating persistence across multiple scales. *Journal of Climate, 24*(14), 3609–3623.

Knutti, R., Furrer, R., Tebaldi, C., Cermak, J., & Meehl, G. A. (2010). Challenges in combining projections from multiple climate models. *Journal of Climate, 23,* 2739–2758.

Knutti, R., Masson, D., & Gettelman, A. (2013). Climate model genealogy: Generation CMIP5 and how we got there. *Geophysical Research Letters, 40,* 1194–1199.

Kwon, H.-H., Lall, U., & Khalil A. F. (2007). Stochastic simulation model for nonstationary time series using an autoregressive wavelet decomposition: Applications to rainfall and temperature. *Water Resources Research, 43,* W05407. https://doi.org/10.1029/2006WR005258.

Labadie, J., Baldo, M., & Larson, R. (2000). *MODSIM: Decision support system for river basin management: Documentation and user manual.* Ft. Collins, CO: Dept. of Civil Eng., Colo. State Univ.

Lopez, A., et al. (2006). Two approaches to quantifying uncertainty in global temperature changes. *Journal of Climate, 19,* 4785–4796.

Manning, L., Hall, J., Fowler, H., Kilsby, C., & Tebaldi, C. Using probabilistic climate change information from a multimodel ensemble for water resources assessment. *Water Resources Research, 45.*

Masson, D., & Knutti, R. (2011). Climate model genealogy. *Geophysical Research Letters, 38.*

McCabe, G. J., Betancourt, J. L., & Hidalgo, H. G. (2007). Associations of decadal to multi-decadal sea-surface temperature variability with upper colorado river flow. *JAWRA Journal of the American Water Resources Association, 43,* 183–192. https://doi.org/10.1111/j.1752-1688.2007.00015.

McCabe, G. J., Palecki, M., & Betacourt, J. (2004). Pacific and Atlantic Ocean influences on multidecadal drought frequency in the United States. *Proceedings of the National Academy of Sciences of the United States of America, 101*(12), 4136–4141.

Meinshausen, M., et al. (2011). The RCP Greenhouse Gas Concentrations and their extensions from 1765 to 2300. *Climatic Change, 109*, 213–241.

Moody, P., & Brown, C. (2012). Modeling stakeholder-defined climate risk on the Upper Great Lakes. *Water Resources Research, 48*, W10524. https://doi.org/10.1029/2012WR012497.

Moody, P., & Brown, C. (2013). Robustness indicators for evaluation under climate change: Application to the upper Great Lakes. *Water Resources Research, 49*, 357. https://doi.org/10.1002/wrcr.20228.

Murphy, J. M., Sexton, D. M. H., Barnett, D. N., Jones, G. S., Webb, M. J., Collins, M., & Stainforth D. A. (2004, August). Quantification of modelling uncertainties in a large ensemble of climate change simulations. *Nature, 430*, 12.

Nowak, K., Hoerling, M., Rajagopalan, B., & Zagona, E. (2012). Colorado river basin hydroclimatic variability. *Journal of Climate, 25*, 4389–4403. http://dx.doi.org/10.1175/JCLI-D-11-00406.1

Pennell, C., & Reichler, T. (2011). On the effective number of climate models. *Journal of Climate, 24*, 2358–2367.

Poff, L., Brown, C., et al. (2015). Eco-engineering decision scaling for sustainable water management under future hydrologic uncertainty. *Nature Climate Change, 6*, 25. https://doi.org/10.1038/nclimate2765.

Ray, P., & Brown, C. (2015). *Confronting climate uncertainty in water resources planning and project design—The decision tree framework* (p. 128). Washington, DC.: World Bank Group Press.

Rocheta, E., Sugiyanto, M., Johnson, F., Evans, J., & Sharma, A. (2014). How well do general circulation models represent low-frequency rainfall variability? *Water Resources Research, 50*, 2108–2123.

Schlaifer, R., & Raiffa, H. (1961). *Applied statistical decision theory* (p. 356). Boston: Clinton Press Inc.

Sexton, D. M. H., Murphy, J. M., Collins, M., & Webb M. J. (2012). Multivariate probability projections using imperfect climate models part I: Outline of methodology. *Climate Dynamics, 38* (11–12), 2513–2542.

Smith, R. L., Tebaldi, C., Nychka, D., & Mearns, L. O. (2009). Bayesian modeling of uncertainty in ensembles of climate models. *Journal of the American Statistical Association, 104*, 97–116.

Stainforth, D. A., et al. (2007). Issues in the interpretation of climate model ensembles to inform decisions. *Philosophical Transactions of the Royal Society of London A, 365*, 2163–2177.

Steinschneider, S., & Brown, C. (2013). A semiparametric multivariate, multi-site weather generator with low-frequency variability for use in climate risk assessments. *Water Resources Research, 49*, 7205.

Steinschneider, S., McCrary, R., Mearns, L., & Brown, C. (2015a). The effects of climate model similarity on probabilistic climate projections and the implications for local, risk-based adaptation planning. *Geophysical Research Letters, 42*, 5014–5022. https://doi.org/10.1002/2015GL064529.

Steinschneider, S., McCrary, R., Wi, S., Mulligan, K., Mearns, L. O., & Brown, C. (2015b). Expanded decision-scaling framework to select robust long-term water-system plans under hydroclimatic uncertainties. *Journal of Water Resources Planning and Management, 141*, 04015023.

Tallaksen, L. M., & Stahl K. (2014). Spatial and temporal patterns of large-scale droughts in Europe: Model dispersion and performance. *Geophysical Research Letters, 41*, 429–434. https://doi.org/10.1002/2013GL058573.

Tebaldi, C., & Sansó, B. (2009). Joint projections of temperature and precipitation change from multiple climate models: A hierarchical Bayesian approach. *Journal of the Royal Statistical Society: Series A (Statistics in Society), 172*, 83–106

Tebaldi, C., Smith, R. L., Nychka, D., & Mearns, L. O. (2005). Quantifying uncertainty in projections of regional climate change: A Bayesian approach to the analysis of multimodel ensembles. *Journal of Climate, 18*, 1524–1540.

Thornton, P. E., Thornton, M. M., Mayer, B. W., Wilhelmi, N., Wei, Y., Devarakonda, R., & Cook, R. B. (2014). Daymet: Daily surface weather data on a 1-km grid for North America, Version 2.

ORNL DAAC, Oak Ridge, Tennessee, USA. Accessed July 07, 2013. http://dx.doi.org/10.3334/ORNLDAAC/1219

Whateley, S., Steinschneider, S., & Brown, C. (2014). A climate change range-based method for estimating robustness for water resources supply. *Water Resources Research, 50,* 8944. https://doi.org/10.1002/2014wr015956.

Winston, W. L., & Goldberg, J. B. (2004). *Operations research: Applications and algorithms* (Vol. 3). Belmont: Thomson Brooks/Cole.

Wise, E. K., Wrzesien, M. L., Dannenberg, M. P., & McGinnis, D. L. (2015). Cool-season precipitation patterns associated with teleconnection interactions in the United States. *Journal of Applied Meteorology and Climatology, 54,* 494–505. http://dx.doi.org/10.1175/JAMC-D-14-0040.1

Yang, Y.-C., Brown, C., Yu, W., Wescoat, J., & Ringler, C. (2014). Water governance and adaptation to climate change in the Indus River Basin. *Journal of Hydrology, 519,* 2527. https://doi.org/10.1016/j.jhydrol.2014.08.055.

Yates, D., Sieber, J., Purkey, D., & Huber-Lee, A. (2005). WEAP21—A demand-, priority-, and preference-driven water planning model: part 1: model characteristics. *Water International, 30,* 487–500.

Dr. Casey Brown is an internationally recognized expert in water resources systems analysis and climate risk assessment. He is Professor of Civil and Environmental Engineering at the University of Massachusetts at Amherst and Adjunct Associate Research Scientist at Columbia University. He has a Ph.D. in Environmental Engineering from Harvard University and headed the water team at the International Research Institute (IRI) for Climate and Society at Columbia University. His research focuses on the resilience of infrastructure systems, especially the effects of climate variability and change, and water resources, and is funded by NSF, Rockefeller Foundation, NOAA, DoD, WRF, among others. He consults for the World Bank, the private sector, state agencies, and municipalities. He has a number of awards to his credit, including the Presidential Early Career Award for Science and Engineering, the National Science Foundation CAREER award, the Huber Research Prize from the American Society of Civil Engineers, and the Climate Science Award from the California Department of Water Resources.

Dr. Scott Steinschneider is an Assistant Professor in the Department of Biological and Environmental Engineering at Cornell University. His primary expertise is in statistical hydroclimatology for decision support in water resources systems. His work has focused on water systems across the USA and globally and has been sponsored by the .S Army Corps of Engineers, National Atmospheric and Oceanic Administration, and National Science Foundation. He earned his B.A. in Mathematics from Tufts University and his M.S. and Ph.D. in Civil and Environmental Engineering from the University of Massachusetts, Amherst. Prior to arriving at Cornell, he was a postdoctoral research fellow at the Columbia Water Center at Columbia University.

Dr. Patrick Ray is an Assistant Professor of Environmental Engineering at the University of Cincinnati, and a member of the Hydrosystems Research Group at the University of Massachusetts, Amherst. His research focuses on increasing the resilience of water systems to climate variability and change through the use of advanced climate science and coupled hydrologic-human system simulation, in combination with innovative water resources management techniques and methods for decision making under uncertainty. His Ph.D. is from Tufts University, and he studied as a postdoc at the University of Massachusetts, Amherst. His academic work has been supported by grants and awards from the National Science Foundation, Fulbright, the World Bank, Army Corps of Engineers, National Center for Atmospheric Research, and California Department of Water Resources, among others. Because of his role in the development of the World Bank's Decision Tree for Confronting Climate Uncertainty, he shared in the World Bank's 2015 KNOWbel Prize for 'Understanding the Impact of Climate Change and Other Risks on Hydropower,' and in 2018 he was named the A. Ivan Johnson Outstanding Young Professional of the American Water Resources Association.

Dr. Sungwook Wi is a Research Assistant Professor in the Department of Civil and Environmental Engineering at the University of Massachusetts Amherst and a Chief Hydrologist at the UMASS Hydrosystems Research Group. His research focuses on the intersection between hydrologic, climatic, and anthropogenic systems with an emphasis on sustainable water resources management. He specializes in developing human-hydrologic systems models and applying the models to assess the impact of climate change and variability as well as human activities on water resources planning and management. His expertise in hydrology has played a critical role in addressing water issues for various watershed systems all over the world, including USA, Mexico, Africa, East Asia, and the Himalayas.

Dr. Leon Basdekas has 25 years of experience in a wide range of water resources engineering projects working in federal government, academic research, private consulting, and municipal government. His experiences include from field engineering, water quantity and quality modeling, multi-objective optimization, and systems analysis. He currently works as an Integrated Water Supply Planning lead for Black and Veatch, where his work includes research projects under the NOAA SARP program and the Water Research Foundation. He earned a B.S. in Geology from the University of Maryland, a B.S. in Civil Engineering from Northern Arizona University, and a Ph.D. in Civil and Environmental Engineering from Utah State University. He also has a graduate school faculty appointment at the University of Colorado at Boulder, where he has advised Ph.D. and Masters-level students in the area of water resources management using advanced planning techniques.

Dr. David Yates is a Scientist in the Research Applications Laboratory at the National Center for Atmospheric Research and an Associate of the Stockholm Environment Institute's US Center. His research has focused both on local-scale hydrologic problems (flash floods, land use-land cover, climate change), as well as climate change impacts, adaptation, and mitigation strategies in the water, agriculture, and energy sectors. Dr. Yates has been a part of the development team of SEI's Water Evaluation and Planning (WEAP) system that seamlessly couples watershed hydrology and water and agricultural systems; and the Long-range Energy Alternatives Planning system (LEAP), which is used to explore energy policy analysis and climate change mitigation assessment strategies. He has trained university, NGO, and government agencies around the world on water management, climate change, and the application of WEAP and LEAP.

Open Access This chapter is licensed under the terms of the Creative Commons Attribution 4.0 International License (http://creativecommons.org/licenses/by/4.0/), which permits use, sharing, adaptation, distribution and reproduction in any medium or format, as long as you give appropriate credit to the original author(s) and the source, provide a link to the Creative Commons licence and indicate if changes were made.

The images or other third party material in this chapter are included in the chapter's Creative Commons licence, unless indicated otherwise in a credit line to the material. If material is not included in the chapter's Creative Commons licence and your intended use is not permitted by statutory regulation or exceeds the permitted use, you will need to obtain permission directly from the copyright holder.

Chapter 13
A Conceptual Model of Planned Adaptation (PA)

Jesse Sowell

Abstract

- Technical and operational systems, especially large complex socio-technical systems, must continuously adapt to a changing context.
- Conventional rulemaking systems have, historically, been ponderous. Low clockspeed rulemaking systems are rarely equipped to keep pace when applied to govern higher clockspeed socio-technical systems.
- This chapter presents and characterizes four organizational complexes that have effectively implemented 'planned adaptation'.
- Based on these cases, an initial general model of planned adaptation is presented: first, in terms of the first principles of where in the rulemaking system adaptation may be introduced; second, in terms of categories of events and triggers; and finally, in terms of the evaluative capabilities and capacities within and across organizations necessary to systematically plan to adapt.
- This general model (which can more generally be termed *systematic adaptation*) characterizes ad hoc adaptation, *planned* adaptation, as well as *ideal forms* of adaptation. These are used as comparators in what is referred to here as the 'systematic adaptation space'.
- Discussion of four cases based on this model highlight instances of critical paths from ad hoc to *planned* adaptation, offering early insights that can lead to effectively designing organizational complexes that can effectively plan to adapt.
- The chapter concludes with a discussion of ongoing work—in particular, how this model will be used for theory-based sampling to identify cases of adaptation that can be used to further refine the model and its application to systematic adaptation by design.

J. Sowell (✉)
Department of International Affairs, The Bush School of Government & Public Service, Texas A&M University, College Station, TX, USA
e-mail: jsowell@tamu.edu

© The Author(s) 2019
V. A. W. J. Marchau et al. (eds.), *Decision Making under Deep Uncertainty*,
https://doi.org/10.1007/978-3-030-05252-2_13

13.1 Introduction

Historically, rulemaking has been a ponderous process. In contemporary societies, rulemaking must cope with combinations of uncertainties that not only arise when dealing with the natural environment, but also those uncertainties that arise from the built, technical environment. Rules—more accurately *rulemaking systems*—must be able to keep pace, incorporating credible knowledge assessments from relevant experts apace with the impact of technology on industry and society. In effect, rulemaking systems must be able to adapt. Rulemaking does not happen in a vacuum; rulemaking happens in organizations with vastly different resources, processes, capabilities, and capacities. The work described in this chapter builds on the existing literature on planned adaptation to offer a conceptual model, based on empirical cases, of the organizational apparatus necessary to execute adaptive policy recommendations.

The literature on decision analysis and its progeny—scenario planning, adaptive policymaking, Robust Decision Making, strategic options—have been the bread and butter of policy analysts since the 1950s. The notion of planned adaptation is a new and understudied aspect of rulemaking. The term was established by McCray et al. (2010), characterizing programs that "both revise rules when relevant new knowledge appears, and take steps to produce such improved knowledge".

In contrast to work on decisionmaking in general and decisionmaking under deep uncertainty in particular, the object of planned adaptation is the character and context of the programs acting on new knowledge, *not* the underlying policy analysis. This is a subtle, but potentially important conceptual distinction, between the substance of the knowledge that comprises policy *recommendations* and the elements of the organization that must implement that recommendation. (This distinction is made clear in Chap. 14, which describes the implementation of Adaptive Delta Management in the Netherlands.)

In this chapter, McCray's notion of planned adaptation is elaborated to explicitly characterize the apparatus necessary to systematically *plan for* adaptation. *To plan* focuses on resources and processes necessary to support the development of, recognize, and/or act on new knowledge:

1. What rules and mechanisms are in place to act on, elicit, consume, or refine policy-relevant information?
2. Who generates this new knowledge?
3. What is the organizational relationship of those generating new knowledge to those adapting the rules?

The process of adaptation comprises the set of resources, both economic and political, necessary to engender the recommended change. Taken together, planned adaptation speaks not only to the act of revising, but elaborates the permutations of organizational configurations necessary to plan for and implement stable, sustainable, systematic adaptation.

Studies of decisionmaking under uncertainty provide a rich analytic framework for evaluating information gaps, but rarely speak to the organizational and political context in which triggers for evaluative action occur. To understand this context, concepts from the theory of rulemaking (Hart 1994) are used as a common conceptual vernacular for understanding adaptation in terms of the rules supporting that process. Hart's categories of rules facilitate disentangling substantive (primary) rules governing the *behavior* of a system from the categories of rules (secondary) used to *identify, evaluate, and change* rules when the rule system no longer effectively sustains system integrity. In general, primary rules represent the current understanding of how to govern the system—in effect a snapshot of a static policy analysis. Secondary rules describe (1) the evaluative capabilities necessary to transform new information about the system into prescriptive knowledge about the (mis)alignment of primary rules with system function and (2) the rules for identifying, creating, and maintaining the capacity necessary to change both primary and secondary rules as necessary, based on prescriptive knowledge.

Moving from abstractions to empirical mechanisms, based on the cases presented in this chapter, an initial set of variables is established for comparing instances of planned adaptation in terms of how secondary capabilities and capacities are used to adapt primary rules. Variables identified in this model characterize the factors that affect the development, implementation, and application of secondary rules. Variables fall into two categories: those that characterize sources of new information and those that characterize the rough organizational structure(s) supporting evaluative capabilities and capacities. Sources of new information are characterized in terms of *triggers* that signal that adaptation may be warranted, and the character of *events* producing the new information. Evaluative variables characterize the *timing* of evaluation relative to triggers and events; the *loci* of evaluative capabilities and capacities in the organizational complex; and how *coupled* the rulemaking principals are to the evaluative agents.

This chapter argues that the four cases presented here offer sufficient diversity in triggers, event context, and organizational configurations to lay the conceptual foundations for an initial model of systematic adaptation. In terms of the case studies presented here, these variables provide the foundation for concisely articulating adaptive capabilities and capabilities of empirical organizational complexes. The model also admits "ideal" adaptive organizational complexes, such as Weber's notion of a high-functioning bureaucracy. Taken together, this model represents the "systematic adaptation space". Mapping these variables facilitates comparing instances with little to no adaptive capabilities, ad hoc adaptation, and planned adaptation. More pragmatically, the analysis argues that, with additional cases, this model could provide the foundation for intentionally designing adaptive organizations.

Section 13.2 summarizes cases on delta management, particulate matter, air transportation safety, and Internet number delegation to provide comparative instances of organizations that have not only planned to adapt, but for which there are empirical instances of successful implementation of the adaptation. In Sect. 13.3, concepts from adaptive policymaking—in particular the character of triggers that signal an adaptation should take place—are synthesized with concepts from the theory of rulemaking to contextualize the factors affecting resource allocation necessary to plan for adaptation. In Sect. 13.4, the variables identified are used to analytically compare these cases and to identify ideal forms, effectively elaborating the range of configurations and contexts that characterize the systematic adaptation space. This systematic adaptation space provides an initial set of bounds that explicitly distinguish organizations with the capabilities and capacity to systematically and intentionally plan for adaptation from those that adapt ad hoc. Given these bounds, the chapter concludes by refining the notion of a systematically adaptive organization, offers a nascent framework for designing such an organizational configuration, and presents a sketch of future studies on planned adaptation.

13.2 Planned Adaptation Cases

Cases of successful, planned, systematic adaptation are not as common as one would hope. Four cases are discussed as the foundations of a conceptual model of systematic adaptation. In contrast to McCray's work, the conceptual unit of analysis in these cases is the rulemaking system, comprising one or more organizations collaborating to plan for adaptation. This unit of analysis includes state-based *and* non-state organizational complexes. The three state-based cases are: (1) a comparison of particulate matter standards development in the European Union (EU) and USA (Sect. 13.2.1); (2) flood risk safety in the Netherlands; and (Sect. 13.2.2) (3) civil air transportation safety standards development in the US (Sect. 13.2.3). The non-state case describes the private transnational regime that manages Internet protocol (IP) address delegation (Sect. 13.2.4).

Across these cases, key variables in systemic adaptation are introduced:

1. The degree of coupling of adaptive processes with substantive rules,
2. The provenance of events requiring change,
3. The timing of triggers signaling the potential need for policy evaluation and adaptation,
4. The loci of monitoring and evaluative capabilities, and
5. Social, political, and organizational incentives for investing in monitoring and evaluative resources.

In the cases, these variables are introduced in context; the variables are defined more formally in the section "Generalizing Elements of Planned Adaptation" (Sect. 13.3).

13.2.1 Particulate Matter Standards

Particulate matter (PM) standards in the EU and the USA comprise a canonical instance of planned adaptation. In both cases, the characteristic element of planning is a *periodic trigger*. In short, regulations limiting the emissions of particulate matter were introduced, with stipulations that combinations of modeling and monitoring would be used to re-evaluate these regulations after a given period of time. At a high level, there are two classes of particulate matter considered in these discussions: PM_{10} and $PM_{2.5}$. The difference is in the size of the particles and their effects on air quality in terms of health impacts. PM_{10} establishes thresholds around particles with a diameter up to 10 micrometers; $PM_{2.5}$ establishes thresholds at a diameter of 2.5 μm. As an instance of planned adaption, the periodic trigger demands a re-evaluation of the substance of the standard (or, very simply, whether the threshold is PM_{10} or $PM_{2.5}$) based on monitoring of particulate matter levels and reconsideration of the regulation based on additional scientific research on the effects of particulate matter.

One of the key findings stressed by Petersen et al. (2006) is the difference in the interpretation of science and monitoring data available to both EU and US standards bodies. In the case of PM standards, the USA, based on common scientific data, was quicker to introduce PM standards and was more precautionary.[1] Although the EU and USA "both use highly transparent, participatory process to inform policy decisions", Petersen et al. (2006) attribute part of the nine-year lag to the review process. It should be stressed that the lag is not intended to say one organizational and regulatory process is better or more efficient. Rather it is a point of comparison; this work will build on to understand the loci of organizational, knowledge creation, and knowledge assessment (evaluation) capabilities in the adaptation of rule systems. This distinction speaks to what will be developed here as the capability to create, consume, and apply knowledge to anticipating change, and the need for subsequent adaptation based on established criteria. In the case of PM standards, Petersen et al. (2006) highlight that not only is the lag due to the review process, but that some of the misanticipation was rooted in misunderstanding (1) the impact of knowledge assessments themselves, such as the costs of detailed modeling in contrast to the observable benefits from these assessments and the impact on overall processes, and (2) how strictly standards would be enforced by the courts. In both of these instances, the character and loci of how knowledge and standards are evaluated and consumed affected both the adaptive process and the application of the standards themselves. In later discussion, this distinction will be elaborated in terms of the interaction between secondary rules (those used to change rules) and the application of primary rules (the substance of a standard or a regulation).

In the EU standards development process, the loci of adaptive capabilities are rather distributed, starting in the European Commission (EC), with implementation of standards by EU states. Petersen et al. (2006) describe the Clean Air for Europe (CAFE) program as the program that drove the development of the 2005 Thematic Strategy on Air Pollution. CAFE provided policy guidance. Technical expertise was contributed through the Technical Analysis Group (TAG), comprising guidance from

---

[1] Typically the roles are reversed. The US regulatory system is considered generally less precautionary than EU regulation. US regulation is considered to be normatively liberal: regulations are typically driven by harms observed (and typically litigated) in a given industry, leading to particular rules safeguarding against those harms. In contrast, EU state regulations are considered to be socially protective: harms are anticipated and the attendant behaviors regulated, with derogations introduced when this precautionary approach is shown to create market inefficiencies in particular cases.

the CAFE secretariat with "consultants carrying out the technical analyses under specific contracts". As noted by Petersen et al., "[t]he organization of this process can be seen as being top-down". This science–policy interface is framed as "top-down"; in terms of organizational complexes, this will be referred to as a top-down configuration. In contrast, US regulation was driven by lawsuits—a more "bottom-up" configuration, injecting new knowledge into the adaptive process by those experiencing the costs of inefficient policies. As this work moves through the remaining cases and elaborates triggers driven by exogenous events and operational experience, it will become evident that this interface may be driven by feedback loops in top-down, bottom-up, or relatively lateral configurations within a given bureaucracy. Petersen et al. (2006) describe this interface in terms of "boundary work". The distinction between actors engaging in risk assessment (attributed to science) and risk management (attributed to policy) is seemingly confounded when the distinction between stakeholders and experts is blurred. This distinction is rather fluid in the cases presented here—in particular the cases on regional Internet registries (RIRs), where the participants in an adaptive consensus process are stakeholders *and* operational experts with a knowledge base distinct from academic network engineers (science).

13.2.2 Delta Management in the Netherlands[2]

In the Netherlands, 55% of the country is at risk of flooding. Dutch present-day flood risk management is an instance of planned adaptation, whose roots stretch back to the coastal flooding disaster of 1953. During the multi-decade project that followed the flooding, substantive "knowledge about the necessary heights and strengths of dunes and dikes was generated ... [but] no system was put in place to systematically refine and update this knowledge and the resulting 'hydraulic requirements', engineering standards that correspond to the set safety standards for the different locations along the coast" (Petersen and Bloemen 2015). In terms of adaptation, the multi-decade response to the 1953 disaster was a systematic effort to improve flood safety, but, in terms of planning, it was an ad hoc response to an event.

Upon completion of the Delta Works project in 1989, planned adaptation was introduced to review safety standards and fund research necessary for generating new knowledge of the systems, risks, and investments in monitoring and data collection. The history of this project provides insight into the evolution of planned adaptation driven by periodic triggers. After the Zuiderzee flooding in 1916, the Rijkswaterstaat (the Directorate-General for Public Works and Water Management) evaluated whether flooding could happen again, and, based on the state of the art at the time, determined that it could not. In 1938, a civil engineer in the Rijkswaterstaat, Johan van Veen, challenged the assumption that no new knowledge would make a re-assessment worthwhile, initiating a study that ultimately led to the introduction of

---

[2] This case is described in detail in Chap. 14.

a plan with higher and newer dikes based on examination of local conditions. Unfortunately, this was too little too late, and "the country was 'slapped in the face' by the 1953 flood known in the Netherlands as 'The Disaster'" (Petersen and Bloemen 2015).

Two factors contributed to the investment of political and organizational capital necessary to yield the Delta Works Plan. First, during World War II, some portions of the Netherlands were intentionally flooded as a military defense against the Germans. "After the war, the Dutch were able to quickly regain the land (sometimes after surmounting huge technical difficulties) and they started to believe more and more that many desired physical and social changes in the Netherlands could be effected by choosing the right government policies and engineering approach ('*maakbaarheid*')" (Petersen and Bloemen 2015).

Experience with rebuilding provided evidence that, with a concerted political and engineering effort, substantive progress could be made. The second factor was the fallout from "The Disaster": "Though the number of casualties was not very high as compared with earlier disasters (1,836 people drowned as compared to tens of thousands in earlier floods in the 1500s for instance), the effects were traumatic—at the individual level, for the Netherlands as a country, and for the coastal engineering profession" (Petersen and Bloemen 2015).[3]

Taken together, these two historical events contributed social capital, and perhaps a new impetus for the coastal engineering profession to improve its engineering and technical capabilities to better serve the public good. In effect, it sets the stage for investing in the organizational, knowledge creation, and monitoring capabilities that ultimately produced the Delta Works with, among others, the storm surge barriers in the Oosterschelde and the Nieuwe Waterweg.

In December 1995, the *Wet op de waterkering* (Water Defense Act) introduced statutory requirements for flood safety levels and "a five-year review cycle of the hydraulic requirements associated with those *fixed* safety levels" (Petersen and Bloemen 2015; emphasis added here). This is a canonical instance of periodic planned adaptation. In the first iteration, the review period was five years. Note above the emphasis added to *fixed*—during the review period, standards levels remain fixed. Upon (periodic) review, these standards, in particular local standards, may be updated to reflect new knowledge of the system.

In terms of the model of systematic adaptation being developed here, the experiences driving adaptation are a combination of flood events in 1916 and 1953 and the experience with both rebuilding after World War II and ad hoc adaptation in the first Delta Plan. To be clear, experience is the aggregate knowledge produced over *a series of events*; experience should not be mistaken as an analytic construct for scoping events or triggers. The trigger is the periodic review. Adaptation comprises the series of local updates that, after review, apply new knowledge to safety standards and, subsequently, their implementation.

The Dutch government constituted a second Delta Commission in 2007. The reasons: outdated flood safety standards (reflecting changes in population and value

---

[3]Petersen and Bloemen (2015) attribute this to Bijker (2007).

of investments in the early 1960s), a backlog in maintenance of dikes, and growing concerns about the possible consequences of the changing climate and sea level rise. In line with the advice of the Commission, the Delta Programme was established. The Delta Programme combined existing insights into water management and in dealing with uncertainty into an approach named Adaptive Delta Management (see Chap. 14).

Adaptive Delta Management (ADM) is based in part on adaptation pathways (see Chaps. 4 and 9). An important aspect of working with adaptation pathways is that the moment of adaptation of existing plans is not fixed, but is the result of an analysis of monitoring and modeling results on both the effectiveness of the measures being implemented and to (actual and expected) changing external conditions. In addition, two pre-fixed adaptation moments are built into the Delta Programme: (1) the overarching Delta Decisions (policy frameworks) and regional strategies are systematically reviewed every six years; and (2) the flood protection standards are reviewed every 12 years. Research programs like the Dutch National Knowledge and Innovation Program "Water and Climate" (NKWK) have a major role in generating the needed information. So the locus of this knowledge generation is the combination of research and lessons from adaptation.

In terms of the social and political capital necessary to sustain planned adaptation, the Netherlands is unique in that it has a distinct social awareness of the importance of water defense: This has been a part of their culture for centuries. Contrast this to the accrual of political capital for other cases: particulate matter and civil aviation safety are comprehensible by the politicians, but are not nearly as prominent a fact of life to individuals as water defense is to the Dutch. In terms of Internet numbers, most users are at best vaguely aware of what an IP address is, much less the impacts of their delegation, which is many layers of abstraction away from their day-to-day navigation of the World Wide Web (WWW—the application layer). It is easy to categorize this form of flood safety management as a somewhat common instance of periodic planned adaptation for rare events, but the difference in the sources of social and political capital make it a valuable instance that hints at an ideal form, where one finds a harmonious confluence of public social, political, and organizational capital necessary to invest in planned adaptation.[4]

13.2.3 Air Transportation Safety

In the USA, the Federal Aviation Administration (FAA) is responsible for "regulat[ing] safety in commercial air transportation and other modes of civil transportation" (McCray et al. 2010). In addition to a "static" license certifying airworthiness,

---

[4]Harmonious is used in the sense of Keohane (1984). In the ideal form, incentives are already aligned. Clearly in the Dutch case, these aligned incentives exist, but planned adaptation serves as a coordination mechanism that fosters the development of capabilities, and the deployment of capacity necessary for periodic review and the subsequent updating of standards.

the FAA has invested in a body with (1) the technical and operational capabilities necessary to evaluate accidents, (2) the capacity to investigate incidents as they occur, and (3) an established channel to provide feedback to update safety regulations. In terms of the language established here, this body has the necessary deep knowledge of airplane mechanics and broader aviation system operations. In contrast to the particulate matter case, where the majority of the knowledge brought to bear is based on science, in this case, the majority of knowledge brought to bear is technical and operational. This body of knowledge builds on scientific knowledge, but the utility of this evaluative and investigative body is its knowledge of how these systems operate on the ground, in situ.

This investigative body, the National Transportation Safety Board (NTSB), is a case of planned adaptation that is, in contrast to the previous two cases, driven by exogenous, stochastic events (accidents), not a periodic review driven by information and knowledge generated by an explicit event monitoring and/or research programs rooted in conventional scientific research. The NTSB deploys "go-teams" that are quickly mobilized to examine every civil aviation accident shortly after those incidents occur. In the terms established here, this is an investment of organizational capital to sustain an investigative capacity. The capability comprises the knowledge necessary to investigate an incident, evaluate the causes in the context of existing safety regulations, and recommend necessary, yet tractable, changes to safety regulations. The NTSB is an investment in capabilities, complemented by the appropriate investment in the capacity necessary to leverage those capabilities.

The result is a relatively tight feedback loop that leverages this combination of capabilities and capacity. As per McCray et al. (2010): "'planning' ... is the provision, in advance, of ample investigatory capacity ... to enable diagnosis of problems when they arise". The adaptation feedback loop is animated by a body that advises, but is independent of, the FAA. The NTSB's capabilities are triggered by the observation of exogenous events. Investigations, facilitated by preexisting capacity, generate the information and knowledge necessary to recommend adaptations.

The history of the NTSB speaks to the distinction between stakeholders and experts. Early on (until 1974), the NTSB was housed within the Department of Transportation (DOT). After an incident in which its recommendations were resisted, Congress ended the NTSB's dependence on the FAA, making it an independent body. "While FAA personnel do participate in some NTSB investigations, they are pointedly excluded from NTSB deliberations to determine the 'probable cause' of civil aviation accidents" (McCray et al. 2010). In effect, establishing the independence of the NTSB created a distinct line between experts recommending regulatory adaptation and the stakeholders that will be affected by that change.

Since then, the NTSB has extended its scope beyond "mere disaster response". The NTSB maintains a database of non-military incidents, supplementing individual investigations with efforts to identify patterns that may not be evident in individual cases. Under this remit, it investigates "near misses" and provides a list of "most wanted" safety improvements, effectively focusing public attention on those areas where technical or policy innovation is needed. This is another instance where we see a distinction in capabilities (research capabilities) and an investment in the capacity necessary to support those application of these capabilities (accident database and supporting human resources).

As an interface in this organizational complex, this is an instance of lateral knowledge sharing between nominally peer agencies. The DOT/FAA is the stakeholder consuming new knowledge and updating rules. The NTSB is the expert community. When Congress made the NTSB independent of the DOT/FAA, it invested political capital (based on public demand), effectively imbuing the NTSB with the ability to enhance its capabilities (expert knowledge, investigative processes, understanding of near misses) and capacity (resources supporting 'go-teams', accident database), as it sees fit, in service of its investigative and evaluative functions.

13.2.4 Internet Number Delegation

The previous three cases describe the organizational capabilities supporting adaptation in conventional government agencies. Internet number delegation is managed by a transnational collection of non-profit organizations, referred to as the Regional Internet Registry (RIR) system.[5] The RIR system comprises five regional registries (RIRs).[6] Each RIR is a non-profit, membership-based organization tasked with delegating Internet Protocol (IP) addresses used for routing traffic from one Internet host to another.

IPv4 addresses are a finite, scarce resource, provisioned as a contiguous series of 2^{32} integers. The community often refers to these simply as "numbers". IP addresses are used to uniquely identify Internet hosts in the Internet's routing system. An Internet host is any device capable of Internet communication, comprising desktop computers, laptops, tablets, mobile phones, elements of the infrastructure such as routers and switches, servers used to host cloud computing services. Early on, the Internet operations community recognized the need to develop policy to order the delegation of this scarce resource, adapting variants of the consensus process used by the Internet Engineering Task Force (IETF).[7]

---

[5] The majority of this section is based on Chap. 4 of Sowell (2015), a study on the governance of RIRs and the consensus-based knowledge assessment processes these organizations use to develop resource delegation policy.

[6] The five RIRs are:

- American Registry of Internet Numbers (ARIN), covering North America and parts of the Caribbean
- Latin America and Caribbean Network Information Center (LACNIC), covering Latin America and parts of the Caribbean
- Réseaux IP Européens (RIPE) Network Coordination Centre (NCC), covering Europe, the Middle East, and Russia
- Africa Network Information Center (AFRINIC), covering Africa
- Asia Pacific Network Information Center (APNIC), covering Asia Pacific

[7] See Resnick (2014) for the modern articulation of the IETF consensus process.

In terms of policy development, (most) of the stakeholders in the RIR community are also numbers and routing experts. This community comprises network operators[8] with deep expertise in the day-to-day operations of the technical systems and protocols that facilitate Internet communication. Network operators are an instance of what Haas (1992) refers to as an epistemic community: "An epistemic community is a network of professionals with recognized expertise and competence in a particular domain and an authoritative claim to policy-relevant knowledge within that domain or issue area."

Conventional epistemic communities comprise professions such as lawyers, physicians, and academics, all of whom derive a substantive portion of their initial training, credibility, and foundational knowledge base from formal schooling. Operator communities differ; the training, credibility, and knowledge base is largely derived from operational experience managing a production Internet infrastructure. The immediate stakeholders—those whose firms depend on number delegations for their value proposition—form a superset of the experts actively evaluating and adapting number policy. Although there are certainly varying degrees of expertise in the larger numbers community, the collective body of numbers stakeholders has the necessary depth of knowledge regarding the role of these resources and how they are used in the operational environment to evaluate and adapt number delegation policy. As such, while those that animate the policymaking process are typically at the upper end of this spectrum, the stakeholder collective is itself an epistemic community.

Consensus-based knowledge assessment is at the heart of number resource policy development processes. Here, consensus-based decision making is used to evaluate and adapt (common) resource management rules to cope with uncertainties stemming from changing resource demands and patterns of use in the broader Internet industry. Resource policy is created by the community as a response to changes in operational use that warrants adaptation to align policy to operational realities. A resource policy proposal is crafted by one or more community members, drawing on their operational expertise, but acting in their role as stakeholders. That resource policy proposal is then evaluated by community experts and RIR staff for fairness; technical, operational, and economic feasibility; and operational implications for the RIR. This evaluation takes place as part of the consensus process. The four phases of the RIR's consensus process are described below.[9]

Phase 1: Problem Identification

When members of the numbers community identify a policy inefficiency, the community must determine if it is legitimately within the scope of RIR number resource

---

[8] Here, "network operator" refers to an individual engineer working for a particular firm, not the firm itself.

[9] The following four phases describe the core tenets of consensus-based decisionmaking across the five RIRs. It should be noted that the generalization presented here does not elaborate the process-specific nuances in each RIR.

policy.[10] Distinguished members of the policy development process (PDP),[11] referred to colloquially as "policy shepherds," evaluate the problem statement to determine legitimacy. Problem identification involves an evaluation of the event—whether the implications warrant adaptation. Here, an event is a change in patterns of number use that warrants a change in delegation policy. Such events driving change are exogenous (typically driven by broader changes and innovation in Internet infrastructure) that are, in turn, a response to the demands from emerging patterns of use at the application layer.[12] The stakeholder that brings this event and its (perceived) implications to the attention of the broader community[13] is the trigger for event evaluation and policy adaptation. In this sense, the event is exogenous, experienced by stakeholders that make up the numbers community, one or more of whom trigger a policy change.

Phase 2: Active Consensus

Where problem identification identifies whether a problem is legitimate and in scope, active consensus evaluates the proposal in terms of demand among operators experiencing inefficiencies, the quality of the proposal and the appropriateness of the proposed solution. Active consensus is the primary locus of policy adaptation. Within a given policy proposal, "adaptation step 0" is the initial proposal itself. Subsequent adaptations (steps 1 … n) occur as members of the numbers community make utterances that suggest credible modifications to the proposed solution. In that sense, active consensus is a process of knowledge assessment evaluating adaptations within the context of a given proposal.

Within the numbers community, politically fungible voting is eschewed as a knowledge assessment process.[14] Under the family of consensus processes, asserting (voting) "No, I do not agree," is insufficient. Credible contributions require a rationale that, through evaluative policy dialogues that encourage constructive conflict, either fit or update the operational epistemic communities' authoritative image of number resource dynamics. Shallow agreement via a marginal "51–49 victory" is not consensus. In this transnational resource management space, there is no hegemon to enforce the rules. Rather, mutual enforcement requires credible commitments from participants that, as both architects of the rules and subjects of those rules, must both reason about how to adapt, adhere to, and enforce resource policy. As such, consensus often demands that >70% of credible contributors agree, building com-

---

[10]In ARIN, the Advisory Council (AC) evaluates initial policy proposals for scope and technical merit. In the RIPE and APNIC regions, the specific working group (WG) chairs evaluate policies based on the WG mandate.

[11]AC members in ARIN; policy forum chairs in other regions.

[12]The canonical OSI model has seven layers, but pragmatically speaking only the infrastructure layers (1–4) and the application layer (7) are used in operational engineering environments.

[13]To be more explicit, this policy entrepreneur is likely one of a number of community members that has recognized the problem, but, as a policy entrepreneur, is incentivized to act on this by engaging the policy development process.

[14]Recall that the RIR's consensus process is derived from the IETF consensus process. Voting is eschewed in that epistemic community, as well.

mitment by reconciling technical and operational critiques with dissenting minority contributors.

Iterative adaptation continues until all contestations have been addressed. Contestations may be integrated or dismissed. A contestation is integrated if the modification of the solution is itself not contested. If the modification suggested by a (credible) contestation is contested, the process of exploring the solution space continues until an uncontested solution is identified. When all contestations have been resolved, active consensus has been achieved and the process moves on to passive consensus.

Phase 3: Passive Consensus

Consensus does not require participation by every member of the numbers community. Passive consensus provides the opportunity for those that may not have had the time to follow each incremental change to weigh in on the product of active consensus. Passive consensus serves as a consistency check—a penultimate evaluation of the proposal that ensures uncertainties and/or inconsistencies that may have crept in during an iterative active consensus are not overlooked.

The character of passive consensus differs from active consensus. Active consensus requires both "active" support of a solution and the resolution of contestations. In contrast, silence on the proposal (the absence of contestation after a designated review period) is sufficient for achieving passive consensus. Over the course of the active consensus process, a policy proposal may have experienced a number of major and/or minor revisions. A minor contestation in the passive consensus process, such as a clarification of language, may be dealt with immediately in the passive consensus process. "Easy" cases, such as language cleanup or minor modifications, rarely see substantive change during the passive consensus process. In contrast, "hard" proposals are typically instances of larger policy issues, such as routing security, transfers, or anti-abuse that defy modularization and may have inconsistencies even after substantive efforts at active consensus. If a substantive contestation arises during passive consensus, the proposal may return to the active consensus phase for additional evaluation and revision.

Phase 4: Evaluation of the Consensus Process

A review of the consensus process itself is the last step of the PDP. Active and passive consensus is "called" by the shepherds of that particular proposal, indicating that phase is complete. This is a point of discretion by the consensus facilitator. The review phase is a check on that discretion.

Evaluating the substance of a proposal *is not* the objective of the review phase.[15] The objective of the review is to ensure:

---

[15]In all but the RIPE region, the membership elected board of the RIR reviews policy proposals that have reached consensus. In the RIPE region, although issue-specific WG chairs determine consensus for policies developed in their WG, the collective of WG chairs evaluates whether the PDP was followed.

1. the RIR's PDP was followed, and
2. the policy does not create undue legal risk or obligations for the RIR.

For instance, in the ARIN region, the Board of Trustees of ARIN evaluates the draft policy in terms of "fiduciary risk, liability risk, conformity to law, development in accordance with the ARIN PDP, and adherence to the ARIN Articles of Incorporation or Bylaws," (ARIN 2009). Review is intended to be final check on the consensus decisionmaking process and feasibility of the policy before moving to implementation. In terms of credible knowledge assessment, with the exception of instances where the board identifies a substantive risk to the RIR itself, the board does not subvert the credible knowledge assessment of the community.

Loci of Number Delegation Capabilities

RIR's and numbers community's investment in sustaining the consensus process is an investment in adaptive capabilities and supporting organizational capacity. There are two loci:

1. The members of the number community that evaluates of the substance of numbers policy constitute adaptive capability.
2. RIR staff supporting the infrastructure facilitating this process, performing impact analyses for RIR implementation, provides organizational capacity.

In terms of the organizational complex, this is a bottom-up interface in which an epistemic community is the locus of evaluative capabilities and the RIR (as an organization) provides organizational capacity (including coordination resources and process support) in which these evaluations can play out. Systematic adaptation through the consensus process sustains and extends the knowledge base of the operational epistemic community, in part by codifying that knowledge into resource policy.

13.3 Generalizing Elements of Planned Adaptation

As noted in the introduction, McCray, Oye, and Petersen's notion of planned adaptation is fundamentally about updating rule systems. There is subtle difference between (1) the knowledge necessary to make substantive policy decisions, and (2) the rulemaking apparatus, capabilities, and authority, necessary to change potentially entrenched rules. Starting with Hart's notion of primary and secondary rules, this conceptual model generalizes and explains the factors at play when adapting the rule systems presented here. Rules are ultimately evaluated in terms of their fit—how new knowledge of the system can be used to evaluate whether a set of rules continues to satisfy obligations to system integrity.

In the abstract, such a model is at best an elegant thought experiment. To make this model pragmatic, the notions of triggers, events, and loci of evaluation in terms of monitoring and knowledge assessment are formalized. Triggers are a known class of mechanisms in adaptive rulemaking. Here, the notion of a trigger is the mechanism linking *events* identified as potentially warranting adaptation to the adaptive process itself: an evaluation of the substantive rules governing system behavior and the potential rules that provide guidance for how the adaptation should proceed. Categories of triggers serve as the bridge between an abstract model of adaptation and observed instances of adaptation discussed in the previous section. Events provide information that, upon assessment, offer an opportunity to adapt. Assessing an event demonstrates an evaluative capability to generate new knowledge of the system from information gleaned from an event, but not necessarily the capacity and capabilities necessary *to adapt the attendant rules*. Note the distinction between the capability to evaluate the substance of the event and the capacity and capabilities necessary to change the rules. The question for a model that aspires to be both analytic and evaluative then becomes: what are the necessary organizational and political capabilities necessary to consistently and reliably act on triggers and the substance of policy analyses?

Cases selected for this chapter are a comparatively interesting mix of decision-making processes, organizational structures, and incentive structures that provide a first view into the efficacy of the conceptual vernacular established in this section. That said, this is a relatively small convenience sample. The conceptual model presented here is a canonical case of contingent theory building.[16] While limited, the concepts established here provide the basic building blocks for a first articulation of the systematic adaptation space—the range of configurations and contexts that sustainably facilitate systematic adaptation.

13.3.1 Disentangling Primary and Secondary Rules

Hart (1994) describes rule systems as the union of primary and secondary rules. *Primary rules* are those "thought important because they are believed to be necessary to the maintenance of social life or some highly prized feature of it". Primary rules describe what behaviors are appropriate and what behaviors are prohibited. For instance, a prohibition on stealing is a primary rule. It is a rule that maintains social order, it ensures that the system (in this case a society) functions as one expects and that property rights are respected.

Much of the work on adaptive policymaking is about establishing primary rules that describe how to sustain system function and integrity. In the cases presented here, primary rules describe the appropriate (safe) levels of particulate matter, what is necessary to sustain appropriate flood protections, the body of flight safety rules, and the rules that articulate number delegation rights and how those rights are conferred. These are the rules that determine either how a system should be monitored and sustained, or how the system itself is to function. The approaches and tools of policy analysis described in the previous parts of this book are one conventional means by

---

[16]Contingent theory building is used in the sense of George and Bennett (2005).

which an analyst may systematically evaluate and develop the knowledge informing primary rules for maintaining a given system.

In contrast, *secondary rules* are about how to recognize, create, maintain, and adjudicate primary rules. Secondary rules are "all about [primary rules]; in the sense that while primary rules are concerned with the actions that individuals must or must not do, these secondary rules are all concerned with the primary rules themselves. They specify the ways in which the primary rules may be conclusively ascertained, introduced, eliminated, varied, and the fact of their violation conclusively determined" (Hart 1994).

Adaptive processes must strike a balance between the canonical objective of regulation (creating stability) and the potential for chaos if feedback loops trigger changes too frequently. Conceptually, primary and secondary rules give us a starting point for decomposing adaptive rulemaking processes. Consider the function of secondary rules above in terms of how primary rules may be "ascertained, introduced, eliminated, varied, and the fact of their violation conclusively determined" (Hart 1994). In terms of adaptation, ascertaining a rule's fitness can be framed as part of the path selection process: given a particular trigger, what rule is most appropriate to cope with the conditions at hand? If no rule is fit, the introduction and elimination of rules represent the action that is taken to update a particular rule system—the act of actually adapting to introduce rules more appropriate, or better fit, to the better understanding of a system's context, as highlighted by the evaluation of an observed event.

Hart also provides a notion of obligation. In this work, the obligation is to the efficiency and efficacy of the system at hand. In that sense, one could say that a rule will not be recognized (it will be determined to have poor fitness for the context) when it no longer fulfills its obligation to sustaining system integrity.

The secondary rules function and role provide the foundation for analyzing how decoupled primary and secondary rules are in a particular adaptive rulemaking system, and the implications of that decoupling for the performance of the system. In the next subsections, Hart's three types of secondary rules (recognition, change, and adjudication) are presented in terms of how they help explain systematic adaptation.

Rules of Recognition

Hart (1994) argues that "the simplest form of remedy for the uncertainty of the regime of primary rules is the introduction of what we shall call a 'rule of recognition'… what is crucial is the acknowledgment of reference to the writing or inscription as authoritative, i.e. as the proper way of disposing of doubts as to the existence of the rule".

Simple rules of recognition may refer to an *authoritative* set of rules—in early societies an inscription on a monument, or in later systems, a particular constitutional document. In more sophisticated systems of rules, rules of recognition provide some "general characteristic possessed by the primary rules" such as "having been enacted by a specific body, or their long customary practice, or their relation to judicial decisions".

Operationalizing recognition in terms of adaptation makes points of contention faced by both the creation of, and sustained application of, adaptation apparent. First, rules of recognition reduce uncertainty by highlighting precisely what characterizes a primary rule. In the context of an adaptive process, rules of recognition can also serve as a means to articulate scope—general characteristics can be framed in terms of the system being adapted. "Authoritative" speaks to the source of the rule. In this context, the "specific body" enacting the rule may well be a form of expert committee or epistemic community that has the necessary access to the system, knowledge, and capabilities to effectively develop rules that shape and/or sustain system behavior. Each of the cases in the previous section calls out those expert communities in terms of their evaluative capabilities.

Another operationalization of a characteristic that may be used in conjunction with the source of the rule is the *state of the art* of the rule. State-of-the-art speaks to how up-to-date the information informing a given rule is. Returning to McCray's definition of planned adaptation, rules of recognition are an assertion that primary rules should be based on the best knowledge available on a given system. This seems like an obvious statement, but it serves an evaluative function that will serve as the foundation of a trigger for change. The criteria set by rules of recognition contribute to the broader objective of ascertaining whether a given rule should be in effect. A consequence of that evaluation is the choice to eliminate a given rule, introduce a new rule, or adapt the rule to reflect updated information.

It is important to stress that rules of recognition establish authoritative, not evaluative, criteria. Such criteria establish what it means for a rule to have an authoritative source—whether a rule is within the scope of the system. Recognition does not indicate *how* the rules should change, nor do they speak to the *substance* of primary rules. Speaking to the substance is more traditionally the domain of substantive policy analysis. The question then becomes, how do we systematically adapt the system when we recognize primary rules are no longer appropriate?

Rules of Change

Secondary rules of change "remedy ... the static quality of the regime of primary rules ... [by] empower[ing] an individual or body of persons to introduce new primary rules for the conduct of the life of the group, or of some class within it, and to eliminate old rules" (Hart 1994).

Rules of change provide the *mechanics* for adding new rules, updating existing rules, or removing rules that no longer satisfy obligations to system integrity. Depending on the rule system, those mechanics may be specialized to the regime in which the rule system is embedded or may be part of a more general rulemaking or legislative process. For instance, in the cases of particulate matter and flood risk, evaluative criteria are updated periodically, but the changes themselves are still part of existing regulatory processes. In contrast, in the case of airline safety, the NTSB is based on an established process of feeding information into the FAA to update safety standards. Moreover, the NTSB was explicitly given greater independence after allegations that the FAA and the White House may have undermined recommendations. Number policy in the RIR system illustrates an instance where rule change is almost completely endogenous to the system: evaluation of primary rules is triggered by the constituency as they recognize failures of primary rules to the obligation to system integrity. Returning to McCray's initial definition, not only is the trigger endogenous, but the knowledge generation process is the factor driving change.

Rules of Adjudication

Hart frames the last set of secondary rules in terms of efficiency—in particular, how to move from primary rules enforced by social pressure, to vesting the authority for adjudicating primary rules into particular actors. Rules of adjudication comprise "secondary rules empowering individuals to make authoritative determinants of the question whether, on a particular occasion, a primary rule has been broken" (Hart 1994).

Like the earlier secondary rules, adjudication hinges on the notion of what constitutes authoritative (here "authoritative determinants"). Rules of recognition designate the authoritative bodies, as noted above, but rules of adjudication provide evaluative procedures.

Hart (1994) elaborates the linkages among secondary rules, as follows: "Besides … resemblances to other secondary rules, rules of adjudication have intimate connections with them. Indeed, a system which has rules of adjudication is necessarily also committed to a rule of recognition of an elementary and imperfect sort. This is so because, if courts are empowered to make authoritative determination of the fact that a rule has been broken, these cannot avoid being taken as authoritative determinations of what the rules are".

Here, Hart elaborates the link between rules of adjudication and recognition relative to courts that this work makes for the authoritative experts on the complex system to which policies must adapt. Conceptually, these are similar—an organization is established to systematize these processes. The pragmatic evaluative procedures are where these will differ, trading legalistic analysis for the delegation of decision-making authority to actors that hold epistemic authority in a particular domain of complex systems. In particular, adjudication differs from recognition of scope. Adjudication evaluates the performance of the rule, whether it is up-to-date, and, ultimately, whether it should be in effect under the current conditions.

Rules of adjudication speak to the criteria delineating *whether* one should change the rules or not, whether the change is warranted by system performance, or the implications for stakeholders (costs, efficiency, distributional effect, etc.). In the next subsection (Sect. 13.3.2), the entry point for recognizing and adjudicating the efficacy of a rule will be operationalized in terms of triggers and event evaluation. In effect, a trigger is a signal—a recognition of an event. That event provides new information about the performance of a rule system. For example, a periodic trigger is established by an authoritative (recognized) body to evaluate event monitoring information (particulate matter). The trigger is the entry point to an analysis under the adjudication function.

Taken together, secondary rules of change are one indicator of adaptive capability. They are durable procedures for systematically evaluating whether a policy or set of policies, continue to meet obligations to system integrity. For each of the systems evaluated here, this organizational construct is the locus of permanent or semi-permanent authority to adjust rules to suit changes in the environment (like a judiciary in conventional governmental arrangements). Moreover, adjudication procedures will differ substantively. For instance, in conventional bureaucracies, domain experts contribute to the evaluation of the efficacy of rules (such as the EPA's periodic evaluation of particulate matter policy, or the event-driven evaluation of airline safety regulations). Expert adjudication is not limited to dedicated analysts, though. In the RIRs, the relevant experts are the engineers that enjoy the efficiencies, or suffer under the deficiencies, of number delegation policy; these actors serve as both monitor and change agent through RIRs' policy development processes. Each of these instances highlights not only the actors, but the triggers for policy change (the topic of the next subsection).

13.3.2 Triggers and Events

Adaptation here means processing system feedback—leveraging organizational capacity to transform information about the system into knowledge that informs the rules an organization uses to sustain system integrity and function. This subsection elaborates a conceptual model of triggers, events, and evaluative information processing and knowledge assessment that (ideally) culminates in effective adaptation of a set of rules. Definitions presented here bridge the substance of policy analysis and the adaptation of Hart's rules to policy adaptation. As noted earlier, rules of change describe how to change primary and secondary rules. They are the mechanics of change within an organizational process, but do not speak to the evaluative component. Triggers, events, and information processing map to the organizational complex's capabilities and capacity created by investing political capital in bodies with the expertise and authority necessary to effect rules of recognition and rules of adjudication. Rules of recognition authoritatively designate those that should monitor triggers and the structure of those bodies. Rules of adjudication authoritatively designate two elements of the evaluative process:

1. When a behavior deviates from the current definition of system integrity (i.e., when the current rules are no longer fulfilling their obligation to system integrity);
2. The process experts should use to evaluate the information inherent in an event to either select a more appropriate rule or, in some cases, create a new one.

In what follows, we elaborate these processes through systematic definitions of the types of triggers, events, and evaluation observed in the cases.

Triggers

A trigger *signals* when a change that potentially affects obligations to integrity has occurred; i.e., a trigger is the means of recognizing the potential need to adapt. (This is practically the same definition of trigger (or Adaptation Tipping Point) that is used in the various DMDU approaches elsewhere in the book.) In the many uses of the term, a trigger is caused by an event, but that is not necessarily always the case. A trigger may also be the consequence of ongoing evaluations of events—this distinguishes reactive triggers from proactive triggers. Triggers activate an evaluative (adjudicative) process that informs whether rules of change should be invoked. *Proactive* triggers initiate an evaluation based on information that has already been collected by some monitoring process. Reactive triggers are more akin to the prevailing use of the term. *Reactive* triggers activate both an information gathering process and evaluative process in the face of a previously unobserved, unanticipated, or ongoing yet unmonitored, event. Three types of triggers are observed in the cases above: periodic, stochastic, and tacit. Periodic and tacit triggers are proactive; stochastic triggers are reactive.

Periodic triggers are proactively clock-driven. Both the particulate matter and flood risk cases are instances of proactive, clock-driven triggers. In both cases, monitoring of system processes for events with significance to system integrity, and the subsequent evaluation of new information occurs before (in preparation for) scheduled (periodic) rule evaluation. In terms of planning for adaptation, this is probably the most intuitive of the triggering mechanisms. It is assumed that the system is continuing to change, but not at a rate that warrants an active rule adaptation scheme.

McCray et al. (2010) argue that "[t]he EPA program for ambient air standards is likely the most fully developed existing program of Planned Adaptation". Stability of rule systems is the primary objective of conventional bureaucracies. Periodic triggers are arguably the most stable form of adaptation. Periodic triggers are predictable and fit nicely into an organization's budgeting cycle. Periodic triggers allow managers to set aside resources (invest in capacity) for an upcoming evaluation, and provide technical analysts with the lead time to perform the kinds of rigorous, systematic analyses they prefer. The evolution of flood risk management, culminating in the Water Defense Act in 1995 and the Adaptive Delta Management framework in 2010, is an instance of learning from ad hoc adaptation to develop a more stable model of (periodic) adaptation.

Stochastic triggers are the direct product of active, continuous monitoring of a system for events that may impact (or have already impacted) the integrity of the system. Unlike known events monitored under a periodic triggering model, the stochastic model is monitoring for classes of events that may occur at any time. The stochastic trigger is a signal that an event has occurred, and that information from that event should be evaluated. Among the cases presented here, airline safety is the canonical instance of a stochastic trigger. Accidents are the events that are monitored by the NTSB. When an accident occurs, an investigation ensues. In this case, a stochastic event occurs, monitoring apparatus recognizes the event and invokes resources necessary to first cope with the impact of the event (mitigation), but, ideally, with appropriate investment in capacity, in short order, to process the implications of these events for the rule system in order to remediate.

Stochastic triggers pose a problem for a conventional organization's capability and capacity planning. In contrast to periodic triggers, the investment is in monitoring the system, and the readiness to act on events that may not happen. Rules of recognition provide an authoritative mapping of event types and the bodies that should evaluate those types. Rules of adjudication document the process for evaluation or, in some cases, in conjunction with rules of recognition, delineate the scope of discretion of the evaluating body. Rules of recognition here are, as may be obvious, not static, but are constantly updated, by rules of change *based on* the evaluative process. With the exception of very well-understood systems, the intersection of rules of recognition and adjudication will be concerned with the *scope* of expert discretion (recognition) and the distributional consequences of evaluation under that discretion (adjudication). This is especially important for high clockspeed stochastic events, where a series of independent evaluations, potentially by different actors, may lead to inconsistent recommendations. Trigger frequency is an important factor. In the simple case, one may assume that trigger frequency is one to one with event frequency. In the case of high clockspeed stochastic events, an aggregation rule may need to be in place to make the evaluative process more efficient and avoid unintended inconsistencies or worse yet, conflicting recommendations. The character of the evaluation process will be discussed shortly.

Tacit triggers are unique in that the trigger is based on a preliminary evaluation of the event. In this chapter, tacit triggers drive policy development in the RIR system. A policy change in the RIR system requires a rationale (a preliminary evaluation) before active consensus is invoked. In this case, the event is a change in industry practice, the numbers market, or the demand for numbers that warrants a change in policy. Tacit triggers are not monitored by the organization (the RIR and PDP shepherds) but rather by the affected stakeholders. In this case, the preliminary evaluation, not the event itself, triggers the process for changing policy. Hence, the trigger is tacit. Evaluating the tacit character of triggers in numbers policy is subtle: as noted in the case on RIRs, the ultimate source of an event is exogenous, but the recognition of the event is endogenous; adaptation is triggered internally by those experiencing the effects of those changes. From outside the system, because tacit triggers are driven by internal events, they may seem stochastic. The next subsection distinguishes among endogenous, exogenous, and boundary events.

In the abstract, tacit triggering is a process in which those actors bound by the rules have the option to propose change at any time. Returning to the issue of the clockspeed of triggers, the option to change policy is continuously available, but the organization has the option to provide feedback on feasibility, cost, and implementation timelines. This review process is also part of the adjudicative process, although, like in the RIRs, it may be authoritatively assigned to a different group of experts than those that evaluate the substance of the event itself. Tacit triggers, like stochastic triggers, may occur in rapid succession. In contrast to stochastic triggers, especially those driven by safety, tacit triggers do not have the immediacy of stochastic events. Tacit triggers still require an investment in on-demand capacity to deploy evaluative capabilities, but there is more flexibility in the deployment of those resources, especially within the timeline of the PDP.

Events

Events have been alluded to in the discussion of triggers, in terms of whether the trigger is a consequence of an event or an evaluation. Another important conceptual distinction is whether the cause of the event is endogenous to the system, exogenous to the system, or whether it is a product of processes at the system boundary. These are referred to as endogenous, exogenous, or boundary events, respectively. In the cases discussed here, events are primarily labeled as belonging to one of these three categories, but this does not mean that systems cannot be affected by multiple categories.

The notions of endogenous and exogenous events are ideal types; no instance fits their criteria perfectly, but the rulemaking process may treat the event as one or the other as a way to scope secondary rules. *Endogenous events* are scoped to the system at hand: the system is shaped almost exclusively by actors managing the system, and the effects of the events are largely experienced by those actors. *Exogenous events* are scoped to the context in which the system being managed (and adapted) operates: these events affect the performance of the rules governing the system, but are not necessarily under the control of actors involved in governing either the system or the rulemaking process. *Boundary events* combine the two. Boundary events are exogenous in the sense that the cause is outside the scope of the rules, but those *causes* may be within the scope of influence of the rule system.

Numbers policy adaptation in the Internet case is largely driven by endogenous events. Application layer changes do affect number utilization, but tacit triggers activate an evaluation process that updates delegation rules to adapt to new conditions. The event is the internal consequence of a mismatch in policy with utilization demands. The effects of that event are borne largely by the numbers community. That said, the numbers system, whose function is defined by the rules dictating how it is managed, is within the control of the stakeholder collective affected by those rules. In this sense, both the events and the tacit triggers are endogenous to that system.

In the case of flood risk and airline safety, the events are exogenous. The events are driven by external factors that vary in the degree to which changes in the rules can control the manifestation of the event. In the case of flood risk, the event is typically a natural event, which is out the control of the rulemaking system. Policy adaptation is a process of monitoring natural events and developing mechanisms for countering the effects of these events. The event is not eliminated, as in the case of the numbers system, but the effects are mitigated. In the case of airline safety, systemic flaws are triggered by exogenous events such as weather, unexpected system conditions, or design flaws. Again, none of these events can be controlled or predicted, but the effects can be mitigated by evaluating the impact of a previously unobserved event and adapting rules appropriately. Note that while the particular accidents (events) are typically previously unobserved, the classes of events are not. The research efforts noted in the NTSB case are an effort to better characterize the classes of these exogenous events and identify patterns.

The particulate matter case is an instance of coping with long-running boundary events that are evaluated periodically. Particulate matter emissions are governed by EU and EPA standards in Europe and the US, respectively. That said, exogenous factors, such as innovations in industry, changes in industrial processes, and changes in the factors affecting industry outputs can all challenge the efficacy of EU and EPA rules. In that sense, the interaction among these exogenous variables and the rules governing particulate matter levels occur at the boundary. Experts evaluate these events and make recommendations to the EU or EPA to update rules that will mitigate if the source of the event is out of the scope of regulation. If, on the other hand, the event is within the scope of standards, such as industry emission levels, remediation makes endogenous some or all of the root causes of the event itself.

For each of these classes of events, the role of evaluation is to understand the characteristics of these events and how susceptible they are to the system's rules. Evaluation is discussed in the next subsection.

13.3.3 Evaluation

Evaluation takes place at various points in the rule adaptation process. Organizational factors affecting evaluation are characterized in terms of the timing of evaluation, the composition of the evaluating body, and how coupled that body is with the actors adversely affected by inefficient rules.

Timing of Evaluation

Timing of the evaluation has been alluded to in terms of both the type of event and, more specifically, the type of trigger. The timing of the evaluation refers to both when the evaluation is activated and the duration of the evaluation. Evaluation may comprise monitoring and/or investigative capabilities. This discussion will distinguish two classes of evaluation: active assessment capabilities and vested assessment capabilities. Like the characterization of events, these are ideal types. Some cases may combine elements of active and vested assessment capabilities.

Active assessment capabilities are those that continuously monitor status of an event (or class of events), collecting information necessary for expert knowledge assessment of the state-of-the event. Active assessment is applied to events that are known to be in progress, but whose effects have yet to manifest. Flood risk management is the canonical instance among the cases presented here. Water defense is an ongoing aggregate event whose status is actively and continuously monitored to mitigate adverse effects. Particulate matter is another instance of active assessment. PM levels and effects are continuously monitored.

Vested assessment capabilities are an investment in the resources (capacity) necessary to respond to unanticipated exogenous events. Vested assessment capabilities comprise the investigative and analytic capabilities necessary to evaluate the effects of an unanticipated event. Vested capabilities are only initiated when an event occurs. Ideally, vested assessment capabilities can quickly offer a means to first mitigate the adverse effects of an event, then formulate a remediation if the effects of the event are within the scope of influence of the rule system. Airline safety is an instance of vested assessment capabilities. The NTSB "go-teams" are an investment in investigative capacity that can be deployed quickly after an accident (event).

The consensus process in numbers delegation policy development comprises elements of both active and vested assessment capabilities. Evaluation is active, in the sense that those that experience inefficiencies in the rules are also those that create and update the rules. In the course of their operations, they monitor the efficacy of the rules. It is known and expected that patterns of delegation will change, but it is unclear precisely when. In this sense, events are unanticipated. The capacity necessary to support the consensus process is a vested assessment capability that can be activated by a member of the numbers community when inefficiencies cross thresholds agreed upon by the community. Although the trigger is tacit, as opposed to the stochastic trigger in the airline case, it signals an event that warrants evaluation to determine if adaptation is necessary. Together these capabilities signal an event should be considered, evaluates the effects of the event, and provides a forum for proposing solutions as a means to remediate the problem by updating delegation rules.

Loci of Assessment

Assessment processes may be concentrated within the same organization that makes the rules, distributed across one or more organizations, or (as in the case of numbers policy) distributed within an epistemic community. The *degree of concentration* is a function of the observability of events, and by whom. In the ideal concentrated form, the rulemaking system has the event observation and assessment capabilities and capacity necessary to generate sufficient triggers and perform event evaluation on its own. In the first distributed form, adaptive capabilities are located in an organization or distributed across organizations, separate from the organization making the rules. In the second distributed form, adaptive capabilities are distributed among individual experts that contribute to monitoring and evaluation. These ideal forms create a spectrum of loci of assessment.

The ideal concentrated form is included as a reference point. Harkening back to Hart's analysis, in historical governmental rulemaking systems, especially legal systems, the knowledge necessary to update rules is part of the knowledge base related to prosecutorial and judicial knowledge, rooted in that epistemic community. In the cases presented here, nearly all of the adaptive processes are distributed to some degree, requiring the rulemaking process to consume knowledge from a separate organization generating that knowledge. Two of the cases have a historical concentrated form: flood risk and the NTSB. Ad hoc adaptation in the Rijkswaterstaat was conducted by its own civil engineers. In the case of airline safety, the NTSB was initially managed by the FAA. In both cases, planned adaptation now follows a distributed form.

Of the cases as they stand now, among these distributed forms, the case of the NTSB is the most concentrated. In this case, there are two organizations: the FAA and the NTSB. The NTSB comprises experts in aviation operations and safety. As indicated by McCray et al. (2010), the NTSB occasionally enlists experts from the FAA. In this sense, along with the NTSB history, the NTSB is independent of the FAA, but its expertise is in certain cases coupled with the FAA.

The next most concentrated of these adaptive systems is the case of Adaptive Delta Management. In this instance, active assessment is performed by a number of expert organizations, but is funded by the government. This distributed form comprises a single organization consuming new information and knowledge, but produced by a diverse set of monitoring and research organizations. Under this model, there is the potential for contention over the findings of these distributed monitors and assessors. One interpretation of this contention would be that it confounds creating stable, consistent rules. Another interpretation is that contention facilitates more effective analysis of risk and uncertainties. Contention highlights precisely where leading knowledge and analysis of events conflict—in effect, where uncertainties in the understanding of the event and its implications remain. Rather than rely on a single, potentially flawed analysis, this contention highlights where rules should incorporate contingencies to cope with unanticipated consequences, or at least to know which events it can mitigate and/or remediate, and which will require more ad hoc measures.

The standards for developing particulate matter are even more distributed. While the effects of PM are relatively localized, both rulemaking and the assessment capabilities necessary to inform these rulemaking processes are distributed. These two standards development processes draw on common bodies of (contending) scientific knowledge. In this case, there is variance in the kinds of triggers that drive adaptation, how evaluations are interpreted, and the evaluations themselves. Petersen et al. (2006) indicate that the EU's lag in adopting $PM_{2.5}$ standards is explained in part by what has been characterized here as the variance in a distributed locus of assessment.

Lastly, number policy adaptation is the most distributed, but is scoped to a rather mutable system. The body of evaluators comprises individuals from a variety of firms that depend on IP addresses for their value proposition. In this sense, the loci of assessment are highly distributed. That said, recall that events are endogenous and the system is in effect a set of delegation rules and the supporting infrastructure for managing this delegation. As such, while highly distributed, the events that create demand for adaptation are well within the scope of the rulemaking process. The result is a highly decentralized evaluative body, but one that has all the necessary organizational levers to mitigate and remediate largely endogenous events.

Coupling

Closely related to, but conceptually independent of the loci of assessment is how coupled the loci of assessment are to rulemakers and those affected by events. Monitors and knowledge assessors range from tightly coupled to loosely coupled. Coupling has two factors: overlap and influence. *Overlap* means that some subset of the actors performing assessment is also the actors responsible for updating rules. *Influence* means that some subset of the actors performing assessment is under the influence of actors making the rules, typically because of a funding or employment relationship. Tightly coupled means that there is substantive overlap in the actors performing assessments and/or there is substantive influence over assessors by the organization responsible for rulemaking. Loose coupling is the other end of the spectrum: there is little to no overlap and/or influence.

Among the cases presented here, numbers policy is the most tightly coupled. Under the consensus process, there is a high degree of overlap: the assessors are effectively the rulemakers. That said, in most of the RIRs, the evaluation *of the process* is performed by the board or executive committee of the RIR. This introduces a check on the discretion of the assessors and policy shepherds. While the numbers case is tightly coupled in terms of overlap, it is loosely coupled in terms of influence. The default, tacit mode of influence is that the numbers community must comply with numbers policy. That said, numbers community members are loosely coupled because they are employed by firms independent of the rulemaker.

The cases of flood risk and particulate matter have a similarly structured coupling. Both draw assessments from actors engaged in scientific research on the corresponding events. In terms of influence, there is greater influence in flood risk given that monitors are (partly) employed by the Rijkswaterstaat, supplemented by researchers funded (partly) by the Delta Programme. In the particulate matter case, monitors are similarly employed by US and EU agencies, but there is much more diversely funded research on particulate matter in both the USA and the EU. As such, there is similarly moderate coupling of the monitoring component of assessment, but the influence over particulate matter research is lower in particulate matter than in flood risk. In both cases, the relative magnitude of influence is based on two factors: first, the diversity of inputs and sources of scientific research; second, what proportion of this is funded by the rulemaking organization. As will be discussed in the next section, if reduced coupling is perceived to improve the quality of assessments and attendant adaptations, influence can be reduced by incorporating knowledge sources that are funded by some other authority, or (ideally) funded independently of any political or rulemaking organization.

Of these cases, the least coupled is the NTSB. The NTSB is also an instance of decoupling to improve the quality of the assessment process. Recall that the NTSB was at one time tightly coupled with the FAA, but, after public backlash over an instance of political influence, the Congress made the NTSB independent of the FAA. Although now loosely coupled in terms of influence, there is some overlap coupling. FAA employees are occasionally incorporated into NTSB investigations, but the NTSB reduces influence by ensuring that the analysis and recommendations are performed by the NTSB itself.

13.4 Conclusions and Ongoing Work

This chapter has reviewed four instances of planned adaptation representing diverse institutional make-up: state-agencies, international organizations, and a private transnational regime. While instances of planned adaptation are rare overall, within the four cases reviewed there is substantive diversity in the mechanisms that animate monitoring, knowledge assessment, triggers, events, and the organizational context in which these processes play out. In this concluding section, the range of configurations that give rise to adaptive capabilities and capacity are compared. This range of configurations highlights observed configurations, configurations that are considered feasible, and those that are considered infeasible. Given this range, the implications for planning and designing for adaptation are presented. The chapter concludes with implications for future studies—in particular, the value of identifying issues that may be ripe for a planned adaptation approach, and the value of analyzing instances of ad hoc adaptation (as the foundations of planned adaptation) and failed adaptation.

13.4.1 Combinations of Adaptive Capabilities

To illustrate the adaptive capabilities space delineated by the cases, Table 13.1 summarizes the characterizations from Sect. 13.3.

A number of these characteristics are interdependent. Across the state-based cases, there is not an instance of tacit triggers. That said, as noted earlier, in conventional rulemaking, as articulated by Hart, the locus of adaptation is the state. In the conceptual vernacular established here, the closest to this form is the numbers community, a private transnational regime. In cases in which a government agency is the rulemaker, adaptation is driven by exogenous or boundary events—there is little overlap with the evaluator. This seems to indicate that these cases illustrate the need for some decoupling of monitoring, knowledge creation, and knowledge assessment from a potentially politicized rulemaking organization. In the case of PM and flood risk, the evaluator is a combination of the state and scientific research. The NTSB is an interesting case because, while the evaluator is loosely coupled, it is another agent of the state. Numbers policy is the counterexample in terms of coupling, but is substantively different because the rulemakers and evaluators are effectively the same body.

At one extreme, the government instances could be explained by simple economic specialization. Rulemaking organizations need some fluency in the system they are regulating, but do not necessarily have direct day-to-day experience with the management of that system. The job of the rulemaking organization is moni-

Table 13.1 Adaptive capabilities configurations[a]

Aspect/Case→	PM	Flood risk	NTSB	Numbers
Organization	State, IGO	State + region	State	Private transnational
Trigger	Periodic	Periodic	Stochastic	Tacit
Event	Boundary	Exogenous	Exogenous	Endogenous
Clockspeed	4–7 years	6, 12 years	1–6 months	6–18 months
Evaluator	State, Research	State, Research	State	Industry
Timing	Active	Active	Vested	Active + Vested
Concentration	3	2	1	4
Overlap	Loose	Loose	Loose	Tight
Influence	Loose	Moderate	Loose	Loose

[a]**Timing** refers to assessment timing; **Overlap** and **Influence** refer to types of coupling; **Concentration** refers to the concentration in the loci of assessment, following the relative ranking described earlier, where 1 is the most concentrated and 4 the least

toring and enforcing rules, which is conceptually different from monitoring events and designing triggers. This view explains the scarcity of planned adaptation among government agencies, but is a bit too simple.

A better explanation is that these agencies' coupling represents variants of the knowledge–policy interface. Among the state-based instances, two are combinations of state-based monitors, whose data are evaluated by both the state and external scientific research organizations. Further, the two that combine state and research have moderate concentration relative to the high concentration in the NTSB and the extremely low concentration in the numbers case. The character of the trigger and the event offer additional insights. In both cases, events are known, low clockspeed events that are monitored and evaluated over longer timescales (order of years) than numbers policy and aviation accidents (order of months). Neither presents an immediately observable hazard event; rather, these events represent changes that culminate in an adverse effect.

In both cases in which the evaluator is a combination of the state and research, the trigger is periodic. Monitoring data from long clockspeed events are fed to external researchers who can, within the bounds of the review schedule, perform evaluations as they see fit. The potential trade-off is in when friction in the knowledge–policy interface manifests. During the trigger period, the loose coupling may reduce friction. During the review period, friction in the knowledge–policy interface may increase. This friction will be revisited in the discussion of "Planning and Designing for Adaptation" (Sect. 13.4.2), and then again in the "Implications for Future Study" (Sect. 13.4.3).

In these cases, the timing of the evaluation is contingent on the clockspeed of the event, but not necessarily the source of the event. For both long clockspeed events, the timing is active. For the shorter clockspeed events, the timing has a vested element. In these latter events, evaluation is triggered by the event, focused on mitigation and remediation in both the case of the NTSB and numbers policy. Of particular interest here is the capability to invest in evaluative capacity by both the state and private actors. Conventionally, states are considered more far-sighted than industry when it comes to managing risk and uncertainty. As illustrated by these cases, they may not have the capabilities and capacity to act on that foresight, but they do have the social and political incentives. Here we see two counterexamples to the more conventionally intuitive cases of planned adaptation (PM and flood risk). In the NTSB, political factors have aligned to invest in "go-teams" necessary to respond to high clockspeed events. In the instance of numbers policy, endogenous economic incentives have led to investment in a consensus process that accounts for obligations to system integrity necessary for function and continued economic benefits.

13.4.2 Planning and Designing for Adaptation

The configurations highlighted thus far are those observed in the cases. One contribution of this conceptual model is the potential to reason about unobserved configu-

13 A Conceptual Model of Planned Adaptation (PA)

rations and under what conditions they may be feasible. While Table 13.1 illustrates clearly feasible configurations, it also hints at those *not* depicted, and that warrant discussion.

The ideal form of government-based adaptation that is tacit, endogenous, active, and vested, with high concentration and both tightly coupled overlap and influence noted earlier, is clearly not evident here. In terms of design and planning, it serves as an ideal form to start from. Which of these elements need to be relaxed to create a more feasible form? The closest instance here is numbers policy. While this instance is successful, it is a constrained case. Part of its success is that the stakeholders and the rulemakers are experts in a very narrowly scoped issue area. Further, unlike government-based rulemaking, which is often confounded by issue tying, the private transnational numbers regime has been historically decoupled from conventional political issues. It does not have to compete for resources with other issue areas, and it has the benefit of rulemakers who have a deep understanding of the implications of both primary and secondary rules. This is not necessarily the case in government-based rulemaking.

Another configuration that is not depicted in Table 13.1 is the case of high clockspeed, exogenous events with periodic triggers based on active assessment timing. Earlier, it was highlighted that high clockspeed feedback loops may create more harm than good. Even if it is assumed that rules of change can keep pace, high-speed adaptation may be ultimately detrimental for system integrity. As also noted earlier, a prized element of both rulemaking systems and the obligation to system integrity is stability.

For such a configuration the question becomes: are the effects of high clockspeed events sufficiently low impact to wait for periodic review? PM and flood risk, while not following this configuration, seem to make this choice. In the case of flood risk, the events driving active, continuous monitoring may be high impact (cost) events, such as "The Disaster" and the political implications of being "slapped in the face" again. In the case of PM, the high impact (cost) event is the cumulative effects. In both cases, cost-benefit analyses balanced these to select the active model over a vested, non-periodic model of adaptation. In the next subsection, this will be revisited in terms of how future studies can incorporate cost-benefits in terms of not only the substance of the policy analysis, but the costs of the attendant organizational structures.

Yet another configuration to consider (a variant of the ideal form) is the case in which the evaluator is the state, but the overlap and influence coupling is tight. In effect, this configuration is the original NTSB, before it was made independent of the FAA. Like the ideal form, this configuration invites issue tying. It may also be limited to domains in which the state has access to sufficient event data and experts necessary to evaluate the status and implications of a given event. Modulo issue tying, the NTSB under the FAA did have access to such data. Many technical systems are influenced more by natural events (flood risk) and industry innovation (PM, numbers). In these cases, an active assessment model requires events be observable by evaluators; PM and flood risk are such cases, but numbers is not. Further, the diversity of research perspectives demonstrated in PM and flood risk is also necessary, but missing.

The scarcity of successful instances of planned adaptation may require a planning and design strategy that not only builds on successful cases, but requires the close inspection of instances of ad hoc and failed adaptation. The instances of unobserved configurations mentioned above illustrate the analytic beginnings of such a modeling strategy. The conceptual model presented here offers an initial set of building blocks and the analytic framing to systematically speculate on unobserved, yet potential, configurations. That said, these are speculations. In the next (concluding) section, this approach will be framed as the foundation for a theory-driven sampling of instances of successful planned adaptation, ad hoc adaptation, and failed adaptation.

13.4.3 Implications for Future Study

The model presented here is a first pass at a framework for intentionally designing for planned adaptation. As noted in the introduction (Sect. 13.1), the substance of policy adaptation—domain-specific knowledge applied in conventional policy analyses—has a substantive literature dating back to the 1950s. This chapter contributes a complementary model that, building on the existing planned adaptation literature, considers policy adaptation in the context of the rulemaking system, and where elements of adaptation play out in the organizational complex that generates knowledge and applies it the process of policy adaptation. This concluding subsection outlines three research directions for the future study of planned adaptation:

1. A survey to identify not only additional cases of planned adaptation, but also borderline cases, instances of ad hoc adaptation, and failed attempts at planned adaptation;
2. Case studies and (where possible) interviews informing the processes that contributed to cases identified in the survey above, to refine the model presented here;
3. Integration of this model into conventional, substantive policy analyses.

An effective model is not only capable of expressing successful outcomes, but must also be capable of expressing instances of failures. More precisely still, it must be capable of expressing the criteria for success and the range of success and failure. The survey proposed would use the model presented here to develop a theoretically informed sample. Building on the unobserved configurations identified above, this effort would first further elaborate this initial set of unobserved configurations, then perform a survey of the literature on science and technology policy to identify cases that fit particular configurations or hint at unidentified configurations.

Given this body of "natural experiments," the next step is to perform a deep dive into these cases to precisely elaborate the mechanics of how efforts at adaptation either succeeded or failed. These case analyses would characterize each case in terms of the model presented here. That said, recall that this model was noted in the introduction as the first iteration in contingent theory building. As instances of successes and failures are characterized, the conceptual model will itself be refined to better represent both successes and failures. Pragmatically, given sufficient refinement, such a model would not only identify the mechanics of individual instances of adaptation, but the critical paths from one configuration to another. Even in this limited sample, the case histories for both flood risk and the NTSB provide early illustrations of such a critical path: flood risk followed a path from an ad hoc configuration to a planned configuration; the NTSB moved from a configuration with tightly coupled influence to a more independent configuration that preserved necessary overlap. Ultimately, this model would move from a planned adaptation space comprising relatively independent configuration points to a family of planned adaptation configurations, replete with observed critical paths, annotated with necessary resource endowments and investments, from one configuration to another.

Finally, this chapter recommends integrating this conceptual model into conventional policy analyses. Like the discussion of the unexpected costs of PM standards by Petersen et al. (2006), policy analyses usually incorporate cost-benefit analyses, but generally do not account for the organizational costs of capabilities and capacity to adapt. This model provides the foundations for systematically identifying the various loci of organizational costs. In particular, the variables describing loci of assessment and coupling can provide a starting point for understanding the cost of friction at the knowledge–policy interface. Such an integration would contribute to moving conventional policy analyses from mathematical models to more immediately applicable prescriptions that would appeal to both those concerned with obligations to system integrity and bureaucrats charged with evaluating the tractability of a solution. This latter may create its own feedback loop, reducing the friction at the knowledge–policy interface for expert policy analysts confronted by organizational barriers to effective implementation of planned adaptation models.

References

ARIN. (2009). *ARIN Policy Development Process*. Retrieved from https://www.arin.net/policy/pdp.html.
Bijker, W. E. (2007). American and Dutch coastal engineering: Differences in risk conception and differences in technological culture. *Social Studies of Science, 37*(1), 143–151.
George, A. L., & Bennett, A. (2005). *Case studies and theory development in the social sciences*. Cambridge, MA: The MIT Press.
Haas, P. M. (1992). Introduction: Epistemic communities and international policy coordination. *International Organization, 46*(1), 1–35.
Hart, H. L. A. (1994). *The concept of law*. Oxford, UK: Oxford University Press.
Keohane, R. O. (1984). *After hegemony: Cooperation and discord in the world political economy*. Princeton, New Jersey: Princeton University Press.
McCray, L. E., Oye, K. A., & Petersen, A. C. (2010). Planned adaptation in risk regulation: An initial survey of US environmental, health, and safety regulation. *Technological Forecasting and Social Change, 77*(6), 951–959.
Petersen, A. C., & Bloemen, P. (2015). Planned adaptation in design and testing of critical infrastructure: The case of flood safety in The Netherlands. In *International Symposium for Next Generation Infrastructure Conference Proceedings,* 30 September–1 October 2014. Schloss Laxenburg, Vienna, Austria: International Institute of Applied Systems Analysis (IIASA).

Petersen, A. C., van der Sluijs, J., Tuinstra, W., & Martin, K. C. (2006). *Anticipation and adaptation in particulate matter policy: The European Union, the Netherlands, and United States*, 10–11 October 2006. Washington, DC: TransAtlantic Uncertainty Colloquium (TAUC) Workshop.

Resnick, P. (2014). On Consensus and Humming in the IETF, RFC7282. https://datatracker.ietf.org/doc/draft-resnick-on-consensus/.

Sowell, J. H. (2015). *Finding order in a contentious internet* (Ph.D. thesis). Cambridge, MA: Massachusetts Institute of Technology.

Dr. Jesse Sowell (Texas A&M University, Bush School of Government and Public Service) studies the capabilities and capacities of the transnational institutions that support Internet and cybersecurity operations. In particular, Dr. Sowell is interested in the role of consensus-based decision-making processes as a form of credible knowledge assessment necessary to adaptation in large, complex engineering systems. Dr. Sowell completed a Ph.D. in Technology, Management, and Policy at the Massachusetts Institute of Technology in 2015. From October 2016 to August 2018, Dr. Sowell was a Postdoctoral Cybersecurity Fellow at Stanford's Center for International Security and Cooperation. In August 2018, Dr. Sowell joined the Department of International Affairs at Texas A&M's Bush School of Government & Public Service.

Open Access This chapter is licensed under the terms of the Creative Commons Attribution 4.0 International License (http://creativecommons.org/licenses/by/4.0/), which permits use, sharing, adaptation, distribution and reproduction in any medium or format, as long as you give appropriate credit to the original author(s) and the source, provide a link to the Creative Commons licence and indicate if changes were made.

The images or other third party material in this chapter are included in the chapter's Creative Commons licence, unless indicated otherwise in a credit line to the material. If material is not included in the chapter's Creative Commons licence and your intended use is not permitted by statutory regulation or exceeds the permitted use, you will need to obtain permission directly from the copyright holder.

Chapter 14
DMDU into Practice: Adaptive Delta Management in The Netherlands

Pieter J. T. M. Bloemen, Floris Hammer, Maarten J. van der Vlist, Pieter Grinwis and Jos van Alphen

Abstract

- Generic rules on how to organize the process of putting DMDU approaches and tools into actual practice are currently lacking.
- Lessons can be drawn from the Adaptive Delta Management (ADM) approach used in the Dutch Delta programme on flood risk management, freshwater availability, and spatial adaptation.
- In the context of putting a DMDU approach into practice, three consecutive phases can be distinguished, with specific aspects that require extra attention: (I) Directly following the political decision to actually start a long-term programme that will deal with the deeply uncertain issue, the programme itself should be designed to keep political involvement "at arm's length". (II) Strategy development requires a narrative that explains how uncertainty is dealt with; that narrative should match the specific societal and political context of the moment. (III) Implementing adaptive strategies requires organizational arrangements for systematically accommodating adjustments, a monitoring system for timely detecting of signals, and a decision-making process that links directly to its output.
- On a more general level, it can be concluded that DMDU approaches, such as described in the scientific literature, can profit from feedback—feedback from other researchers and feedback from practitioners. Organizing the latter, an

---

P. J. T. M. Bloemen (✉) · J. van Alphen
Staff Delta Programme Commissioner, Ministry of Infrastructure and Water Management, The Hague, The Netherlands
e-mail: pieter.bloemen@deltacommissaris.nl

F. Hammer
Independent consultant, specializing in executing technical programmes and projects in complex political environments, The Hague, The Netherlands

M. J. van der Vlist
Rijkswaterstaat, Ministry of Infrastructure and Water Management, The Hague, The Netherlands

P. Grinwis
Member of the Municipal Council of The Hague and Policy Officer for the ChristenUnie in the Dutch Senate and the House of Representatives, The Hague, The Netherlands

© The Author(s) 2019
V. A. W. J. Marchau et al. (eds.), *Decision Making under Deep Uncertainty*,
https://doi.org/10.1007/978-3-030-05252-2_14

instrumental element of "coproduction of knowledge", might be more time consuming, but is likely to be very effective.

14.1 Organizational Aspects of Putting a DMDU Approach into Practice

It is implicitly assumed by DMDU scholars that their approaches will be welcomed—that they will be embraced politically and accepted institutionally—suggesting that implementing a DMDU approach is mainly a technical and intellectual challenge. Practice, however, shows that in real-life decisionmaking, organizational aspects play a major role in determining the willingness and ultimate success in applying approaches for dealing with deep uncertainty. This chapter explores how the notion of deep uncertainty influences the organization of processes involved in setting up and running an execution-oriented policy programme.

In a typical execution-oriented policy programme, three consecutive and sometimes partly overlapping phases can be distinguished: (I) the phase prior to the actual start of the programme, in which political commitment is mobilized; (II) the phase of strategy development and decisionmaking, focusing on the analysis of the challenges and on inventorying and selecting options for intervening; and (III) the phase of elaboration and implementation of the strategies, in which monitoring and adjustment of strategies are central. These phases may overlap, and it may be necessary to pass through them more than once before producing a stable output.

This chapter distinguishes, for each of these three phases, one aspect that is typical for that phase (and that requires a different way of organization in the case of deep uncertainty compared with the traditional way):

- In Phase I (prior to the actual start of the programme), the aspect that is singled out is the balance between politicization and de-politicization. In the case of deep uncertainty, there is a high risk of under- and overspending. This heightens the threshold for political commitment to take the lead. Actually, addressing this type of issue often requires a deep breath and consistency over long periods of time (implying politics "at arm's length").
- In Phase II (strategy development and decisionmaking), the aspect that is analyzed concerns the way uncertainties are addressed in strategy development. In a situation of deep uncertainty, the statistics necessary for calculating an optimal solution are not available. Several approaches are available for dealing with deep uncertainty, often developed for specific contexts.
- In Phase III (elaborating, implementing, and adjusting the strategy), the design of a monitoring and evaluation system that permits timely adjustments is essential. Choosing an adaptive strategy implies higher requirements for generating and interpreting data on actual and possible future changes in system[1] conditions and

---

[1] Different categories of systems can be distinguished (physical, societal, and cyber, among others), and combinations. In the context of ADM, reference is made to physical systems.

system integrity.[2] The dynamic character of the strategies, and the high stakes involved, require continuous alertness of decision makers.

In Sects. 14.3–14.5, these phases, aspects, and organizational implications are illustrated using the experiences in the Netherlands with respect to climate adaptation. The final section (Sect. 14.6) reflects on the results of the way in which uncertainty has been dealt with so far in "climate proofing" of the Netherlands. Section 14.2 introduces the case study.

14.2 The Case Study: Adaptive Delta Management

The Netherlands, with a land surface of 34,000 km^2, is situated on the North Sea in the combined delta of the rivers Rhine, Meuse, Scheldt, and Ems. With about 17 million inhabitants and a gross domestic product of about € 693 billion, the Netherlands is the eighteenth largest economy worldwide (International Monetary Fund 2018). The fact that nearly 60% of the country is flood prone makes flood risk management an existential issue for the Netherlands.

The Delta programme plays a central role in climate proofing the Netherlands. The programme has its legislative foundation in the Delta Act, and has a Delta Fund with a budget of € 1 billion per year. This yearly budget is reserved until 2029—and the lifetime of the Delta Fund is prolonged by another year every year. The programme started in 2010, and presently unites the central government, provinces, municipalities, and waterboards on the improvement of flood risk management, reduction of vulnerability to water scarcity, and spatial adaptation (Delta programme Commissioner 2017; van Alphen 2016).

Involving social organizations and the business community, the public authorities that are united in the Delta programme prepared (in the period 2010–2014) five "Delta Decisions". These over-arching (or key) decisions form the basis of the work that the Netherlands will perform over the next 35 years (with a planning horizon of 2100). The decisions concern new flood safety standards, sustainable freshwater provision, climate-resilient design and construction of urban and rural areas across the Netherlands, and regionally structuring choices for flood risk management and freshwater supply in two critical regions: the IJsselmeer region and the Rhine-Meuse delta. Six regional strategies were developed iteratively, consisting of goals, measures, and a tentative timeline. These regional strategies were developed in the regional subprogrammes of the Delta programme, in which the national government, provinces, municipalities, and waterboards work together, involving the scientific community, NGO's, and the private sector.

---

[2]In line with the terminology used in chapter 13 'system integrity' is defined as "A system's state where its intended functions are being performed without degradation or being impaired by other changes or disruptions to its environments." (From 'The Law Dictionary'). Changes in system integrity may signal degradation in integrity, and may represent a failure mode.

Four "Delta Scenarios" were developed to guide the process of formulating the Delta Decisions and constructing the regional strategies (Bruggeman et al. 2011; Bruggeman and Dammers 2013; KNMI 2014). These scenarios combine the two main sources of uncertainty that determine the future water challenges: climate change and socio-economic conditions.

The existing flood safety standards were based on the size of the population and value of investments in the early 1960s. The new standards, which came into effect on January 1, 2017, take into consideration the "high end" of the four Delta Scenarios. For 2050, the date on which the new protection level has to be realized, they assume considerable climate change (average temperature +2 degrees, sea level rise +35 cm, and winter precipitation +14%), an increase in population (to 20 million people), and an increase in the value of investments (ongoing economic growth of 2.5% a year).

The Delta Plan on flood risk management and the Delta Plan on freshwater, both financed from the Delta Fund, comprise the measures from the regional strategies. The Delta Decisions, regional strategies, and two Delta Plans formed the central elements of the proposal sent to Parliament in September 2014. The proposal contains a total of 14 adaptation pathways, developed with a planning horizon of 2100. The proposal was accepted and the necessary annual budget of over € 1 billion until the end of 2028 was allocated (Delta programme Commissioner 2014).

From discussions in the year the Delta programme was set up (2009), the choice evolved to tailor "our own" approach for dealing with deep uncertainty, an approach that matches the specific characteristics and context of the Dutch Delta programme. Elements from available methods would be "cherry-picked" to build an approach that would fit well with the mission and tasks of the programme, would easily be explainable to policymakers from regional and local public authorities that were to develop the regional strategies, and would offer both structure (for consistency) and flexibility (for tailoring to theme-specific and region-specific characteristics).

The approach was named "Adaptive Delta Management" (ADM). The process of developing ADM was inspired by the practical experience of the UK Thames Estuary 2100 programme (Environment Agency UK 2012) and the development of the DAPP approach by the knowledge institute Deltares (Haasnoot et al., Chap. 4 of this book; Lawrence et al., Chap. 9 of this book). ADM follows an adaptive and integrated approach: *adaptive* in order to be able to speed up or temporize efforts or to change strategy if the actual or expected rate of climatic and socio-economic developments indicate this might be necessary (Dessai and Van der Sluijs 2007; Van Buuren et al. 2013), and *integrated* in order to address the highly interconnected fields of water management and physical developments that characterize dynamic and densely populated deltas. The value of combining an adaptive and integrated approach in water management had been shown by in-depth research on water management regimes across Europe, Africa, and Asia (Huntjens et al. 2011).

14.3 Phase I: Prior to the Start of ADM (Politicization and De-politicization)

Addressing a deeply uncertain issue such as climate adaptation in a policy programme often requires a "deep breath" and consistency over long periods of time. These characteristics need to be built into the programme. That programme needs to be built shortly after the "founding decision," in order to make maximum use of the political momentum. This implies a swift change in the character of the central process. Agenda-setting and political debates are often dominated by sweeping statements and firm standpoints; in contrast, setting up a programme requires structure and sobriety. A legislative basis for intervening, and a guaranteed budget, need to be organized. They are necessary conditions for keeping political involvement "at arm's length": well-informed on the main results of the programme, but not incentivized or compelled to actively intervene in every day operational issues. Hence the central question in this phase is: How can the necessary political commitment for starting up a programme of interventions that adequately addresses the issue of a deeply uncertain future be created, and how can stable political commitment in the implementation phase be organized on the main lines of the programme, while keeping more detailed decisions outside the everyday political arena? As stated above, generic rules are lacking, but it can be informative to describe how this was done in the process of climate proofing the Netherlands.

14.3.1 Build a Constituency for Change that Will Allow Political Commitments to Be Made

Building a constituency for change that will allow political commitments to be made is not a singular action deliberately planned and executed by an individual organization. It is an evolving process. In the case of the Delta programme, that process was fed by many organizations and events, including formal advice on the subject of climate adaptation by National Planning Bureaus and knowledge institutes (e.g., Netherlands Environment Assessment Agency (MNP) 2005); and The National Institute for Public Health and the Environment (RIVM) 2004; and National Advisory Committees (e.g., The Netherlands Scientific Council for Government Policy (WRR) 2006; Council for Housing, Spatial Planning, and Environment (VROM-raad) 2007). They stimulated the debate on what was to be expected in terms of changing climatic conditions, and what could be adequate interventions.

Political interest for climate adaptation in general was growing, but there was no consensus yet on where to begin and how to approach this issue. Following the March 2005 motion of Senator Lemstra on the necessity of an investment strategy that anticipates long-term developments like climate change and sea level rise, the Ministry of Housing, Spatial Planning, and Environment took the initiative, in May

2005, to set up the national programme *Climate Adaptation and Spatial Planning* (ARK).

Creating awareness about the necessity for climate adaptation was one of the major challenges at the start of the ARK programme. On that subject, ARK got a major boost when Hurricane Katrina caused New Orleans to flood in August 2005. Already for a long time there had been discontent with the way flood risk was managed in the Netherlands, and the (near) flooding of the Rhine and Meuse river system in 1993 and 1995 had already resulted in a decision to prepare for unprecedented future peak discharges. The flooding of New Orleans added to the pressure building up that something needed to be done. It was a wake-up call in the sense that, also in a highly developed Western country, disastrous floods are possible. That event, and the movie *An Inconvenient Truth*,[3] helped get the adaptation process under way in the Netherlands. Along that line, the Netherlands Environmental Assessment Agency wrote in the foreword of their December 2005 report on the effects of climate change in the Netherlands: "The scientifically indisputable changes in the climate and extreme weather conditions elsewhere raise questions as to the possible effects for the Netherlands" (Milieu- en Natuur Planbureau 2005, p. 3). The arguments used in the Working Plan for the ARK programme (Ministries of Housing, Spatial Planning, and Environment et al. March 2006) for mobilizing political commitment are given in Table 14.1. The columns of the table highlight respectively the priors, the updates, and the new understandings. The rows are more or less in chronological order, often overlapping in time. The organizations that contributed to the new insights range from the IPCC and the Royal Netherlands Meteorological Institute (especially rows 1–3), planning bureaus and advisory councils (especially rows 4–6), knowledge institutes, specialized consultancies, and nongovernmental organizations (especially rows 5–7), and the private sector (especially row 7).

The ARK programme, uniting the four levels of government (national, provinces, municipalities, and waterboards) and the relevant knowledge institutes, produced the first Dutch *National Adaptation Strategy* (NAS) (Ministries of Housing, Spatial Planning, and Environment et al. 2007). The NAS was formally approved by the Government in 2007. In the NAS, flood safety was identified as one of the most pressing of four issues that had to be addressed.[4] The combination of the attention for climate change from dramatic events abroad, and the processes set up by the ARK programme for drafting the NAS, further increased the public and political attention.

Developing a constituency for change might be a pre-condition for political commitment—but it does not automatically lead to it. The constituency for change was growing, but a substantial political commitment to establishing a major climate adaptation programme apparently required more than that. The grand politically mobilizing perspective was still missing. The initiative for generating that perspective would not come from the Ministry of Housing, Spatial Planning, and Environment, which

---

[3] May 2006; released in the Netherlands October 2006.

[4] The other three were: living conditions (heat stress and cloudbursts), biodiversity (shifts in ecosystems, salinization), and economy (vulnerability of vital infrastructure; transport, energy, communication).

Table 14.1 Why start <u>now</u> with a national adaptation programme (Ministries of Housing, Spatial Planning, and Environment, et al. 2006)

At first we thought …	But now …	So …
1. We will prevent climate change by reducing greenhouse gas emissions following agreements made in the UN IPCC context	… we see that the agreements are not met, that the measures that are taken are not sufficient, and that the climate will continue to change, at least in the coming decades	… parallel to the mitigation-track (reducing greenhouse gas emissions) an adaptation track is needed (reducing the negative consequences of the changing climate)
2. Climate change is a gradual process that might cause problems in the far future	… we know that climate change is also about an increase in the likelihood of extremes; and that these extremes might happen tomorrow	… time for a loud wake-up call; a broader awareness of possible consequences is indispensable
3. We should concentrate on model calculations of global weather systems	… we notice that translations to the local context require concrete information at a lower scale	… we need to connect generic signs with local experience
4. In the far future we will have to include climate change in our investment decisions	… we realize that cost efficiency of investments (e.g., in infrastructure, flood protection, etc.) will already be influenced by the climatic conditions in the coming decades	… the policy instruments that are used to compare different investment options (e.g., benefit-cost analysis) need to be re-calibrated in order to better include costs and benefits in the long term
5. The different national departments are reasonably in control of their policy field and can add a paragraph on climate adaptation	… the interaction among the different possible effects contributes to the urgency of the problem; a series of sectoral interventions will not be sufficient	… we now choose for an integrated cross-cutting approach
6. Climate change is an environmental issue that is mainly about flood protection, so certainly the responsibility of government	… climate change is a broad societal issue with large social and economic consequences for all parties	… climate adaptation requires close cooperation among governments, knowledge institutes, the private sector, and NGOs
7. Climate change is a problem now, and will be a big problem in the future	… climate change also offers opportunities	… it is important to explicate, at national, regional, and local scale, the opportunities of a changing climate, both for the private and public sectors

had the lead in the climate adaptation dossier with the ARK programme producing the NAS, but from the "competing" Ministry of Transport, Public Works and Water Management.

14.3.2 Develop Attractive and Plausible Perspectives: The Second Delta Committee

In June 2007, as the NAS was being discussed in the Cabinet, the Ministry of Transport, Public Works and Water Management published the *Water Vision* (Ministry of Transport, Public Works and Water Management 2007). The background for this vision was the backlog in maintenance of the dike system, the fact that the flood safety standards were outdated, and the feeling that climate change would sooner or later significantly increase chances of flooding if nothing were done. One of the most important suggestions of the Water Vision was to install an independent, high-level advisory committee on the subject of long-term planning and climate adaptation. The Water Vision was formally approved by the national government on September 7, 2007, and that same day a high-level *ad hoc* advisory council was installed: the Second Delta Committee.

The Second Delta Committee, combining expertise in water management with expertise in the fields of spatial planning, sustainability, civil engineering, food and agriculture, public finance, communication, and private sector involvement, was asked: "*How can we guarantee that our country will remain, for many generations to come, an attractive place to live, work, invest, and recreate?*" (Second Delta Committee 2008).

After one year, the Second Delta Committee concluded in its formal advice that the Netherlands would remain an international nexus for goods, services, and knowledge, and that it would not be necessary to displace these to higher grounds. The credo of the Committee's advice was: the situation (the challenges related to flood protection and freshwater supply) is not acute, but it is urgent to begin preparations to deal with them. The Committee concluded that flood protection and freshwater supply in the Netherlands could be guaranteed for the coming centuries, even under worst-case climate change conditions, but the investigations and preparations regarding the necessary measures should start soon (Second Delta Committee 2008). The advice was to build more robustness into the flood safety domain—both in terms of the policy processes that are necessary to speed up investments in flood safety, and in terms of the physical system of rivers, lakes, and sea, and of dunes, dams, and dikes. After delivering its formal advice, the Second Delta Committee was dissolved.

Thanks to the effective cooperation between the Second Delta Committee and the Ministry of Transport, Public Works and Watermanagement, the advice was accepted quickly. Within one week after the advice of the Second Delta Committee was published, the Cabinet published its positive reaction to the advice. Within four months, most of the Committee's recommendations were formally integrated into the

National Water Plan (Ministry of Transport, Public Works and Water Management, December 2008).

The climate adaptation policy processes were to be stabilized by installing a government commissioner, assigned for periods of seven years, a Delta Fund with a considerable yearly budget, a Delta programme for coordinating measures, and a Delta Act to consolidate this process. The advice also included suggestions for dramatically increasing the robustness of the physical system. Examples: sealing off the Rotterdam area with a ring of floodgates and sluices, constructing new canals, replacing the existing flood protection standards by standards that are 10 times stricter, and raising the water level of the central IJsselmeer Lake by 1.5 m. These were bold measures that would secure protection against floods for centuries. Framing the issue of climate change in this way helped put climate adaptation high on the political agenda. This advice had the "grandeur," the sweeping storylines, and the spectacular interventions that politics had been waiting for.

14.3.3 Enhance Public Awareness and Political Commitment

At the peak of public and political attention, generated by the advice of the Second Delta Committee, the "grand decision" was taken to actually start a large-scale climate adaptation programme. Now was the time to organize political commitment on the main lines of the programme, and at the same time build into the programme that politics are kept "at arm's length" in the details of running that programme.

A preparatory team consisting of civil servants from four Ministries[5] involved in climate adaptation was assigned the task to set up, in 2009, the Delta programme. In the context of programmeming concrete measures, there is more need for nuance than in the context of agenda-setting. Also, to distinguish the role of the Second Delta Committee (the *ad hoc* advisory council which existed for only the one year that was necessary to formulate the advice) from the role of the Delta programme Commissioner (a permanent high position laid down in the Delta Act), it was decided that the tone of voice of the Delta programme should not be one of looming disasters, but one of sobriety. That switch led to shifting from the "worst-case" futures used by the Second Delta Committee, to the "plausible" Delta Scenarios of the Delta programme Commissioner, covering a wide range of future uncertainty.

The 2014 OECD Study *Water Governance in the Netherlands—fit for the future?* concludes that the success of the Delta Works following the 1953 flooding came at the price of "a striking 'awareness gap' among Dutch citizens related to key water management functions, how they are performed, and by whom. Similarly, the perception of water risks is low." But the trust of the Dutch citizen in the capacities of the government to manage flood risks is high. And there is no real political debate about the relevance for the Netherlands of a reliable defense system against flooding.

---

[5]The Ministries of Housing Spatial Planning, and Environment; Economic Affairs; Agriculture Nature and Fishery; and Transport, Public Works and Water Management.

14.3.4 Stabilize Processes; Build Trust and Continuity into the Structure of the programme

An important ingredient of the advice of the Second Delta Committee is to install a Delta Fund especially assigned for flood risk management and freshwater availability "at distance from the regular national budgets and other funds that have a wider field of application. (…) Funds are available on the moment that they are needed; expenditure follows the realization of the necessary measures. This secures the financial resources necessary for flood risk management and freshwater availability, and prevents competition with the short-term agenda."

As observed in the formal evaluation of the Delta programme in 2016 (Office for the Senior Civil Service (ABD) 2016), the Delta programme Commissioner "operates separately from 'the issues of the day', contributes to depoliticization and continuity …". It is concluded that "The Delta Act has led to remarkable results in a period of economic setbacks and (in its first period) political instability. The de-politicization—meaning the decoupling of the heat of the day from long-term perspectives—of the water policy domain has contributed to stability in the Parliament around the Delta programme." The Delta Act guarantees "calm on the short-term (depoliticization, continuity), and (financial) security on both the short- and long-term, all conditional for a successful system as a whole."

The introduction of the new position of the government commissioner for the Delta programme (hereafter called the Delta programme Commissioner) did not provoke much political debate, and there was little media coverage. On a more generic level, introducing a government commissioner implies political recognition of the theme: a relatively free role at a certain distance from ministerial responsibilities and political interference (van Twist et al. 2013, p. 5).

In the decades preceding the disastrous 1953 flood, the risk of dikes breaching during storms was well known, but political and administrative processes consistently prioritized other themes. Deciding to install a Delta programme Commissioner with an influential role at a certain distance from politics resembles the decision of Odysseus ordering his men to bind him to the mast of the ship, and his men to plug their ears, so that the ship could safely pass the Sirens; he wanted his men *not* to follow his commands in that dangerous part[6] of the journey.[7]

14.4 Phase II: Developing Strategies and Decisionmaking

In a situation of deep uncertainty, there are no statistics available for calculating an optimal solution. As the previous chapters of the book show, there are several

---

[6]Pushing for serious action in calm times defies the immediate demands of political actors. In the case of anticipatory climate adaptation, the narrative of tying to the mast flips from times of turmoil to "calm times".

[7]In political science this is a well known strategy of credibly committing by "tying one's hands".

approaches available for dealing with deep uncertainty. It is difficult to determine up front which approach fits best with your specific issue. But: "Rather than arguing over whether to apply the approach Robust Decision Making or the Dynamic Adaptive Policy Pathways, the discussion should be which combination of deep uncertainty tools are appropriate to use given the nature of the problem situation" (Chap. 15).

14.4.1 Create a Narrative that Mobilizes Administrative and Political Decisionmakers

It was clear for all parties involved that the notion of uncertainty should have an explicit place in the process of developing strategies. By choosing for an adaptive approach, the Delta programme Commissioner gave a clear signal that resonated with both the believers and the deniers of man-induced climate change: in developing adaptive strategies and proposing costly measures he would not only pay attention to the risk of *under* investing, but also to the risk of *over*investing. Inherent to this approach for dealing with uncertainty is that adaptivity is key, and (therefore) monitoring is essential.

It also made clear that the new Delta Plans would not automatically contain the same type of gigantic structures as the Delta Plans that was drafted after the flooding of 1953. The near-floods of 1993 and 1995 already changed the flood risk management scene in the Netherlands. Climate change, uncertainty, and the adaptive approach had made their entrance (Zevenbergen et al. 2015). In the wake of Hurricane Katrina a new type of challenge arose. This time it was not about channeling political and societal commitment, mobilized by a flood or near-flood in the Netherlands. In situations in which floods or near-floods form the starting point, these conditions typically "outline" the dimensions of the type of situation that will have to be addressed. In the case of the Delta programme, these "outlining" events had *not* occurred; now it was about dealing with *deep* uncertainty about preparing for *multiple* future developments for which, at the outset, *no reliable statistical data are available*[8] and that might become disastrous.

This new situation called for a broader set of interventions, and at the same time allowed more time for designing integrated measures. There was more time and opportunity for tailoring solutions to local challenges and preferences, for combining traditional flood risk management interventions like dike reinforcement with measures providing room for the river, and for spatial measures increasing the flood resilience of built-up areas and for developing and testing emergency plans. In other words: there was time, but also a need for an integrated and innovative approach that would combine the challenge of dealing with uncertainty in future developments

---

[8]Depending on the characteristics of the challenge, appropriate monitoring may be a means to effectively collect data, and transform the data into an increased understanding of the system, thus contributing to a reduction of the uncertainty. Monitoring is critical to the narrative of adaptive strategies.

with the ambition to match solutions with ambitions in other policy fields. As shown by Funfgeld et al. (2018) "organizations working effectively with decision-support tools to adapt to climate change will need to feel ownership of them and have confidence in modifying them to suit their particular adaptation needs and organizational goals."

The narrative of Adaptive Delta Management is best summarized by its four elements (Delta programme Commissioner 2013):

- Connect short-term decisions in the wide field of spatial planning (housing, nature, infrastructure, recreation, etc.) with long-term objectives in the (more narrow) field of flood risk management, freshwater availability, and spatial planning. A typical instance is the construction of a river bypass close to the city of Nijmegen. The bypass is expected to be necessary to accommodate an expected increase in peak river discharges in the coming decades. Although not necessary for dealing with the present-day climatic conditions, it is constructed now to be sure that future urban developments will not sprawl over the allocated area.
- Develop adaptation pathways that visualize what measures address what physical conditions, and estimate when these conditions could occur under what scenario.
- In choosing strategies, look for and "rate" flexibility. The high uncertainty on the possible increase in sea level rise makes sand suppletion an attractive alternative for heightening sea dikes.
- Link Delta programme measures with other investment agendas (e.g., aging infrastructure, urban development, nature, shipping, and recreation). A typical example would be the Prins Hendrik dike in Texel. The flood safety-oriented works foreseen for the dike were adjusted to accommodate ambitions in nature conservation. The additional costs were covered by the regional, nature-oriented Wadden Fund.

14.4.2 Involve All Parties in Developing an Approach for Dealing with Deep Uncertainty

The working process for developing the envisioned approach was designed to (a) involve experts from both research institutes and responsible departments in order to arrive at (b) a state-of-the-art approach that (c) can be applied relatively easy in practice and (d) can count on commitment at the level of policymakers. The working process that was constructed in the Delta programme consisted of an organizational complex with the following three categories of organizations contributing.

Coregroup for ADM, Strategy Working Group, and Research Coordinator

A Coregroup for ADM was assigned the task of developing an approach for dealing with uncertainty that would meet the design criteria (a) through (d), described above. This group consisted of three members of the Delta programme Commissioner's staff and one principal expert from the division 'Rijkswaterstaat' of the Ministry of Infrastructure and Water Management, Rijkswaterstaat division. (These are the first four authors of this chapter.) The Coregroup coordinated all activities related to developing the ADM approach and applying it in practice. For the latter, the Strategy Working Group was added to the existing organizational complex of the Delta programme. This group consisted of the senior policy advisors of the subprogrammes of the Delta programme. Depending on the need for support in strategy development, the frequency of meetings varied from once per week to once per month. It was soon concluded that additional support was needed in the form of focused research and model development. The responsibility for that task was assigned to the Research Coordinator of the Committee's Staff (the fifth co-author of this chapter).

Ministerial Working Group for Adaptive Delta Management

A Ministerial Working Group for ADM was established in which different sections of the Ministries of Infrastructure and Environment, Economic Affairs, Agriculture Nature and Food Quality, and Finance participated. The task of this group was to scrutinize (with a specific focus on juridical, administrative, and financial aspects) the principles underlying ADM, and to discuss possible consequences—both positive and negative—of applying ADM. This process can be viewed as an *ex-ante* evaluation. The group consisted of eight senior experts with in-depth knowledge of the policy processes in their departments, and of the corresponding requirements for the substantiation of policy options. The discussions contributed to the rigor of the ADM approach, and to support from other departments for applying it in preparing decisionmaking. The group's conclusions were made public in the Delta programme 2012 (Delta programme Commissioner 2011). They concluded that the approach can contribute to reducing chances of over- and underinvesting. They also signaled the risk that keeping long-term options open might result in lost income. No legislative barriers that hinder mutual fine-tuning of measures and ambitions in other policy fields were found. Last but not least, the Ministerial Working Group urged that attention should be paid that further development of the approach should not make it more complex.

ADM Research Network and Specialized Consultancies

The ADM Research Network was closely linked to the Coregroup. The Dutch knowledge institutes with expertise relevant to the development and application of this approach (Deltares, Delft University of Technology, Wageningen University Research, Royal Netherlands Meteorological Institute, Netherlands Environmental Assessment Agency) were represented in this group. The Research Network made specialized contributions to the process and reflected on intermediate results. In addition, specialized consultancies (including Stratelligence, Blueconomy, and Royal Haskoning DHV-SMC) were actively involved both in developing the ADM approach and in assisting the regional and thematic subprogrammes of the Delta programme in applying it.

ADM was described (in its ideal form) in an extensive manual (van Rhee 2012). But all involved were aware that applying the approach to themes and regions as diverse as in the Delta programme would not be a straightforward matter of following a cookbook recipe. It was decided to actively support the tailoring (and sometimes simplification) of the approach to match characteristics of the subprogrammes, such as region-specific interests and data availability. Such further elaboration was performed "on-site" in concert with the national and regional teams responsible for developing the strategies, supported by a team of external experts from knowledge institutes and specialized consultancies.

In terms of organizational interfaces, the Coregroup for Adaptive Delta Management was the central entity. In that group, the findings and questions of the Ministerial Working Group for Adaptive Delta Management, the ADM Research Network, and the Strategy Working Group were discussed and used in the process of developing the ADM manual.

Results from, and experiences in developing and applying, ADM were presented and discussed in national and international conferences with both scientists and practitioners in, among others, the National Delta Conference in 2011 (Amsterdam), the Knowledge Conferences of the Dutch Delta programme in 2012 (Delft), 2013 (Wageningen), and 2014 (Almere), the European Conference on Climate Adaptation in 2013 (Hamburg) and 2015 (Copenhagen), the international conference Deltas in Times of Climate Change in 2014 (Rotterdam), and the annual workshops of the Society for Decision Making under Deep Uncertainty (DMDU) in 2014 (Santa Monica) and 2015 (Delft).

14.4.3 Evaluate and Upgrade the Approach Regularly

It is recommended to organize regular evaluations, looking at the approach from different angles, because they contribute to the quality of and commitment to the newly developed approach.

A large-scale survey (645 respondents) done in 2013 by the Erasmus University in Rotterdam (van Buuren and Teisman 2014) on the functioning of the Delta programme showed that 72% of all respondents agreed that the Delta programme was successful in connecting short-term decisions with long-term objectives—one of the four elements of Adaptive Delta Management. The survey indicated that Adaptive Delta Management and thinking in adaptation pathways are seen as important outcomes of the Delta programme. It was concluded that the three "core qualities" of the Delta programme that need to be preserved in the next phase are "shared ownership," "coherence," and "adaptivity."[9]

A more focused online survey was performed to gain insight into the experience in the Delta programme of applying in practice the specific aspect of ADM of adaptation pathways (Rijke 2014). The survey addressed the people that actually worked on the formulation of the Delta Decisions and regional strategies—typically policymakers from different levels of government, consultants, and practice-oriented researchers. They assigned the highest added value of applying adaptation pathways to the way

---

[9] In terms of the conceptual elements of Planned Adaptation, described in more detail in Chap. 13, the first two are elements of 'organizational interfaces and engagement'; the last is the 'outcome'.

it helps to incorporate long-term objectives (in flood safety, freshwater supply) into short-term decisions (in a broad range of sectors, including water management, urban planning, nature, aging infrastructure, recreation, and shipping). The added value on this aspect was rated 4.6 out of a maximum of 5 points. The most difficult aspects of DAPP to apply in ADM were the determination of Adaptation Tipping Points (ATPs) and the quantification of the added value of flexibility.

The formal statutory evaluation of the Delta programme, executed in 2016, stressed the importance of maintaining adaptivity in the Delta programme, but stated that it is too early to judge its added value in the phase of elaboration and implementation. Maintaining adaptivity was considered as crucial. At the same time, it ranked high in the top five of major challenges for the future. The evaluation commission urged all parties involved in the Delta programme to acknowledge the importance of maintaining this specific aspect for the successful functioning of the Delta programme.

14.4.4 Operationalize the DMDU Approach

Practical experience with applying the adaptation pathway approach in the UK (UK Thames Estuary E2100 project) and the Netherlands (Delta programme) was compared with literature studies in order to formulate challenges for its further development (Bloemen et al. 2017). Two of the most prominent *challenges* are briefly described below. The overall results of this study are summarized in Table 14.2.

Determining tipping points, in the absence of precise policy goals, for intrinsically flexible strategies, and in situations with large natural variability

An implicit assumption in the adaptation pathways approach (APA) is that some physical parameter, whether it is climatic conditions influencing probabilities of a flood, or socio-economic developments influencing possible consequences of a flooding, changes gradually—thus, slowly but surely forcing society to react, and ultimately switch to a different strategy. This approach seems to work best in the case of gradual trend-dominated developments like sea level rise, forcing a clear-cut decision on, for instance, the upgrading or replacement of a flood surge barrier.

Determining ATPs proved challenging in other contexts. Attempts to operationalize the adaptation pathways developed for freshwater availability in terms of defining when the next generation of measures should be implemented have temporarily been put on hold, as it has become clear that the policy objectives in that field were not precise enough to determine an approximate timing of the tipping points under the different climatological and socio-economic scenarios (Haasnoot et al. 2017).

The strategy chosen in the Delta programme for the threat of flooding from the sea is "beach nourishment." In this strategy, sand is supplied in the sea close to the coast, thus reducing wave erosion. This strategy is intrinsically flexible: every year the volume of sand supplied can be increased or decreased depending on the rate of

Table 14.2 Lessons learned on the application of the adaptation pathways approach

Lessons learned from the application of the adaptation pathways approach (APA)	Source	Phase
Added value of applying APA		
APA is effective in informing and mobilizing decisionmakers, and in keeping decision processes going forward; it helps to gain approval and buy-into the plan with key stakeholders	Analysis experience UK, NL Literature review	Design
Added value is highest if pathways are focused at a strategic level of decisionmaking, e.g., regarding a concrete investment decision on water-related infrastructure. It provides political support for keeping long-term options open, and motivates decisionmakers to modify their plans to better accommodate future conditions	Analysis experience UK, NL Literature review	Design
APA helps to increase awareness about uncertainties. It helps to incorporate long-term objectives into short-term decisions, offers visualization of multiple alternatives, and helps positioning measures in a physical context and in an indicative timeframe	Analysis experience UK, NL	Design
Recommendations for organizing the design of adaptation pathways in practice		
Foster "free thinking space" for the consideration of actions that may not be politically or financially acceptable in the short term	Analysis experience UK, NL	Design
A test version of adaptation pathway diagrams early in the development process is beneficial for communicating concepts and garnering stakeholder support	Analysis experience UK, NL	Design
Incorporate local information in the design of the adaptation pathways. Organize stakeholder participation in pathway development	Literature review; in line with experience UK, NL	Design

(continued)

Table 14.2 (continued)

Lessons learned from the application of the adaptation pathways approach (APA)	Source	Phase
Organize coordination at a higher level to ensure consistency	Literature review; in line with experience UK, NL	Design
Make periodic updates of the adaptation pathways	Literature review; in line with experience UK, NL	Design, Implementation
Challenges for the further development of APA		
Determining tipping points in the absence of precise policy goals, for intrinsically flexible strategies, and in situations of large natural variability	Analysis experience UK, NL	Design, Implementation
Unraveling the relations among parallel strategies implemented simultaneously	Analysis experience UK, NL	Design
Maximizing broad commitment in situations of low predictability	Analysis experience UK, NL	Implementation
Preparing a switch to transformational strategies	Analysis experience UK, NL Literature review	Design, Implementation

observed sea level rise. In this case, it is not possible (and not necessary) to determine tipping points—as long as there is enough sand available, there are no tipping points.

The monitoring of the changes in the frequency of storms, droughts, and heat waves remains difficult, due to the lack of observations of extreme events, which are by definition rare. In the case of climate change induced changes in peaks of river discharge, research combining monitoring data with model calculations shows that the natural variability in river discharge is so high that, even when rapid (but not extreme) climate change is assumed, it can take 3–4 decades before the climate change signal can actually be distilled in a statistically sound way from monitoring data on river discharges (Diermanse et al. 2010; Klijn et al. 2012).

From a practical point of view, research is needed to find alternative approaches and/or parameters for distilling the climate change signals from river discharge measurements. This could be achieved through combining data-based detection of changes in observed events with exploration of possible future events through scenarios and modeling (Hall et al. 2014). Accordingly, Haasnoot et al. (2015) identified a possible signaling role for "decreasing summer river discharge" as an indicator for changes in peak river discharge in the River Rhine.

Preparing the switch from incremental to transformational strategies

While in theory the pathways approach is "neutral" to the choice of the type and order of measures, practice shows that the selected pathway or the preferred strategy often contains incremental measures in the short term, firmer measures in the middle term, and (options for) system-changing interventions or transformational measures in the long term. The rationale behind this seems obvious: the longer the time horizon, the larger the climatic challenges, thus the heavier the required interventions.

Incremental measures are "protective," in the sense that they can be considered as investments in a further gradual improvement of the resilience of the present system. The flipside is that this may increase the transfer costs to a new or significantly modified system. Increasing the resilience of the present system may also lead to an increase in sunk costs, further increasing the threshold for switching from an incremental strategy to a transformational strategy. It is often stated that there are many plans for transformational measures; but these measures are implemented only as a reaction to extreme events. It is tempting "in calm times" to postpone highly demanding decisions that have their justification only in long-term, uncertain challenges. Comparable to England's TE 2100 Plan, the Delta programme aims "to stay ahead of major flooding." Due to climate change, transformational measures are inevitable in the long term. So at some point in time, the transition from incremental strategies to transformational strategies will have to be made. Though several authors have addressed the difficulty of making a planned shift to transformational strategies (Folke et al. 2010; de Haan et al. 2014; Lonsdale et al. 2015; Rijke et al. 2013), the issue has not yet been tackled adequately.

From this analysis, it also follows that middle-term investments in the resilience of the present system should be re-evaluated in the light of a possible future transition to a significantly modified system. Two options are to adopt shorter depreciation periods, or to consider alternative measures that are specifically designed for relatively short periods.

14.5 Phase III: Elaborating, Implementing, and Adjusting Strategies

14.5.1 Plan the Adaptation

Choosing an adaptive approach implies being able and ready to adjust frameworks, strategies, and plans. These adjustments themselves need to be prepared. Planned Adaptation, as defined by McCray et al. (2010), hints at the bureaucratic apparatus necessary *"to plan for adaptation,"* but does not explicitly address it. programmes that apply Planned Adaptation "both revise rules when relevant new knowledge appears, and take steps to produce such improved knowledge." This notion is described in detail in Chap. 13. The next subsection elaborates on the experience of applying this notion in the Delta programme.

Choosing an adaptive strategy imposes higher requirements for generating and interpreting data on actual changes and possible future changes in external physical conditions (e.g., changes in climate, socio-economic conditions, and land use), in knowledge and innovation (e.g., dike stability, modeling techniques, and technology), and societal preferences (e.g., resistance against ever-heightening dikes, willingness to pay for landscape quality, and market value of waterfront housing).

The dynamic character of adaptive strategies, and the high stakes usually involved (e.g., flooding, shortages of freshwater), require the continuous alertness of decisionmakers. The information generated by monitoring and evaluation has to be analyzed, interpreted, and translated into options for adjusting strategies, and ultimately intervening in the implementation of the strategies. Decisionmaking has to be linked directly to the output of the monitoring and evaluation system.

14.5.2 Organize the Adaptation

As described in Chap. 13, the organizational complex needs to be prepared to execute adaptive policy recommendations. The object of Planned Adaptation is the character and context of the programmes acting on new knowledge.

The Delta programme combines elements of Dynamic Adaptive Policy Pathways (Chaps. 4 and 9) with elements of Planned Adaptation (Petersen and Bloemen 2015; Chap. 13). This subsection analyzes how three generalizing elements of Planned Adaptation have been incorporated into the Delta programme: the notion of primary and secondary rules, the role of triggers and events, and the organizational characteristics of evaluation. The box below illustrates that adapting a plan is not necessarily a case of Planned Adaptation.

> **Unforeseen, ad hoc and reactive adaptation of the plan versus planned, systematic, and proactive adaptation.**
>
> Some of the projects that were part of the first Delta Plan (Delta committee 1960) that was established after the 1953 flooding, such as the Oosterschelde storm surge barrier, were not built as originally designed. The Oosterschelde storm surge barrier was adjusted following changes in societal preferences in the early 1970s. These adjustments were not foreseen. The original design was a complete closure of the Oosterschelde from the North Sea by a dam. The first 5 km of the barrier were built like that. Environmental and local economic considerations led to adjustments in that plan. The remaining 4 km were built with tidal gates that are open, but can be closed when sea levels rise above certain standards. In an interview in the magazine *De Ingenieur* (October 4, 2016) Frank Spaargaren, the head of the responsible Rijkswaterstaat section

at the time, stressed that the engineers that had worked on the original design felt that deviating from it was "a defeat." The adaptation was not planned. The adaptive approach, in contrast, assumes *beforehand* that new insights will lead to adjustments, and prepares for accommodating these adjustments; adaptation is not *ad hoc* and reactive, but is systematic and proactive.

Primary and secondary rules

McCray's notion of Planned Adaptation is fundamentally about updating rulemaking systems. Adaptation in conventional rulemaking systems, in particular legal systems, can be framed in terms of Hart's primary and secondary rules (1994). Historically, adaptations in legal systems have been ponderous—they are "low clockspeed" precisely because they are reflective of changes in societal norms. In contrast, technical systems move at a faster pace, often at a pace orders of magnitude faster than legal processes. "Primary rules describe what behaviors are appropriate and what behaviors are prohibited. In contrast, secondary rules are about how to recognize, create, maintain, and adjudicate primary rules. Adaptive processes must strike a balance between the canonical objective of regulations (creating stability) and the potential for chaos if feedback loops trigger changes too frequently" (Chap. 13). The challenge is to facilitate rulemaking capability and capacity that satisfies the obligation to system integrity while not falling into the trap of creating chaotic processes by attempting to adapt too frequently.

The Delta programme covers the policy fields of flood risk management, freshwater availability, and spatial adaptation. The field of flood risk management is dominated by substantive "primary rules" in the first of the three levels of the multi-level safety approach: the prevention level. The first level is characterized by quantitative standards for flood protection; the second and third level focus on impact reduction by flood proofing land use (level 2), and adequate disaster management to reduce fatalities (level 3). Policy development in these last two levels has not (yet) resulted in quantitative legislative standards. This also goes for the policy fields of freshwater availability and spatial adaptation.

Among recent acts that are considered important in the field of water management are the Delta Act (2012), the Water Act (2009), and the Flood Protection Act (also called the Water Defence Act (1995)). These acts set fixed time periods for reviewing the safety standards, but do not provide criteria for the adjustments themselves.

Rules, be they primary or secondary, are not necessarily of a legislative nature. The notion of agreeing up front on the criteria that should be applied for adjustments has been incorporated into the Delta programme in the form of rules for "sieving" from the observed and calculated external developments the ones that call for reconsidering of the Delta Decisions or regional strategies. Developments that have actually been measured, and for which a logical physical explanation is available and that are expected to have an impact in the near-future, are considered convincing enough; taken together they inform a set of secondary rules for reviewing the Delta Decisions (which comprise primary rules like flood safety standards). Detecting these developments, and using a set of secondary rules to monitor and evaluate, is the task of the Delta programme Signal Group, described in more detail in Sect. 14.5.3.

Triggers and events

Planned Adaptation defines a trigger as a signal that "potentially affects obligations to integrity" of the system (Chap. 13). In Chap. 13, Sowell distinguishes three types of triggers: periodic triggers and tacit triggers (both proactive in nature), and stochastic triggers (reactive in nature):

- *Periodic triggers* are clock-driven. Built into the Delta programme are two clock-driven triggers. The over-arching Delta Decisions and regional strategies are systematically reviewed every six years, and the flood protection standards are reviewed every 12 years.
- In the case of *tacit triggers,* the evaluation of the rules is the starting point, which may result in an adaptation of the policy. In this case, tacit triggers are not monitored by the bureaucracy, but by the constituency itself.[10] With tacit triggers, the evaluation of events, not the events themselves, activate the process for changing policy. Actors bound by the rules have the option to propose changes at any time; the option to change policies is always available. Characteristic of ADM is that strategies can be adjusted yearly. Different types of evaluation contribute to that process, as described in the next subsection.
- *Stochastic triggers* are "the direct product of active, continuous monitoring of a system for events that may impact the integrity of the system," (Chap. 13). Stochastic triggers are reactive.

Organizational characteristics of evaluation

As mentioned earlier in this chapter, monitoring and evaluation play a crucial role in the implementation phase of an adaptive strategy. Three organizational aspects are considered relevant in the context of Planned Adaptation: the timing of evaluation, the composition of the evaluating body, and the degree of coupling between the evaluating body and the actors adversely affected by the rules (Chap. 13).

When it comes to the timing of evaluation, Planned Adaptation distinguishes "ideal types" of assessment capabilities at the two ends of a spectrum of evaluation classes: "active' and "vested" assessment capabilities. Active assessment capabilities monitor continuously; vested assessment capabilities respond to unanticipated events. ADM typically focuses monitoring efforts on the timely detection of changes in developments that are known to be in progress, making it a canonical instance of active assessment. In addition, extreme events are analyzed to improve understanding of the way the water system reacts to changes in climate.

As for the composition of the evaluating body, the evaluative process can be concentrated or distributed. When concentrated, evaluation is performed by a single

---

[10]This is not always the case; triggers may be monitored by the bureaucracy, the constituency, or a combination of the two. See Chap. 13 for a more complete typology of trigger monitoring mechanisms and supporting organizational constructs.

organization. This may be the organization that develops rules or policy frameworks, or an organization consuming expert knowledge of the system. A distributed evaluative body comprises two or more organizations elaborating permutations of the rulemaking body and/or expert communities. In the case of flood risk management, formal evaluation is traditionally concentrated. In the Netherlands, the Rijkswaterstaat used to rely heavily on its own civil engineers to decide when flood safety rules needed to be adjusted. After the near flooding of the Rhine and Meuse rivers in 1993 and 1995, it was decided to give the rivers more room, as an alternative to further strengthening and heightening the existing dikes. This broadened the knowledge community associated with the flood safety domain. The distributed evaluative bodies brought to the decision broadened the epistemic communities in the flood safety domain. A further broadening followed in 2017, as a new type of flood safety standard developed by the Delta programme came into effect that more systematically follows the risk approach, thereby incorporating population and value of investments into the flood safety equation.

The third aspect of Planned Adaptation focuses on the relationship between the organization that assesses the rules and the parties affected by (changes in) the rules. In the case of flood risk management, a strong coupling can be observed: experts from the Rijkswaterstaat and waterboards are heavily involved in developing, applying, and adapting the rules.

14.5.3 Rethink Monitoring and Evaluation

A well-functioning monitoring and evaluation (M&E) system is the *conditio sine qua non* for the implementation of an adaptive strategy. Choosing an adaptive approach implies that a system for the timely detection and interpretation of relevant signals has to be installed, and that the decisionmaking process has to be designed to translate these signals directly into suggestions for adjustments of policy frameworks, strategies, and plans at different levels of government.

The goals for monitoring and evaluation, as set by the National Steering Group Delta programme, are:

1. Generate information on changes in external conditions that might require adjustments of policy frameworks, strategies, and plans;
2. Fuel learning processes by facilitating the sharing of successes and the exchange of opportunities for further improvements;
3. Provide a basis for the external justification of the budget and energy spent.

The last goal is standard for all administrative processes and programmes. The following text concentrates on the first two goals.

Generate information on changes in external conditions

The ambition to adjust plans in a timely manner implies that M&E is about more than the actions (output) and their effects (outcomes). It is also about developments in the surrounding world, about the effects of these developments on the effectiveness and efficiency of the programmed actions and contingent actions, and on the results of ongoing research on the assumptions underlying the plans.

The M&E system for the Delta programme, named monitoring, analyzing, acting (MAA), distinguishes three groups of developments that may require adjustment of choices and plans: "knowledge and innovation," "climatic and socio-economic developments," and "changes in societal preferences." These developments and their implications for the Delta Decisions, regional strategies, and Delta Plans are monitored by the MAA system. Observed and modeled developments can be analyzed in relation to the tipping points in the adaptation pathways. In a broader sense, only developments that are more extreme than the Delta Scenarios need attention, as the present strategies have been designed to be robust for the whole field of future conditions spanned by the Delta scenarios.

For the timely detection of external developments, the Signal Group Delta programme was set up (Haasnoot et al. 2018). It has been operational since 2017. The Signal Group is comprised of experts from knowledge institutes specialized in the domains of climate, water management, and spatial planning. In its first year, it identified and discussed 20 external developments that might require adjustments of the present strategies. Applying the secondary evaluative rules[11] described in the previous subsection, two developments were identified that would justify a systematic review of strategies and plans: an acceleration of the deterioration processes of the Antarctic ice cover (possibly leading to an increase of the sea level rise), and an increase in the frequency and intensity of cloudbursts. These developments are now the subject of in-depth research. Research programmes like the Dutch "National Knowledge and Innovation programme Water and Climate" (NKWK) have a major role in generating the needed information. The evaluative character of the secondary rules indicates active monitoring producing knowledge sufficient to inform a change in (primary) rules. Those changes are governed by (secondary) rules of change that, depending on the discretion imbued upon the evaluative body, may imply direct changes in the rules based exclusively on the findings, or be a part of a political process.

Fuel learning processes; link monitoring and evaluation results directly to decision-making

MMA is designed to facilitate multi-level programmeming in which the programmeming of measures is jointly done by the different levels of government. It focuses M&E efforts on answering four governance questions:

- Are we "on scheme"? Have the measures that were programmed been executed in time?

---

[11]MMA monitors the adaptive process, not the substance of system integrity. Monitoring and goals related to the adaptive process itself may be informed by progress on monitoring and evaluating the substance, but ultimately the objective is to ensure the integrity of the adaptive elements of the system. At a system level, this is one of the components of the obligation to system integrity.

- Are we "on track"? Do the executed and planned measures suffice to reach the goals of the programme[12]?
- Do we follow an integrated approach? Do we reach out to match the design, planning, and location of our interventions with ambitions in other policy domains?
- Do we follow a participative approach? Do we actively involve stakeholders and citizens in our planning process?

To facilitate the adaptation of the policy frameworks, strategies, and plans to new insights, MMA contains two monitoring lines: a blue line and a green line. In the blue line, the elaboration and implementation of the strategies and plans are followed. Monitoring if the measures are executed in time provides the answer to the question if the Delta programme is "on scheme." The blue line is about output. The subprogrammes of the Delta programme have the lead in answering the "on scheme" question.

The green line focuses on a timely detection of external developments that might require adjusting the Delta Decisions, Regional Strategies, or Delta Plans. Here, the specialized knowledge institutes represented in the Signal Group have the lead. The findings of the Signal Group are discussed in the Delta programme Knowledge Network, which includes the subprogrammes of the Delta programme, the responsible authorities, and the most relevant knowledge institutes.

Twice a year, the Signal Group and the Knowledge Network meet to discuss output (blue line) and external developments (green line). These discussions provide input for the answer to the question of whether the Delta programme is "on track" and for distilling statements about outcomes: Are the measures that we are executing, plus the ones that we have planned, still sufficient? Do they continue to fulfill obligations to system integrity, given observed and expected changes in external conditions, based on the goals that were formulated? Or, is it necessary to update the rules governing the adaptive elements of the system and/or the goals that represent the obligation to substantive system integrity?

These statements are then used to formulate advice on the elaboration, implementation, and adjustment of the Delta Decisions, Regional Strategies, and Delta Plans. Depending on their content, the advice targets the appropriate scales of action and government level, distinguishing among the Delta Decisions, the Regional Strategies, and the Delta Plans (see Fig. 14.1).

Figure 14.1 shows that the information coming from the blue line (monitoring the elaboration and implementation of strategies and plans—the three blocks on the left) is discussed in combination with the information from the green line (monitoring external developments—the three blocks at the top). The results of these discussions form the basis for formulating advice for the appropriate level of decisionmaking (the orange blocks on the right), for each of the three abstraction levels (the beams in the middle).

---

[12]Goals considered necessary and enough to safeguard the integrity of the system.

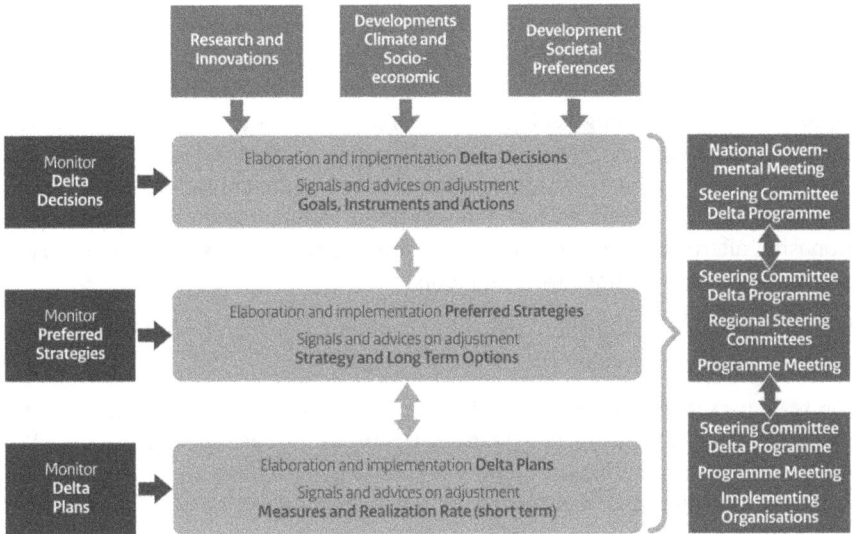

Fig. 14.1 Structure of the monitoring, analyzing, and acting (MAA) system of the Delta programme

14.6 Conclusions, Reflections, and Outlook

The preceding sections of this chapter have shown how deep uncertainty was dealt with in practice, in setting up and running a large-scale climate adaptation programme. Is enough being done and planned to prevent a next flood from happening? Do new insights require further stepping up the efforts? Or is the Netherlands already overinvesting—preparing for climate conditions that will never materialize? Only time can tell for sure. But reflecting on what was done in order to deal with deep uncertainty in the different phases, the following observations are considered worth sharing.

Phase I (prior to the start of the programme): politicization and depoliticization

The ambition to mobilize political interest in starting up an ambitious programme for dealing with an issue characterized by deep uncertainty, and to subsequently position politics at arm's length in the phase of actual programmeming of measures, has been realized to a certain degree. The Parliament and Senate agreed that the water-related challenges of climate proofing the Netherlands have to be addressed in a long-term programme, secured by the underlying Delta Act, the position of a relatively independent Delta programme Commissioner assigned the task to manage the programme, and a guaranteed budget of € 1 billion per year for twenty years.

Already in its first years, this robust structure has proven its added value. Attempts to dismantle the structure by a short-lived government in which one of the partners was not motivated to address climate change failed because of the programme's legislative foundation. The subsequent government, composed of parties that all recognized the urgency of dealing with climate change, actively contributed to keeping the process of setting up the Delta programme moving forward.

In the process of planning the actual implementation of concrete measures, like the heightening and strengthening of dikes and the creation of bypasses in the river system, politics did interfere. In specific situations, local stakeholders managed to mobilize local or national political commitments to significantly adjust or completely obstruct flood safety projects. Interventions of the Delta programme Commissioner, proposing alternative, less invasive, flood protection measures, were necessary to prevent stalemate situations from emerging.[13]

Phase II (developing strategies and decisionmaking): explicitly incorporating uncertainty into strategy development

The regular evaluations contributed to improvements over time and helped secure commitment to the working method that had evolved. Practice showed that the ADM approach for incorporating uncertainty into strategy development is relatively easy to explain to laymen and decisionmakers, and offers enough room for tailoring. The online survey on the use of ADM that addressed the people that actually worked on the formulation of the Delta Decisions and Regional Strategies helped to further improve the approach and tailor assistance to the regional subprogrammes of the Delta programme, where the strategy development was taking place. The results of the study on the challenges for the further development of the Adaptation Pathways Approach is being translated into a "research—design—use agenda."

The process of designing and discussing adaptation pathways has proven to be effective in keeping decision processes moving forward to final approval of the long-term plan. By making transparent how short-term decisions can be related to long-term tasks, it has motivated and facilitated policymakers, politicians, and other decisionmakers to incorporate uncertainty about future conditions into their decisions and plans, and to keep long-term, contingent actions open. It has helped to raise awareness about uncertainties, offered visualization of multiple alternatives, provided political support for taking the measures necessary for securing the possibility to implement contingent actions at a later stage, and motivated decisionmakers to modify their plans to better accommodate future conditions.

On a more general level, it can be concluded that DMDU approaches, such as described in the scientific literature, can profit from feedback—feedback from other researchers, and feedback from practitioners. Organizing the latter might be more time-consuming, but is likely to be very effective.

---

[13]This could also be framed as feedback that facilitated solutions that were amenable to all stakeholders, including those affected by local elements of the system being implemented or modified. Even if the political process could be seen as adversarial, it can also be viewed as constructive if framed correctly. This may be a lesson in the early feedback necessary for Phase I planning to avoid conflicts down the line.

Phase III (elaboration, implementation, and adjusting strategies): rethinking monitoring and evaluation

The notion of combining clock-driven triggers for adaptation with stochastic triggers has helped in structuring monitoring efforts. Rethinking monitoring and evaluation in the context of deep uncertainties and adaptive approaches have resulted in tailoring a monitoring and evaluation system for the Delta programme. That system, labelled "Monitor, Analyze, Act" (MAA) has been applied for 2 years. Lessons learned from these first years are:

- An M&E structure that focuses attention on a small number of central administrative questions helps to keep policymakers and researchers actively involved, and to keep national and regional steering groups interested in the output of the system;
- The processes of collecting information and discussing results unveil discrepancies among different organizations in views on the distribution of responsibilities;
- The way the MAA system is set up stimulates technical and strategic learning. Yearly and 6-yearly reviews directly follow from the processes of gathering information and discussing results.

- Outlook. It is concluded that there is still a lot of work to be done:

- The relevance of the output from the MAA system for steering *at the regional level* needs to be improved. This is crucial because the regional subprogrammes are the major source of information on the implementation process.
- The original planning of the implementation of the MAA system was too optimistic. Parts are still under construction. It has become clear that criteria for determining if the measures that are executed and planned suffice to reach the goals of the Delta programme in time (outcomes) need more reflection, research, and discussion. Also the goals of the strategies still need to be elaborated in more detail and made more concrete in order to be able to measure their outcomes.
- The relation between the criteria that can be measured and the criteria that can be modeled needs additional reflection and discussion.

References

Bloemen, P., Reeder, T., Zevenbergen, C., Rijke, J., & Kingsborough, A. (2017). Lessons learned from applying adaptation pathways in flood risk management and challenges for the further development of this approach. *Mitigation and Adaptation Strategies for Global Change*. https://doi.org/10.1007/s11027-017-9773-9.

Bruggeman, W. A., & Dammers, E. (Eds.). (2013). *Deltascenario's voor 2050 en 2100, nadere uitwerking 2012–2013*. The Hague, The Netherlands: Ministry of Infrastructure and Environment.

Bruggeman, W., Hommes, S., Haasnoot, M., Te Linde, A., & van der Brugge, R. (2011). Deltascenarios: Scenarios for robustness analysis of strategies for fresh water supply and water safety (Deltascenario's: Scenario's voor robuustheidanalyse van maatregelen voor zoetwatervoorziening en waterveiligheid). Technical Report, Deltares (in Dutch).

Council for Housing, Spatial Planning, and Environment (VROM-raad). (2007). De hype voorbij – Klimaatverandering als structureel ruimtelijk vraagstuk (in Dutch).

de Haan, F. J., Ferguson, B. C., Adamowicz, R. C., Johnstone, P., Brown, R. R., & Wong, T. H. F. (2014). The needs of society: A new understanding of transitions, sustainability and liveability. *Technological Forecasting and Social Change, 85*. http://dx.doi.org/10.1016/j.techfore.2013.09.005.

De Ingenieur – interview Spaargaren. (oktober 2016). De Oosterschelde bracht het keerpunt (in Dutch).

Delta Committee. (1960). Final report of the Delta Committee (Rapport Deltacommissie. Eindverslag en interim adviezen). Technical Report (in Dutch).

Delta Programme Commissioner. (2011). The 2012 Delta Programme Working on the delta. Acting today, preparing for tomorrow (English version). Ministry of Transport Public Works and Water Management, Ministry of Agriculture Nature and Food Quality, Ministry of Housing Spatial Planning and the Environment, Dutch national government.

Delta Programme Commissioner. (2013). The 2014 Delta Programme Working on the delta. Promising solutions for tasking and ambitions (English version). Ministry of Transport Public Works and Water Management, Ministry of Agriculture Nature and Food Quality, Ministry of Housing Spatial Planning and the Environment, Dutch national government.

Delta Programme Commissioner. (2014). The 2015 Delta Programme Working on the Delta. The decisions to keep the Netherlands safe and liveable (English version). Ministry of Infrastructure and the Environment, Ministry of Economic Affairs, Dutch national government.

Delta Programme Commissioner. (2017). The 2018 Delta Programme Working on the Delta. Continuing the work on sustainable and safe delta (English version). Ministry of Infrastructure and the Environment, Ministry of Economic Affairs, Dutch national government.

Dessai, S., & van der Sluijs, J. P. (2007). *Uncertainty and climate change adaptation: A scoping study*. Copernicus Institute for Sustainable Development and Innovation, Department of Science Technology and Society.

Diermanse, F. L. M., Kwadijk, J. C. J., Beckers, J. V. L., & Crebas, J. I. (2010). Statistical trend analysis of annual maximum discharges of the Rhine and Meuse rivers. In *British Hydrological Society Third International Symposium*, Newcastle 2010.

Environment Agency United Kingdom. (2012). *Managing flood risk through London and the Thames Estuary TE2100 Plan*. London: Environment Agency.

Folke, C., Carpenter, S. R., Walker, B., Scheffer, M., Chapin, T., & Rockstrom, J. (2010). Resilience thinking: Integrating resilience, adaptability and transformability. *Ecology and Society, 15*(4), 20. http://www.ecologyandsociety.org/vol15/iss4/art20/.

Funfgeld, H., Lonsdale, K., & Bosomworth, K. (2018). Beyond the tools: Supporting adaptation when organisational resources and capacities are in short supply. *Climatic Change*. https://doi.org/10.1007/s10584-018-2238-7.

Haasnoot, M., Schasfoort, F., Eilander, D., Diermanse, F., & Oosterberg, W. (2017). Knikpunten in zicht: een signaleringssysteem voor tijdige adaptatie in het Deltaprogramma Zoetwater; 11200588-003 © Deltares, 2017, B.

Haasnoot, M., Schellekens, J., Beersma, J. J., Middelkoop, H., & Kwadijk, J. C. J. (2015). Transient scenarios for robust climate change adaptation illustrated for water management in the Netherlands. *Environmental Research Letters*. https://doi.org/10.1088/1748-9326/10/10/105008.

Haasnoot, M., Van 't Klooster, S., & van Alphen, J. (2018). Designing a monitoring system to detect signals to uncertain climate change. *Global Environmental Change, 52*, 273–285. https://doi.org/10.1016/j.gloenvcha.2018.08.003.

Hall, J., Arheimer, B., Borga, M., Brazdil, R., Claps, P., et al. (2014). Understanding flood regime changes in Europe: A state of the art assessment. *Hydrology and Earth System Sciences, European Geosciences Union, 18*(7), 2735–2772. https://doi.org/10.5194/hess-18-2735-2014%3e%3chal-01141526.

Hart, H. (1994). *The concept of law* (2nd ed.). Oxford: Oxford University Press.

Huntjens, P., Pahl-Wostl, C., Rihoux, B., Schlüter, M., Flachner, Z., Neto, S., et al. (2011). Adaptive water management and policy learning in a changing climate: A formal comparative analysis of eight water management regimes in Europe, Africa and Asia. *Environmental Policy and Governance, 21,* 145–163.

International Monetary Fund (IMF). (2018). *World Economic Outlook Data Base.*

Klijn, F., de Bruijn, K. M., Knoop, J., & Kwadijk, J. C. J. (2012). Assessment of the Netherlands' flood risk management policy under global change. *Ambio, 41,* 180–192. https://doi.org/10.1007/s13280-011-0193-x.

KNMI. (2014). *KNMI '14 climate scenarios for the Netherlands; A guide for professionals in climate adaptation.* De Bilt, The Netherlands: KNMI.

Kwakkel, J., & Haasnoot, M. (this book, Chapter 15). Supporting decisionmaking under deep uncertainty: A taxonomy of approaches and tools.

Lawrence, J., Haasnoot, M., McKim, L., Atapattu, D., Campbell, G., & Stroombergen, A. (this book, Chapter 9). Dynamic Adaptive Policy Pathways: From Theory to Practice.

Lonsdale, K., Pringle, P., & Turner, B. (2015). Transformative adaptation: What it is, why it matters and what is needed. UK Climate Impacts Programme, University of Oxford, Oxford, UK.

McCray, L. E., Oye, K. A., & Petersen, A. C. (2010). Planned adaptation in risk regulation: An initial survey of US environmental, health, and safety regulation. *Technological Forecasting and Social Change, 77*(6), 951–959. https://doi.org/10.1016/j.techfore.2009.12.001.

Ministries of Housing, Spatial Planning, and Environment, of Economic Affairs, of Agriculture Nature and Fishery, and of Transport, Public Works and Water Management. (2006). Werkplan Nationaal Programma Adaptatie Ruimte en Klimaat (ARK).

Ministries of Housing, Spatial Planning, and Environment, of Economic Affairs, of Agriculture Nature and Fishery, and of Transport, Public Works and Water Management. (2007). The National Adaptation Strategy.

Ministry of Transport, Public Works and Water Management. (2007). The Water Vision (Watervisie. Nederland veroveren op de toekomst).

Ministry of Transport, Public Works and Water Management. (2008). The National Water Plan.

Netherlands Environment Assessment Agency (Milieu- en Natuurplanbureau). (2005). The effects of climate change in the Netherlands.

Netherlands Environment Assessment Agency (Milieu- en Natuurplanbureau) and The National Institute for Public Health and the Environment (RIVM). (2004). Risico's in bedijkte termen - een thematische evaluatie van het Nederlandse veiligheidsbeleid tegen overstromen.

Netherlands Scientific Council for Government Policy (WRR). (2006). Klimaatstrategie – tussen ambitie en realisme.

OECD. (2014). *Water governance in the Netherlands, fit for the future?* OECD studies on Water. Paris: OECD Publishing.

Office for the Senior Civil Service (ABD), Ministry of the Interior and Kingdom Relations. (2016). Statutory ex-post evaluation of the Delta Act.

Petersen, A. C., & Bloemen, P. (2015). Planned adaptation in design and testing of critical infrastructure: The case of flood safety in The Netherlands (Proceedings paper). In *International Symposium for Next Generation Infrastructure Conference Proceedings*. Retrieved 2016-02-07, http://www.ucl.ac.uk/steapp/isngi/proceedings.

Rijke, J. (2014). Adaptief deltamanagement – Ontstaansgeschiedenis en toepassing in het Deltaprogramma (in Dutch).

Rijke, J., Farrely, M., Brown, R., & Zevenbergen, C. (2013). Configuring transformative governance to enhance resilient urban water systems. *Environmental Science & Policy, 25,* 62–72. https://doi.org/10.1016/j.envsci.2012.09.012.

Second Delta Committee. (2008). Working together with water: A living land builds for its future. Findings of the Dutch Delta Committee. Ministerie van Verkeer en Waterstaat, The Hague, The Netherlands. Available from http://www.deltacommissie.com/en/advies.

van Alphen, J. (2016). The Delta Programme and updated flood risk management policies in The Netherlands. *Journal of Flood Risk Management, 9,* 310–319.

van Buuren, A., Driessen, P. J., van Rijswick, M., Rietveld, P., Salet, W., Spit, T., & Teisman, G. (2013). Towards adaptive spatial planning for climate change: Balancing between robustness and flexibility. *Journal for European Environmental and Planning Law, 10*(1), 29–53.

van Buuren, A., & Teisman, G. (2014). Samen verder werken aan de delta. De governance van het Nationaal Deltaprogramma na 2014 (in Dutch).

van Rhee, G. (2012). Handreiking Adaptief Deltamanagement – definitief concept. Stratelligence, Leiden (in Dutch).

van Twist, M., Schulz, M., van der Steen, M., & Ferket, J. (2013). *De Deltacommissaris. Een kroniek van de instelling van een regeringscommissaris voor de Nederlandse delta*. NSOB. ISBN 978-90-75297-33-I.

Zevenbergen, C., Rijke, J., Herk, S., & Bloemen, P. (2015). Room for the river: A stepping stone in adaptive delta management. *International Journal of Water Governance, 3*, 121–114. https://doi.org/10.7564/14-ijwg63.

Drs. Pieter J. T. M. Bloemen (Ministry of Infrastructure and Water Management—Staff Delta Programme Commissioner) is the Chief Strategic Officer of the Dutch Delta programme. He is responsible for the development and application of Adaptive Delta Management. He led the Strategic Environmental Assessment of the Delta Decisions and preferred regional strategies published in 2014. He presently works on the development of a monitoring and evaluation system that matches the adaptive approach of the Delta programme and is responsible for the first six-yearly review of the Delta Decisions and preferred regional strategies, planned for 2020. Bloemen has been Visiting Researcher at IHE Delft Institute of Water Education (Chair Group Flood Resilience) since January 2015 and is working on a Ph.D. thesis on the governance of the adaptive approach.

Floris Hammer, M.Sc. (formerly of Ministry of Infrastructure and Water Management—Staff Delta Programme Commissioner) worked on the initial development of ADM. He researched and described several possible applications of ADM in the Delta programme. Examples are an adaptive planning of shipping infrastructure and water quality/safety measures around the Volkerak Zoommeer region. He trained policy advisors and civil servants in applying ADM in practice. Floris is currently an independent consultant, specializing in executing technical programmes and projects in complex political environments.

Dr. ir. Maarten J. van der Vlist (Ministry of Infrastructure and Water Management, Rijkswaterstaat) is currently principle expert on adaptive water management at the Rijkswaterstaat and has a joint appointment as Associate Professor adaptive water management and land use planning at Wageningen University. He has a Ph.D. in agricultural engineering on the issue of sustainability as a task of planning. His recent work has focused on replacement and renovations (redesigning) of aging water resource infrastructure in the national waterways and national water system in the Netherlands. He was co-founder of Next Generation Infrastructures 2.0, a collaboration among the most important infrastructure managers in the Netherlands, supported by a scientific research programme. He is one of the key players of De Bouwcampus, a center for co-creation between experts on issues such as energy transition, and replacement and renovation. In his research at Wageningen University, he focusses on the impact of climate change and the related uncertainty on renewal of aging infrastructure and the design of alternatives on the system level and the object level.

Pieter Grinwis was closely involved in the establishment of the Delta Programme as a political assistant to the State Secretary for Transport, Public Works, and Water Management (2007–2010). From 2010 to 2015, he worked at the Delta programme Commissioner as a political and financial–economic advisor for the Delta Commissioner on the initial development of ADM. He was particularly concerned with valuing flexibility in investment decisions and with the level of the discount rate in relation to dealing with long-term effects. Pieter is currently a Member of the Municipal Council of The Hague and a Policy Officer for the ChristenUnie in the Dutch Senate and the House of Representatives.

Drs. Jos van Alphen (Ministry of Infrastructure and Water Management—Staff Delta Programme Commissioner) is senior advisor of the Dutch Delta Programme. During his former positions at the Rijkswaterstaat, he became an international expert on flood risk management. Presently, he is responsible for knowledge management in the Delta Programme and is Chairman of the Delta Programme Signal Group. The Signal Group monitors external developments that might influence the implementation of the adaptive strategies. From his experience with the Dutch Delta Programme, van Alphen supports the preparation and implementation of delta plans in Bangladesh and Manila Bay.

Open Access This chapter is licensed under the terms of the Creative Commons Attribution 4.0 International License (http://creativecommons.org/licenses/by/4.0/), which permits use, sharing, adaptation, distribution and reproduction in any medium or format, as long as you give appropriate credit to the original author(s) and the source, provide a link to the Creative Commons licence and indicate if changes were made.

The images or other third party material in this chapter are included in the chapter's Creative Commons licence, unless indicated otherwise in a credit line to the material. If material is not included in the chapter's Creative Commons licence and your intended use is not permitted by statutory regulation or exceeds the permitted use, you will need to obtain permission directly from the copyright holder.

Part IV
DMDU-Synthesis

Chapter 15
Supporting DMDU: A Taxonomy of Approaches and Tools

Jan H. Kwakkel and Marjolijn Haasnoot

Abstract

- A wide variety of tools and approaches for supporting the making of decisions under deep uncertainty have been put forward, but we lack a comparative overview.
- This chapter presents a taxonomy of approaches and tools for supporting decisionmaking under deep uncertainty.
- The taxonomy is based on a decomposition of the tools and approaches into a set of common building blocks.
- Analysts can use the taxonomy for designing context-specific approaches to support DMDU.

15.1 Introduction

Over the last decade, various researchers have put forward approaches for supporting decisionmaking under deep uncertainty (DMDU). For example, Lempert et al. (2006) put forward Robust Decision Making. This was later expanded by Kasprzyk et al. (2013) into Many-objective Robust Decision Making. Other researchers put forward adaptive policymaking (Walker et al. 2001; Kwakkel et al. 2010a) and adaptation pathways (Haasnoot et al. 2012), which were subsequently combined into Dynamic Adaptive Policy Pathways (Haasnoot et al. 2013). At present, a variety of approaches and tools are available, but there is little insight into how the approaches and tools are similar or different, where they overlap, and how they might be meaningfully combined in offering decision support in a specific context. The aim of this chapter is to offer some thoughts on these questions.

J. H. Kwakkel (✉)
Delft University of Technology, Delft, The Netherlands
e-mail: J.H.Kwakkel@tudelft.nl

M. Haasnoot
Water Resources and Delta Management, Deltares, Delft, The Netherlands

Faculty of Physical Geography, Utrecht University, Utrecht, The Netherlands

To provide a tentative, more synthetic view of the DMDU field, this chapter will first discuss the key ideas that underpin the field, followed by a proposed taxonomy. These key ideas are an attempt at articulating the presuppositions that emanate from much of the DMDU literature. They have emerged out of discussions over the last few years with various people. The taxonomy builds on an earlier taxonomy put forward by Herman et al. (2015). It decomposes each approach into its key building blocks. This exercise is particularly useful, for it enables one to move beyond a single approach. By focusing on the building blocks that are available, analysts can start to compose context-specific approaches for supporting DMDU. Moreover, it allows academics to move away from very general statements at the level of comprehensive approaches and instead focus on specific differences and similarities at the level of building blocks.

A challenge one faces when trying to put forward a taxonomy of approaches is that they are subject to interpretation, offer quite some leeway to the user in how they are being used, and are changing over time. A taxonomy pigeonholes approaches and thus runs the risk of failing to do sufficient justice to the intrinsic flexibility that currently exists in practice with respect to DMDU approaches. This risk is somewhat alleviated in the present case, since the focus is on the building blocks that make up the various approaches. By understanding an approach as being composed of building blocks, changes to approaches may be construed as swapping one building block with another one.

The remainder of this chapter is organized as follows. In Sect. 15.2, a broad overview is given of the key ideas that underpin the literature on supporting DMDU. In Sect. 15.3, we present a taxonomy for the various approaches and tools. In Sect. 15.4, we discuss how each of the approaches and tools fit into this taxonomy. Section 15.5 presents our concluding remarks.

15.2 Key Ideas

Decisionmaking with respect to complex systems requires coming to grips with irreducible uncertainty. This uncertainty arises out of intrinsic limits to predictability that occur when dealing with a complex system. Another source of uncertainty is that decisionmaking on complex systems generally involves a variety of stakeholders with different perspectives on what the system is, and what problem one is trying to solve. A third source of uncertainty is that complex systems are subject to dynamic change and are never completely understood.

The intrinsic limits to predictability, the existence of legitimate alternative interpretations of the same data, and the limits to knowability of a system have important implications for decisionmaking. Under the label of 'decisionmaking under deep uncertainty', these are now being explored. Deep uncertainty means that the various parties to a decision do not know or cannot agree on how the system works, how likely various possible future states of the world are, and how important the various outcomes of interest are (Lempert et al. 2003). This suggests that under deep uncertainty, it is possible to enumerate possible representations of the system, to list plausible futures, and to enumerate relevant outcomes of interest without being able to rank order these representations of the system, the list of plausible futures, or the enumeration of outcomes of interest in terms of likelihood or importance (Kwakkel et al. 2010b).

In the literature, there is an emerging consensus that any decision regarding a complex system should be robust with respect to the various uncertainties. Intuitively, a decision is robust if its expected performance is only weakly affected by the actual future states that emerge as a function of the values actually observed among the various deeply uncertain factors. Various operationalizations of this intuition may be found in the literature (Herman et al. 2015; Giuliani and Castelletti 2016; Kwakkel et al. 2016a; McPhail et al. 2018). On the one hand, there are robustness metrics that focus on the performance of individual policy options and assess their performance over a set of plausible futures. Well-known examples include minimax and the domain criterion (Starr 1963; Schneller and Sphicas 1983). On the other hand, there are metrics of the performance of policy options relative to a reference point. The best-known example of this type is Savage's minimax regret (Savage 1951), which uses the best possible option for a given future as the reference point against which all other options are to be evaluated.

Over the last decade, a new strategic planning paradigm, known as 'decisionmaking under deep uncertainty,' has emerged that aims to support the development of robust plans. This paradigm rests on three key ideas: (i) Exploratory Modeling; (ii) adaptive planning; and (iii) decision support.

15.2.1 Exploratory Modeling

The first key idea is Exploratory Modeling (EM). In the face of deep uncertainty, one should explore the consequences of the various presently irreducible uncertainties for decisionmaking. Typically, in the case of complex systems, this involves the use of computational scenario approaches.

The idea to systematically explore the consequences of the various uncertainties that are present is rooted in the idea of what-if scenario thinking. Scenarios are (plausible) descriptions of what the future might look like. Scenario thinking is a means for thinking about possible threats and opportunities that the future might hold and their impacts on an organization, business, or system. Scenario thinking gained prominence in part due to pioneering work by Shell in the late 1960s. One of the scenarios that emerged described a very rapid rise in oil price, forcing Shell to consider futures quite different from business as usual. This was believed to have given Shell a competitive advantage during the oil crisis of 1973. Thinking with scenarios when making decisions may help in choosing options that perform reasonably well under a wide range of conditions.

Why is there such a strong insistence on the use of models? There is ample evidence that human reasoning with respect to complex uncertain systems is intrinsically insufficient. Often, mental models are event based, have an open-loop view of causality, ignore feedback, fail to account for time delays, and are insensitive to nonlinearity (Sterman 1994). In complex systems, the overall dynamics, however, are due to accumulations, feedbacks, and time delays, with nonlinear interactions among them. Thus, mental simulations of complex systems are challenging to the point of infeasibility. This is confirmed empirically in various studies (Sterman 1989; Brehmer 1992; Kleinmuntz 1992; Diehl and Sterman 1995; Atkins et al. 2002; Sastry and Boyd 1998). This strongly suggests that it is worthwhile to support human reasoning on uncertain complex systems with simulation models that are much better at adequately deriving the consequences from sets of hypotheses pertaining to the functioning of these systems (Sterman 2002).

EM is a research method that uses computational experimentation for analyzing complex and uncertain systems (Bankes 1993; Bankes et al. 2013). In the presence of deep uncertainty, the available information enables the development of a set of models, but the uncertainty precludes the possibility of narrowing down this set to a single true representation of the system of interest. A set of models that is plausible or interesting in a given context is generated by the uncertainties associated with the problem of interest and is constrained by available data and knowledge. A single model drawn from the set is not a prediction. Rather, it is a computational what-if experiment that reveals how the real-world system would behave if the specific assumptions about the various uncertainties encapsulated in this model were correct. A single what-if experiment is typically not that informative other than to suggest the plausibility of its outcomes, which in turn may contribute to the substantiation of the necessity to intervene. Instead, EM aims to support reasoning and decisionmaking on the basis of a comprehensive set of such models for the system of interest. In contrast to more traditional scenario planning approaches, EM allows reasoning over a much larger set of cases than a scenario process can generate, while maintaining consistency across the set of cases. The analysis of this set of cases allows humans to infer systematic regularities among subsets of the full ensemble of cases. Thus, EM involves searching through the set of models using (many-objective) optimization algorithms, and sampling over the set of models using computational design of experiments and global sensitivity analysis techniques.

15.2.2 *Adaptive Planning*

The second key idea underpinning many deep uncertainty approaches is the idea of adaptive planning. Adaptive planning means that plans are designed from the outset to be adapted over time in response to how the future may actually unfold. The way a plan is designed to adapt in the face of potential changes in conditions is announced simultaneously with the plan itself rather than taking place in an ad hoc manner *post facto*. The flexibility of adaptive plans is a key means of achieving decision robustness. While the future is unfolding, many deep uncertainties are being resolved. Having an adaptive plan allows decisionmakers to adapt the implementation of the plan in response to this. This means that a wide variety of futures have to be explored during plan design. Insight is needed into which actions are best suited to which futures, as well as what signals from the unfolding future should be monitored to ensure the timely implementation of the appropriate actions. The timing of plan adaptation is not known a priori; it depends on how the future unfolds. In this sense, adaptive planning differs from planned adaptation (see Chapter 13), where changes generally occur at predetermined moments (e.g., every 5 years) and which entails a review of conditions that results in an adaptation of the original plan. Adaptive planning involves a paradigm shift from planning in time to planning conditional on observed developments.

15.2.3 Decision Support

The third key idea is decision support. Decisionmaking on complex and uncertain systems generally involves multiple actors coming to agreement. In such a situation, decisionmaking requires an iterative approach that facilitates learning across alternative framings of the problem, and learning about stakeholder preferences and trade-offs, in a collaborative process of discovering what is possible (Herman et al. 2015). In this iterative approach, the various approaches for decisionmaking under deep uncertainty often put candidate policy decisions into the analysis by stress testing them over a wide range of uncertainties. The uncertainties are then characterized by their effect on the decision. The challenges inherent in such processes are reviewed in depth by Tsoukiàs (2008). He envisions that various decision support approaches and tools are used to enable a constructive learning process among the stakeholders and analysts. Decision support in this conceptualization moves from an a priori agreement on (or imposition of) assumptions about the probability of alternative states of the world and the way in which competing objectives are to be aggregated with the aim of producing a preference ranking of decision alternatives, to an a posteriori exploration of trade-offs among objectives and the robustness of this performance across possible futures. That is, under deep uncertainty decision support should move away from trying to define what is the right choice and instead aim at enabling deliberation and joint sense making among the various parties to a decision.

15.3 A Taxonomy of Approaches and Tools for Supporting Decision Making Under Deep Uncertainty

The availability of a variety of approaches for supporting the making of decisions under deep uncertainty raises a new set of questions. How are the various approaches different? Where do they overlap? Where are they complementary? What computational support tools do they use? Answering these questions may help pave the way for the future harmonization and potential integration of the various approaches. It might also help in assessing if certain approaches are more applicable in certain decisionmaking contexts than others. Both Hall et al. (2012) and Matrosov et al. (2013b) compare Info-Gap Decision Theory and Robust Decision Making. They conclude that, along quite different analytical paths, both approaches arrive at fairly similar but not identical results. Matrosov et al. (2013a) compare Robust Decision Making with an economic optimization approach. In this case, the results from applying both techniques yield different analytical results, suggesting value to efforts seeking to combine both approaches. Roach et al. (2015, 2016) compare Info-Gap Decision Theory and robust optimization. They conclude that there are substantial differences between the plans resulting from these two approaches and argue in favor of mixed methodologies. Gersonius et al. (2015) compare a Real Options Analysis (in detail reported in Gersonius et al. 2013) with an Adaptation Tipping Point analysis (Kwadijk et al. 2010). They highlight the substantial differences in starting points and suggest that both approaches could be applied together. Essentially, the same is argued by Buurman and Babovic (2016), who compare DAPP and Real Options Analysis. Kwakkel et al. (2016b) compare DAPP and RDM. They argue in favor of combining both approaches. DAPP primarily provides a systematic structure for adaptive plans, while RDM provides a clear iterative model-based process for designing adaptive plans.

To move beyond a discussion of the similarities and differences among the various DMDU approaches, a taxonomy of the components that make up the approaches is useful. A first such taxonomy was put forward by Herman et al. (2015). This taxonomy focused on model-based robustness frameworks for supporting decisionmaking under deep uncertainty. In light of the key ideas introduced in the previous section, one could say that the Herman taxonomy focused exclusively on EM. The purpose of this chapter, however, is to cover the broader field and not restrict ourselves to approaches based exclusively on EM. Practically, this means that there is a need to supplement the Herman taxonomy with an additional category related to adaptive planning (the second key idea).

Figure 15.1 shows the proposed taxonomy of the components that make up the various DMDU approaches. It covers five broad categories:

1. *Policy architecture*, which covers the various ways in which adaptive policies may be structured.
2. *Generation of policy alternatives*, which covers how policy alternatives or components thereof are identified, given a specification of the available policy levers.
3. *Generation of scenarios*, which covers how context scenarios are identified given a variety of uncertainties.
4. *Robustness metrics*, which cover the various ways in which policy robustness is operationalized.
5. *Vulnerability analysis*, which covers the various analysis techniques that are used to understand how policy robustness is influenced by both uncertainties and policy levers.

15 Supporting DMDU: A Taxonomy of Approaches and Tools

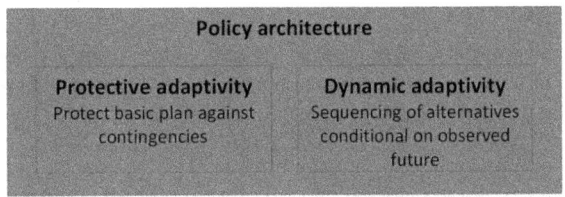
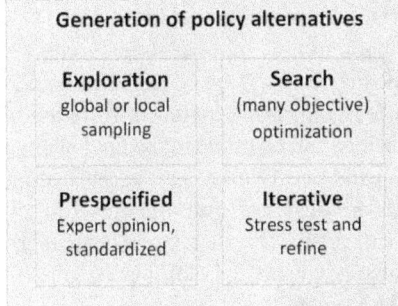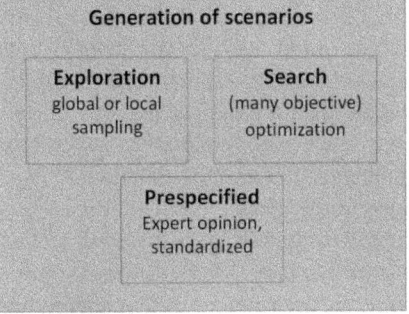
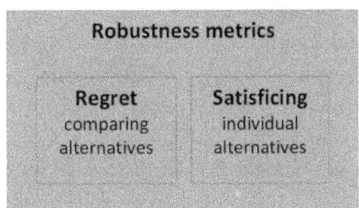
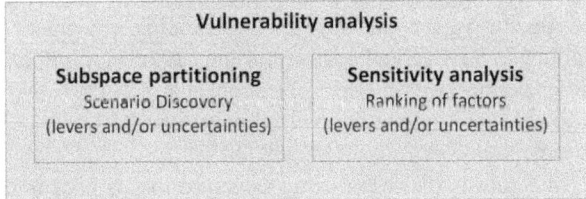

Fig. 15.1 A taxonomy of components that make up the DMDU approaches

In the following subsections, we discuss each of these categories in more detail.

15.3.1 Policy Architecture

In the deep uncertainty literature and before, various ideas on adaptive plans or policies and the importance of flexibility in planning and decisionmaking can be found. Under deep uncertainty, static plans are likely to fail, become overly costly to protect against failure, or are incapable of seizing opportunities. An alternative is to design flexible plans that can be adapted over time. In this way, a policy may be yoked to an evolving knowledge base (McCray et al. 2010). Adaptive actions are implemented only if, in monitoring how the various uncertainties are resolving, there is a clear signal that these actions are needed. Signals can come both from monitoring data and from computer simulations of future developments. Many examples of authors arguing in favor of this paradigm may be found (Albrechts 2004; Schwartz and Trigeorgis 2004; Swanson et al. 2010; Eriksson and Weber 2008; Walker et al. 2001; Kwakkel et al. 2010a; Lempert et al. 2003; de Neufville and Odoni 2003).

The initial seeds for the adaptive planning paradigm were sown almost a century ago. Dewey (1927) put forth an argument proposing that policies be treated as experiments with the aim of promoting continual learning, and adapted in response to experience over time. Early applications of this idea can be found in the field of environmental management (Holling 1978; McLain and Lee 1996) where, because of the uncertainty about system functioning, policies are designed from the outset to test clearly formulated hypotheses about the behavior of an ecosystem being changed by human use (Lee 1993). A similar attitude is also advocated by Collingridge (1980) with respect to the development of new technologies. Given ignorance about the possible side effects of technologies under development, he argues that one should strive for correctability of decisions, extensive monitoring of effects, and flexibility. Policy learning is also major issue in evolutionary economics of innovation (De La Mothe 2006; Faber and Frenken 2009; Mytelka and Smith 2002).

More recently, Walker et al. (2001) developed a structured, stepwise approach for dynamic adaptation. They advocate that policies should be adaptive: One should take only those actions that are non-regret and time-urgent, and postpone other actions to a later stage. They suggest that a monitoring system and a pre-specification of responses when specific trigger values are reached should complement a basic policy. The resulting policy is flexible and adaptive to the future as it unfolds. The idea of adaptive policies was extended further by Haasnoot et al. (2012), who conceptualized a plan as a sequence of actions to be realized over time. In later work, they called this adaptive policy pathways (Haasnoot et al. 2013).

A policy architecture is the overarching structure that is used to design a plan. A range of adaptive policy architectures is possible. At one extreme (left box under 'Policy architecture' in Fig. 15.1), we have a basic plan to be implemented immediately, complemented by a set of contingency actions that are to be implemented if and when necessary (see Fig. 15.2).

This style of policy architecture can be categorized as 'protective adaptivity.' It is implicitly advocated in Assumption-Based Planning (Dewar 2002; Dewar et al. 1993). Dynamic Adaptive Planning (Walker et al. 2001; Kwakkel et al. 2010a; Chap. 3) also sits in this box. The other extreme no longer views a basic policy as a basic plan plus contingency actions but as a series of actions, the implementation of which coevolve with how the future unfolds. This style of policy architecture can be categorized as 'dynamic adaptivity.' Dynamic Adaptive Policy Pathways (Haasnoot et al. 2013; Chap. 4) exemplifies this type of policy architecture (see Fig. 15.3).

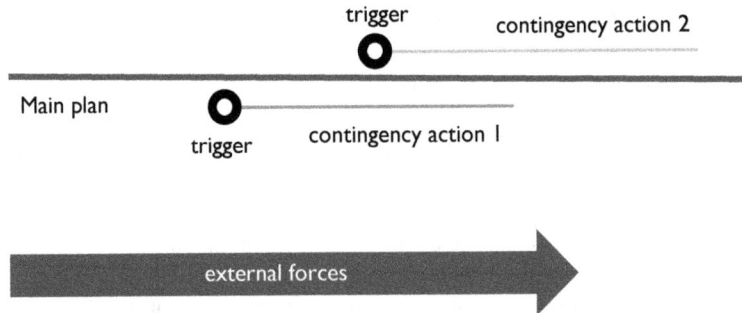

Fig. 15.2 Adaptive policy

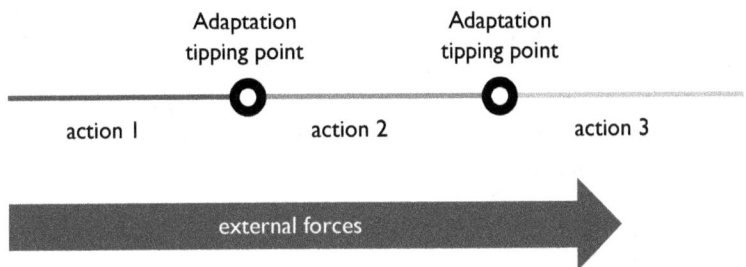

Fig. 15.3 Adaptation pathway

15.3.2 Generation of Policy Alternatives and Generation of Scenarios

Given a wide variety of deeply uncertain factors and a set of policy levers that may be used to steer the system toward more desirable functioning, the analyst has to choose how to investigate the influence of both uncertainties and levers on outcomes. Broadly speaking, a distinction can be drawn between two different strategies: 'exploration' and 'search.' Exploration strategies investigate the properties of the uncertainty space and the policy lever space by systematically sampling points in these spaces and evaluating their consequences. Exploration relies on the careful design of experiments and can use techniques such as Monte Carlo sampling, Latin Hypercube sampling, or factorial methods. Exploration may be used to answer questions such as 'under what circumstances would this policy do well?', 'under what circumstances would it likely fail?', and 'what dynamics could this system exhibit?'. Exploration provides insight into the global properties of the uncertainty space and the policy lever space.

In contrast, search strategies hunt through these spaces in a more directed manner, searching for points with particular properties. Search may be used to answer questions such as 'What is the worst that could happen?', 'What is the best that could happen?', 'How big is the difference in performance between rival policies?', 'What would a good strategy be given one or more scenarios?' Search provides detailed insights into particular points in the uncertainty space (called *scenarios*) or in the policy lever space (called *policy alternatives*). Search relies on the use of (many-objective) optimization techniques.

A third strategy is to have pre-specified scenarios or policies, instead of requiring systematic investigation of the entire uncertainty space and policy lever space. For example, in its original inception, RDM assumed that a set of candidate policies is pre-specified. The performance of these policies is then investigated over a wide range of scenarios. Here, the focus of the exploration is on the impact of the uncertainties on the performance of the pre-specified policies.

In practice, the different strategies for investigating the impacts of uncertainties and policy levers can be combined, as well as executed in an iterative manner. For example, if exploration reveals that there are distinct regions in which a policy fails, search may be employed to identify more precisely where the boundary is located between these distinct regions. (This is the essence of Scenario Discovery.) Another example is to use exploration to identify the conditions under which a policy fails and use this insight to modify the policy. By iterating, a policy can be designed that performs acceptably under a wide range of future conditions. RDM, for example, strongly emphasizes the iterative refinement of candidate policies.

15.3.3 Robustness Metrics

A well-established distinction regarding robustness metrics is the distinction between 'regret' metrics and 'satisficing' metrics (Lempert and Collins 2007). *Regret metrics* are comparative and originate from Savage (1954, p. 21), who defines the regret of a policy option as the difference between the performance of the option in a specific state of the world and the performance of the best possible option in that state of the world. Using Savage's definition, a robust policy is one that minimizes the maximum regret across the states of the world (SOWs). Other regret metrics use some type of baseline performance for a given state of the world instead of the performance of the best policy alternative (Popper et al. 2009; Kasprzyk et al. 2013; Lempert and Collins 2007).

Satisficing metrics build on the work of Simon (1996), who pointed out that decisionmakers often look for a decision that meets one or more requirements, but may not achieve the optimal possible outcomes. Satisficing metrics aim at maximizing the number of states of the world in which the policy alternative under consideration meets minimum performance thresholds. A well-known example of this is the 'domain criterion' (Starr 1963; Schneller and Sphicas 1983), which focuses on the fraction of the space in which a given performance threshold is met; the larger this space, the more robust the policy. Often, this is simplified to looking at the fraction of states of the world, rather than the volume of the space.

Synthesizing the foregoing, in defining robustness an analyst must make two choices. The first choice is about the policy alternatives. Is the analyst interested in the performance of individual policies or in comparing the performance of the

Table 15.1 Conceptual representation of various robustness metrics

		Characterizing performance of policy alternatives	
		Comparing alternatives	Performance of individual policies
Characterizing performance over SOWs	Threshold	Satisficing regret	• Domain criterion (Starr 1963) • Radius of stability, Info-Gap (Ben Haim 2001)
	Descriptive statistics	• Minimax regret (Savage 1954) • 90th percentile baseline regret (Kasprzyk et al. 2013) • 90th percentile best option regret (Herman et al. 2015)	Moments of the distribution (e.g., mean, variance), minimum, maximum, and functions thereof (Hurwicz, signal-to-noise, coefficient of variation)

alternative policies? Out of this choice comes a distribution of performance of each policy alternative over the set of states of the world. The second choice is how to succinctly describe this distribution. Here, the analyst also has two options: imposing some user-specified performance threshold, or characterizing the distribution using descriptive statistics. Table 15.1 presents a simple typology of robustness metrics building on these two choices.

Many of the classic decision analytic robustness metrics belong to the lower right hand corner of Table 15.1; they focus on the performance of individual policies and try to describe the performance of a policy over the set of states of the world using descriptive statistics. Maximin and minimax focus on the best and worst performance over the set of scenarios, while Hurwicz is a function of both. Similarly, Laplace's principle of insufficient reason assigns an equal weight to each of the states of the world and then suggests that the best policy is the one with the best mean performance. More recently, there has been interest in using higher-order moments as well. Hamarat et al. (2014) use a signal-to-noise ratio that considers both the average and the variance. A problem here is that combinations of the mean and variance are not always monotonically increasing (Ray et al. 2013). Moreover, focusing on the variance or standard deviation means that positive and negative deviations from the mean are treated equally (Takriti and Ahmed 2004). This explains why higher-order moments, skewness, and kurtosis have attracted attention (Kwakkel et al. 2016a).

15.3.4 Vulnerability Analysis

Vulnerability analysis is often used in combination with an exploration strategy for the generation of scenarios. That is, exploration is used to generate an ensemble of scenarios. Then, one or more alternative policies are evaluated over those scenarios, with vulnerability analysis being used to discover the influence the various uncertainties have on the success or failure of these policies. However, the vulnerability analysis techniques need not be restricted to understanding the role of the uncertainties; they can be used equally well to investigate the role of the policy levers.

Broadly speaking, two distinct styles of vulnerability analysis are being used. On the one hand, we can use global 'sensitivity analysis' to identify the relative importance of the various uncertainties or policy levers on the outcomes of interest. As Herman et al. (2015) point out, this is underutilized in the deep uncertainty literature. Global sensitivity analysis can serve various functions in the context of a deep uncertainty study. It may be used for factor prioritization—that is, to identify the relative influence of the various uncertainties or policy levers on the outcomes of interest. This helps reduce the dimensionality of the problem by focusing subsequent analyses on the key sources of uncertainties or to search over the most influential policy levers. The results of a sensitivity analysis may also enhance understanding of which uncertainties really matter. This is valuable information for designing policies—for example, by designing policies that are capable of adaptation conditional on how these uncertainties resolve over time, or by designing policies that have a reduced sensitivity to these factors. See Herman et al. (2014) for some examples of the usefulness of global sensitivity analyses for supporting the making of decisions under deep uncertainty.

On the other hand, we can use 'subspace partitioning' to find particular subspaces in either the uncertainty space or the policy lever space that result in a particular class of model outcomes. A well-established tool for this is Scenario Discovery (SD), in which one tries to find subspaces of the uncertainty space in which a candidate policy fails. The Adaptation Tipping Point (ATP) concept, which is central to DAPP, is essentially the same: An ATP specifies the uncertain conditions under which the existing actions on the pathway fail to achieve the stated objectives. Decision Scaling (DS) also employs a similar idea. In all three cases, the analyst tries to partition the uncertainty space into distinct regions based on the success or failure of a candidate policy.

15.4 Application of the Taxonomy

Table 15.2 contains a summary overview of the various approaches and computational support tools for DMDU using the taxonomy presented in Fig. 15.1. RDM is not explicitly based on a policy architecture. In most of its applications, however, it uses a form of protective adaptivity, drawing on ideas from Assumption-Based

Planning (ABP) and adaptive policymaking. For the generation of scenarios, it uses sampling. Policies typically are pre-specified and iteratively refined. RDM uses a domain criterion in the vulnerability phase, while it closes with a regret-based analysis of the leading decision alternatives. For the vulnerability analysis, RDM uses SD. Many-Objective Robust Decision Making (MORDM) uses the RDM structure, but replaces pre-specified policies with a many-objective search for a reference scenario. The advantage of this is that the set of initial decision policy alternatives resulting from this process have good performance and represent a careful exploration of the design space.

Three approaches that clearly fit into the policy architecture are Assumption-Based Planning (ABP), Dynamic Adaptive Planning (DAP), and Dynamic Adaptive Policy Pathways (DAPP). They represent opposing ends of the spectrum of adaptive policy architectures. ABP and DAP focus on the use of adaptivity to protect a basic plan from failing. ABP relies primarily on qualitative judgments. DAP says little with respect to how to use models for designing adaptive polices. Hamarat et al. (2013) use RDM for this. In contrast, DAPP can only be assembled from a given set of actions. These need to be found first. In most applications of DAPP, the actions are assumed to be given. For vulnerability analysis, DAPP uses ATPs, which can be seen as a one-dimensional version of SD.

SD, ATP, DS, and Info-Gap Decision Theory (IG) are all essentially vulnerability analysis tools that may be used to design adaptive plans. Of these, SD is the most general form. ATP analysis is, in essence, a one-dimensional SD. Similarly, DS is a one- or at most two-dimensional form of SD that is exclusively focused on climate information. All three focus on finding subspace(s) of the uncertainty space in which a policy fails. For DS and ATP, this subspace has generally been characterized by climate change information, although the latter can easily be used more generally. SD is agnostic with respect to the uncertainties to be considered. IG is a bit different in this respect, since it requires a reference scenario from which the 'radius of stability' is calculated. As such, it does not produce insights into the subspace in which a policy fails, but instead into how far the future is allowed to deviate from the reference scenario before a policy starts to fail. This means that the choice of the reference scenario and the distance metric used become analytical issues.

Engineering Options Analysis (EOA) is a bit of an outlier in this overview. Its primary focus is on assigning economic value to flexibility. As such, it might be used as part of an RDM study. In this case, SD might be used to identify when the net present value of a given option is lower than its costs. Note also that, in contrast to Real Options Analysis, EOA is easier to combine with both types of policy architectures. Again, EOA might be used to analyze in more detail the value of actions on a given pathway, or across pathways.

Many-Objective Robust Optimization (MORO) is a generic computational tool for designing adaptive plans. It has been used in combination with both a DAP approach (Hamarat et al. 2014) and DAPP (Kwakkel et al. 2015). It extends the argument found in the MORDM literature on the relevance of finding promising designs prior to performing in-depth analyses by bringing robustness considerations into the search itself. Because of this, it does not include a vulnerability analysis.

Table 15.2 Application of the taxonomy to various approaches and tools for supporting the making of decisions under deep uncertainty

	Key references	Policy architecture	Generation of scenarios	Generation of policy alternatives	Robustness metrics	Vulnerability analysis
Assumption-Based Planning (ABP)	Dewar (2002), Dewar et al. (1993)	Protective adaptivity	Not considered: focus is on assumptions that might fail, not on scenarios where they might fail	Pre-specified	Typically focused on satisficing measures	Qualitative judgment
Robust Decision Making (RDM)	Lempert et al. (2006)	Not explicitly considered, but often combined with protective adaptivity	Sampling	Generally pre-specified, and iteratively refined	The SD phase uses a satisficing metric (domain criterion); the trade-off analysis phase relies on regret	Scenario Discovery (Bryant and Lempert 2010; Kwakkel and Jaxa-Rozen 2016)
Many-Objective Robust Decision Making (MORDM)	Kasprzyk et al. (2013), Watson and Kasprzyk (2017)	Not explicitly considered	Sampling	Many-objective search using one or a few scenarios separately	The SD phase uses a satisficing metric (domain criterion); the trade-off analysis phase relies on regret	Scenario Discovery
Dynamic Adaptive Planning (DAP)	Kwakkel et al. (2010a), Walker et al. (2001)	Protective adaptivity	Not explicitly considered	Not explicitly considered	Typically focused on satisficing measures	Not explicitly considered
Dynamic Adaptive Policy Pathways (DAPP)	Haasnoot et al. (2012, 2013)	Dynamic adaptivity	Not explicitly considered, but strong emphasis on the need for transient scenarios	Typically pre-specified	Typically focused on satisficing measures	ATP (Kwadijk et al. 2010)
Info-Gap Decision Theory (IG)	Ben Haim (2001, 2004)	Not considered	Sampling outward from a reference scenario	Pre-specified	Satisficing (Radius of stability)	Not considered

(continued)

Table 15.2 (continued)

	Key references	Policy architecture	Generation of scenarios	Generation of policy alternatives	Robustness metrics	Vulnerability analysis
Engineering Options Analysis (EOA)	de Neufville and Scholtes (2011)		Monte Carlo simulation	Pre-specified	Values at risk	Values at risk
Decision Scaling (DS)	Brown et al. (2012), LeRoy Poff et al. (2015)	Not considered	Pre-specified, or sampling constrained by climate information	Pre-specified	Satisficing (domain criterion)	Climate response function (visual), ANOVA ranking
Scenario Discovery (SD)	Bryant and Lempert (2010), Groves and Lempert (2007), Kwakkel and Jaxa-Rozen (2016)	Not considered	Sampling	Typically pre-specified	Domain criterion (implicit)	Dedicated vulnerability technique in itself
Adaptation Tipping Points (ATP)	Kwadijk et al. (2010)	Not considered	Pre-specified, or sampling	Typically pre-specified	Satisficing (domain criterion)	Dedicated vulnerability technique in itself
Many-Objective Robust Optimization (MORO)	Hamarat et al. (2014), Kwakkel et al. (2015)	Not considered, can be used for both protective and dynamic adaptivity	Pre-specified, or sampling	Many-objective search	Compatible with all satisficing metrics, and reference scenario regret	Not considered

15.5 Concluding Remarks

This chapter has presented and described a taxonomy of the components that make up the various approaches and tools for supporting the making of decisions under deep uncertainty. In short, there are five categories of components: policy architecture, generation of scenarios, generation of alternatives, definition of robustness, and vulnerability analysis. Any given DMDU approach makes choices with respect to these five categories. For some, these choices are primarily or almost exclusively in one category while remaining silent on the others. For others, implicit or explicit choices are made with respect to each category. Table 15.2 summarizes this.

Going forward, it might be useful to use the taxonomy to articulate which choices are made in a given case, rather than having to argue over the exact differences among approaches and the merits of these differences. That is, rather than arguing over whether to apply RDM or DAPP, the discussion should be which combination of tools are appropriate to use given the nature of the problem situation. Different situations warrant different combinations. For example, when designing a new piece of infrastructure (say a large reservoir), it makes sense to use a many-objective search to find promising design alternatives prior to performing a vulnerability analysis. In other cases, a set of policy alternatives might already be available. In such a case, doing additional search might be less relevant.

References

Albrechts, L. (2004). Strategic (spatial) planning reexamined. *Environment and Planning B: Planning and Design, 31,* 743–758.

Atkins, P. W., Wood, R. E., & Rutgers, P. J. (2002). The effects of feedback format on dynamic decision making. *Organizational Behavior and Human Decision Processes, 88*(2), 587–604.

Bankes, S. C. (1993). Exploratory modeling for policy analysis. *Operations Research, 4*(3), 435–449.

Bankes, S. C., Walker, W. E., & Kwakkel, J. H. (2013). Exploratory Modeling and Analysis. In S. Gass & M. C. Fu (Eds.), *Encyclopedia of operations research and management science* (3rd ed.) Berlin, Germany: Springer.

Ben Haim, Y. (2001). *Information-gap decision theory: Decision under severe uncertainty.* London, UK: Academic Press.

Ben Haim, Y. (2004). Uncertainty, probability and information-gaps. *Reliability Engineering and System Safety, 85*(1–3), 249–266.

Brehmer, B. (1992). Dynamic decision making: Human control of complex systems. *Acta Psychologica, 81*(3), 211–241.

Brown, C., Ghile, Y., Laverty, M., & Li, K. (2012). Decision scaling: Linking bottom-up vulnerability analysis with climate projections in the water sector. *Water Resources Research, 48*(9), 1–12. https://doi.org/10.1029/2011WR011212.

Bryant, B. P., & Lempert, R. J. (2010). Thinking inside the box: A participatory computer-assisted approach to scenario discovery. *Technological Forecasting and Social Change, 77*(1), 34–49. https://doi.org/10.1016/j.techfore.2009.08.002.

Buurman, J., & Babovic, V. (2016). Adaptation pathways and real options analysis—An approach to deep uncertainty in climate change adaptation policies. *Policy and Society*. http://www.dx.org/10.1016/ j.polsoc.2016.05.002.

Collingridge, D. (1980). *The social control of technology*. London, UK: Frances Pinter Publisher.

De La Mothe, J. (2006). *Innovation strategies in interdependent states*. Gloucestershire: Edward Elgar Publishing Ltd.

de Neufville, R., & Odoni, A. (2003). *Airport systems: Planning, design, and management*. New York: McGraw-Hill.

de Neufville, R., & Scholtes, S. (2011). *Flexibility in engineering design*. Cambridge, Massachusetts: The MIT Press.

Dewar, J. A. (2002). *Assumption-based planning: A tool for reducing avoidable surprises*. Cambridge: Cambridge University Press.

Dewar, J. A., Builder, C. H., Hix, W. M., & Levin, M. H. (1993). *Assumption-based planning: A planning tool for very uncertain times*. Santa Monica, CA: RAND Report MR-114-A. Available at: http://www.rand.org/pubs/monograph_reports/2005/MR114.pdf.

Dewey, J. (1927). *The public and its problems*. New York: Holt and Company.

Diehl, E., & Sterman, J. D. (1995). Effects of feedback complexity on dynamic decision making. *Organizational Behavior and Human Decision Processes, 62*(2), 198–215.

Eriksson, E. A., & Weber, K. M. (2008). Adaptive foresight: Navigating the complex landscape of policy strategies. *Technological Forecasting and Social Change, 75*(4), 462–482.

Faber, A., & Frenken, K. (2009). Models in evolutionary economics and environmental policy: Towards an evolutionary environmental economics. *Technological Forecasting and Social Change, 76*(4), 462–470. https://doi.org/10.1016/j.techfore.2008.04.009.

Gersonius, B., Ashley, R., Jeuken, A., Pathinara, A., & Zevenbergen, C. (2015). Accounting for uncertainty and flexibility in flood risk management: Comparing real-in-options optimisation and adaptation tipping points. *Journal of Flood Risk Managmeent, 8,* 135–144. https://doi.org/10.1111/jfr3.12083.

Gersonius, B., Ashley, R., Pathirana, A., & Zevenbergen, C. (2013). Climate change uncertainty: building flexibility into water and flood risk infrastructure. *Climatic Change, 116,* 411. https://doi.org/10.1007/s10584-012-0494-5.

Giuliani, M., & Castelletti, A. (2016). Is robustness really robust? How different definitions of robustness impact decision-making under climate change. *Climatic Change, 135*(3–4), 409–424. https://doi.org/10.1007/s10584-015-1586-9.

Groves, D. G., & Lempert, R. J. (2007). A new analytic method for finding policy-relevant scenarios. *Global Environmental Change, 17*(1), 73–85.

Haasnoot, M., Kwakkel, J. H., Walker, W. E., & ter Maat, J. (2013). Dynamic adaptive policy pathways: A method for crafting robust decisions for a deeply uncertain world. *Global Environmental Change, 23*(2), 485–498. https://doi.org/10.1016/j.gloenvcha.2012.12.006.

Haasnoot, M., Middelkoop, H., Offermans, A., van Beek, E., & van Deursen, W. P. A. (2012). Exploring pathways for sustainable water management in river deltas in a changing environment. *Climatic Change, 115*(3–4), 795–819. https://doi.org/10.1007/s10584-012-0444-2.

Hall, J. W., Lempert, R. J., Keller, A., Hackbarth, A., Mijere, C., & McInerney, D. (2012). Robust climate policies under uncertainty: A comparison of robust decision making and info-gap methods. *Risk Analysis, 32*(10), 1527–1672. https://doi.org/10.1111/j.1539-6924.2012.01802.x.

Hamarat, C., Kwakkel, J. H., & Pruyt, E. (2013). Adaptive robust design under deep uncertainty. *Technological Forecasting and Social Change, 80*(3), 408–418. https://doi.org/10.1016/j.techfore.2012.10.004.

Hamarat, C., Kwakkel, J. H., Pruyt, E., & Loonen, E. (2014). An exploratory approach for adaptive policymaking by using multi-objective robust optimization. *Simulation Modelling Practice and Theory, 46,* 25–39. https://doi.org/10.1016/j.simpat.2014.02.008.

Herman, J. D., Reed, P. M., Zeff, H. B., & Characklis, G. W. (2015). How should robustness be defined for water systems planning under change. *Journal of Water Resources Planning and Management*, 141(10). https://doi.org/10.1061/(asce)wr.1943-5452.0000509.

Herman, J. D., Zeff, H. B., Reed, P. M., & Characklis, G. (2014). Beyond optimality: Multistakeholder robustness tradeoffs for regional water portfolio planning under deep uncertainty. *Water Resources Research, 50*(10), 7692–7713. https://doi.org/10.1002/201-4WR015338.

Holling, C. S. (1978). *Adaptive environmental assessment and management*. New York: Wiley.

Kasprzyk, J. R., Nataraj, S., Reed, P. M., & Lempert, R. J. (2013). Many objective robust decision making for complex environmental systems undergoing change. *Environmental Modelling and Software, 42*, 55–71. https://doi.org/10.1016/j.envsoft.2012.007.

Kleinmuntz, B. (1992). Computers as clinicians: An update. *Computers in Biology and Medicine, 22*(4), 227–237.

Kwadijk, J. C. J., Haasnoot, M., Mulder, J. P. M., Hoogvliet, M. M. C., Jeuken, A. B. M., van der Krogt, R. A. A., et al. (2010). Using adaptation tipping points to prepare for climate change and sea level rise: A case study in the Netherlands. *Wiley Interdisciplinary Reviews: Climate Change, 1*(5), 729–740. https://doi.org/10.1002/wcc.64.

Kwakkel, J. H., Walker, W. E., & Marchau, V. A. W. J. (2010a). Adaptive airport strategic planning. *European Journal of Transportation and Infrastructure Research, 10*(3), 227–250.

Kwakkel, J. H., Walker, W. E., & Marchau, V. A. W. J. (2010b). *From predictive modeling to exploratory modeling: How to use non-predictive models for decisionmaking under deep uncertainty*. Paper presented at the Proceedings of the 25th Mini-EURO Conference—Uncertainty and Robustness in Planning and Decision Making, Coimbra, Portugal.

Kwakkel, J. H., Haasnoot, M., & Walker, W. E. (2015). Developing dynamic adaptive policy pathways: A computer-assisted approach for developing adaptive strategies for a deeply uncertain world. *Climatic Change, 132*(3), 373–386. https://doi.org/10.1007/s10584-014-1210-4.

Kwakkel, J. H., & Jaxa-Rozen, M. (2016). Improving scenario discovery for handling heterogeneous uncertainties and multinomial classified outcomes. *Environmental Modelling and Software, 79*, 311–321. https://doi.org/10.1016/.envsoft.2015.11.020.

Kwakkel, J. H., Eker, S., & Pruyt, E. (2016a). How robust is a robust policy? Comparing alternative robustness metrics for robust decision-making. In M. Doumpos, C. Zopounidis, & E. Grigoroudis (Eds.), *Robustness analysis in decision aiding, optimization, and analytics*. Springer.

Kwakkel, J. H., Haasnoot, M., & Walker, W. E. (2016b). Comparing Robust decision-making and dynamic adaptive policy pathways for model-based decision support under deep uncertainty. *Environmental Modelling and Software, 86*, 168–183. https://doi.org/10.1016/j.envsoft.2016.09.017.

Lee, K. (1993). *Compass and gyroscope: Integrating science and politics for the environment*. Washington, DC, USA: Island Press.

Lempert, R. J., & Collins, M. (2007). Managing the risk of uncertain threshold response: Comparison of robust, optimum, and precautionary approaches. *Risk Analysis, 24*(4), 1009–1026. https://doi.org/10.1111/j.1539-6924.2007.00940.x.

Lempert, R. J., Groves, D. G., Popper, S. W., & Bankes, S. C. (2006). A general, analytic method for generating robust strategies and narrative scenarios. *Management Science, 52*(4), 514–528. https://doi.org/10.1287/mnsc.1050.0472.

Lempert, R. J., Popper, S., & Bankes, S. (2003) *Shaping the next one hundred years: New methods for quantitative, long term policy analysis*. Santa Monica, CA: RAND Report MR-1626-RPC.

LeRoy Poff, N., Brown, C., Grantham, T. E., Matthews, J. H., Palmer, M. A., Spence, C. M., et al. (2015). Sustainable water management under future uncertainty with eco-engineering decision scaling. *Nature Climate Change, 6*(1), 25–34. https://doi.org/10.1038/NCLIMATE2765.

Matrosov, E. S., Padula, S., & Harou, J. J. (2013a). Selecting portfolios of water supply and demand management strategies under uncertainty—Contrasting economic optimisation and 'robust decision making' approaches. *Water Resource Management, 27*(4), 1123–1148. https://doi.org/10.1007/s11269-012-0118-x.

Matrosov, E. S., Woords, A. M., & Harou, J. J. (2013b). Robust decision making and info-gap decision theory for water resource system planning. *Journal of Hydrology, 494*(28 June 2013), 43–58.

McCray, L. E., Oye, K. A., & Petersen, A. C. (2010). Planned adaptation in risk regulation: An initial survey of US environmental, health, and safety regulation. *Technological Forecasting and Social Change, 77*(6), 951–959.

McLain, R. J., & Lee, R. G. (1996). Adaptive management: Promises and pitfalls. *Environmental Management, 20*, 437–448.

McPhail, C., Maier, H. R., Kwakkel, J. H., Giuliani, E., Castelletti, A., & Westra, S. (2018). Robustness metrics: How are they calculated, when should they be used and why do they give different results? *Earth's Future*. https://doi.org/10.1002/2017ef000649.

Mytelka, L. K., & Smith, K. (2002). Policy learning and innovation theory: An interactive and co-evolving process. *Research Policy, 31*(8–9), 1467–1479.

Popper, S., Griffin, J., Berrebi, C., Light, T., & Min, E. Y. (2009). *Natural gas and Israel's energy future: A strategic analysis under conditions of deep uncertainty*. Santa Monica, California: RAND.

Ray, P. A., Watkins, D. W., Vogel, R. M., & Kirshen, P. H. (2013). Performance-based evaluation of an improved robust optimization formulation. *Journal of Water Resources Planning and Management, 140*(6). https://doi.org/10.1061/(asce)wr.1943-5452.0000389.

Roach, T., Kapelan, Z., & Ledbetter, R. (2015). Comparison of info-gap and robust optimisation methods for integrated water resource management under severe uncertainty. *Procedia Engineering, 119*, 874–883. https://doi.org/10.1016/j.proeng.2015.08.955.

Roach, T., Kapelan, Z., Ledbetter, R., & ledbetter, M. (2016). Comparison of robust optimization and info-gap methods for water resource management under deep uncertainty. *Journal of Water Resources Planning and Management, 142*(9). https://doi.org/10.1061/(asce)wr.1943-5452.0000660.

Sastry, L., & Boyd, D. R. (1998). Virtual environments for engineering applications. *Virtual Reality, 3*(4), 235–244.

Savage, L. T. (1951). The theory of statistical decisions. *Journal of the American Statistical Association, 46*(253), 55–67.

Savage, L. T. (1954). *The foundations of statistics*. New York: Wiley.

Schneller, G. O. I., & Sphicas, G. P. (1983). Decision making under uncertainty: Starr's domain criterion. *Theory and Decision, 15*(4), 321–336. https://doi.org/10.1007/BF00162111.

Schwartz, E. S., & Trigeorgis, L. (2004). *Real options and investment under uncertainty: Classical readings and recent contributions*. The MIT Press.

Simon, H. A. (1996). *The sciences of the artificial*. Cambridge, Massachusetts: The MIT Press.

Starr, M. K. (1963). *Product design and decision theory*. Englewood Cliffs, NJ: Prentice Hall.

Sterman, J. D. (1989). Modeling managerial behavior: Misperceptions of feedback in a dynamic decision making experiment. *Management Science, 35*(3), 321–339.

Sterman, J. D. (1994). Learning in and about complex systems. *System Dynamics Review, 10*(2–3), 291–330.

Sterman, J. D. (2002). All models are wrong: reflections on becoming a systems scientist. *System Dynamics Review, 18*(4), 501–531.

Swanson, D., Barg, S., Tyler, S., Venema, H., Tomar, S., Bhadwal, S., et al. (2010). Seven tools for creating adaptive policies. *Technological Forecasting and Social Change, 77*(6), 924–939. https://doi.org/10.1016/j.techfore.2010.04.005.

Takriti, S., & Ahmed, S. (2004). On robust optimization of two-stage systems. *Mathematical Programming, 99*(1), 109–126. https://doi.org/10.1007/s10107-003-0373-y.

Tsoukiàs, A. (2008). From decision theory to decision aiding methodology. *European Journal of Operational Research, 187*, 138–161. https://doi.org/10.1016/j.ejor.2007.02.039.

Walker, W. E., Rahman, S. A., & Cave, J. (2001). Adaptive policies, policy analysis, and policymaking. *European Journal of Operational Research, 128*(2), 282–289.

Watson, A. A., & Kasprzyk, J. R. (2017). Incorporating deeply uncertain factors into the many objective search process. *Environmental Modelling and Software, 89*, 159–171. https://doi.org/10.1016/j.envsoft.2016.12.001.

Dr. Jan H. Kwakkel is Associate Professor at the Faculty of Technology, Policy and Management (TPM) of Delft University of Technology. He is also the Vice President of the Society for Decision Making under Deep Uncertainty, and a member of the editorial boards of *Environmental Modeling and Software, Futures, and Foresight*. His research focuses on model-based support for decision making under uncertainty. He is involved in research projects on smart energy systems, high-tech supply chains, transport, and climate adaptation. Jan is the developer of the Exploratory Modeling Workbench, an open-source software implementing a wide variety of state-of-the-art techniques for decisionmaking under uncertainty.

Dr. Marjolijn Haasnoot is a Senior Researcher and Advisor at Deltares and Associate Professor in adaptive delta management at Utrecht University, the Netherlands. Her research focus is on water management, integrated assessment modeling, and decisionmaking under deep uncertainty. She is working and consulting internationally on projects assessing impacts of climate change and socio-economic developments, and alternative management options to develop robust and adaptive plans. Marjolijn developed the Dynamic Adaptive Policy Pathways planning approach. She has a Ph.D. in engineering and a Masters in environmental science.

Open Access This chapter is licensed under the terms of the Creative Commons Attribution 4.0 International License (http://creativecommons.org/licenses/by/4.0/), which permits use, sharing, adaptation, distribution and reproduction in any medium or format, as long as you give appropriate credit to the original author(s) and the source, provide a link to the Creative Commons licence and indicate if changes were made.

The images or other third party material in this chapter are included in the chapter's Creative Commons licence, unless indicated otherwise in a credit line to the material. If material is not included in the chapter's Creative Commons licence and your intended use is not permitted by statutory regulation or exceeds the permitted use, you will need to obtain permission directly from the copyright holder.

Chapter 16
Reflections: DMDU and Public Policy for Uncertain Times

Steven W. Popper

Abstract

- Public policy has always confronted future uncertainties. Projecting likely futures has been viewed as best practice for assessing proposed plans even though few would expect exactly those futures to occur. But in an era of deep uncertainties in which prior rules of thumb are no longer believed likely to hold true in years to come, sufficient diligence for policy analysis demands a different standard.
- DMDU approaches collectively represent an evolving capacity to deal with the challenge of the future by providing a *technology* of complexity, especially in the analysis of problems in public policy.
- New approaches to policy analysis may provide the means to enable policy processes better suited to deep uncertainty and dynamic change. And the recognition by policymakers that there exist (and that they should demand) new means for analysis that comport better to the emerging needs of policy would, in turn, allow more rapid diffusion of technique into realms not yet exposed to means for decisionmaking under deep uncertainty.
- The interaction between analysts and policymakers requires contact between two distinct cultures. DMDU applications collectively present a body of theory and practice with the potential for providing a common vocabulary to the work of both analysts and of those charged with policy design and implementation during uncertain times.
- For a problem requiring treatment by DMDU approaches, "*any job worth doing is worth doing superficially.*" An analysis based on an initial fast and simple exploratory model will frequently elucidate many of the major interactions between policy choices and the problem system. (Exploratory Modeling is discussed extensively in Chaps. 2, 7, and 15.) Those factors that appear most salient may then be examined in more detailed fashion in later model revisions and recalibrations.

S. W. Popper (✉)
Pardee RAND Graduate School, RAND Corporation, Santa Monica, CA, USA
e-mail: swpopper@rand.org

- By developing a basis for planning and operating under conditions of deep uncertainty, complexity, or both, DMDU approaches support governance and decisionmaking in an environment characterized by multiplying centers of authoritative expertise and interests, as well as the diffusion of government responsibility beyond the control of any one agency.

16.1 Introduction

The preceding chapters of this book taken together provide a snapshot of the emerging field of DMDU research and analysis. They have been selected to showcase a broad range of approaches currently at the forefront of the field's development. In doing so, they draw from a field that is itself in motion, as these various approaches all undergo refinement, development, evolution, and even possible combination. Each year broadens the spectrum of policy questions to which they have been applied.

The application chapters were selected to illustrate a growing ability to address problems that had previously been treated either with assumptions of greater certainty than was warranted by the actual state of knowledge, or were found to be intractable using more traditional analytical techniques. They present evidence in support of the contention in Chap. 1 that DMDU approaches may enhance the understanding of our choices and aid decisionmaking to more reliably achieve long-term objectives in the presence of deep uncertainty.

Beyond these specific instances, to what character of problem could DMDU methods be applied more ubiquitously, and in addressing themselves to such questions, what should DMDU practitioners seek to provide? This chapter steps back from the examination of individual approaches to take account of the broader movement underway. In particular, while Chap. 1 spoke of decisionmaking in general, the other chapters make clear that many of the applications lie in the realm of public policy. This is not just an artifact of chance or funding, but rather points toward specific needs that DMDU approaches may satisfy. Shifting the focus from the individual methodological trees of prior chapters to the larger DMDU forest they represent, what role may these approaches and the emerging field play in addressing some of the pressing questions presented to analysts by societies increasingly aware of the dynamic of change?

This chapter seeks to provide a thematic connection between the theory and application chapters that preceded it and the emerging needs of public policy decisionmaking processes. It first takes an excursion into the relationship between science and technology to discuss the emergence of tools for addressing issues of complexity. It then discusses the challenges now faced by institutions for democratic governance and suggests the value of achieving greater integration between the operation of those institutions and the activities of analysts seeking to support their decision processes. The chapter concludes with the contention that the world of analysis and that of policy operate as two different cultures. It lays out those characteristics of DMDU approaches discussed in this book that may bridge the gaps between them.

16.2 Viewing DMDU Approaches as a "Technology of Complexity"

It is a persistent human predisposition to view the present as a time of troubles, while in retrospect, the past appears an era of comparative calm and certainty. Yet, a case can be made that today we are increasingly confronted in our public life by challenges of a depth, breadth, complexity, and rapidity of emergence that are hard to match in the prior record.[1] Two subjective factors may also make our condition unprecedented. The first is that our social and political systems are predicated upon the axiom that management is both possible and expected. We recognize the element of chance, the occurrence of the unexpected, and the potential for our affairs to go awry through *force majeure*, but there is an implicit expectation that the institutions we have put in place will deal with and overcome such exigencies through our skill in understanding and managing the underlying systems. The prospect of not being able to do so is not only dismaying to the individual but collectively undermines the basis of consent within society.

The second subjective factor is that, unlike previous times during which people may have viewed themselves as passing through a temporary crisis before the world would either resume its normal course or settle into a new equilibrium from which new and understandable rules of thumb could be deduced, we have no such expectation. Those elements leading to the deep uncertainties we confront show no signs of abating. Indeed, we can well accept that the next generation may in their turn view our times as being milder versions of their own experience of turbulence and change. The contradiction between these two perceptions gives rise to frustration and anxiety.

In Chap. 1, several sources of deep uncertainty were laid out, including fundamental indeterminacy or unpredictability of system outcomes. That is, beyond just lacking a desired level of knowledge or failing on our part to be sufficiently diligent, a condition of complexity or emergence beyond any means for meaningful anticipation may be hardwired into the systems within which we operate. Seeing emergence, nonlinear response, and unpredictability due to path dependence—in a word, systemic *complexity*—as being an inherent property in some systems themselves (rather than the more colloquial use of the term, which would have it be synonymous with the *complication* arising from keeping track of a very large set of forces, factors, and their interactions,[2]) began with the work of Poincaré in the late nineteenth century but became widely documented in the work of scientists in the last four decades of the

---

[1] Or, one might say that, while other eras may also have witnessed periods of upheaval (1914–1945 comes to mind), we should determine to meet our own challenges armed with more fitting tools than were available to our forebears.

[2] The US space program of the 1960s is an example of a highly complicated system (that is, one with many moving parts) that nonetheless was not complex in the sense that it could be divided into projects directed toward subsystems that upon recombination would successfully succeed in its goal of manned flight to the Moon. An analytically complex system is not susceptible to this degree of predictability and control. For a discussion of complexity and policy, see Mitchell (2009).

twentieth century.[3] By the end of that century, there were even research institutions and groups devoted to elaborating the science of complexity.[4]

This book was constructed to showcase different DMDU approaches and their practical application. There is method in this. The book's twinned chapters on theory and practice reproduce in microcosm the intertwined histories of science with its theories and technology with its tools and outputs. With the one representing the application of a rigorous system for posing questions and evaluating answers, and the other consisting of "activities, directed to the satisfaction of human needs which produce alteration in the material world,"[5] to consider them together is to recognize how dependent each often is on the other. While the popular imagination perceives a linear path, with science opening the way for new discovery, application, and innovation, it is frequently the case that the work of the technologist provides the scientist not only with the tools with which to conduct inquiries, but even with newly revealed phenomenology challenging prior explanations or requiring new ones.[6] The advent of the telescope, microscope, steam engine, and semiconductor all preceded the major developments in the sciences of astronomy, biology, optics, thermodynamics, and solid-state physics.

One can argue about the degree to which the project of working out a science of complexity may have met its objectives or the distance it may have travelled toward achieving them. But to analogize to the more nuanced interaction between science and technology perceptible in other fields, we may ask whether complexity science had available a sufficient "technology of complexity" to support this effort—and whether a need for one was perceived.

A case can be made that efforts by the authors represented in this volume and others constitute such a parallel journey of discovery. To be sure, while aware of the developing field of complexity studies, few of these methodologists and practitioners were motivated principally to contribute to the complexity science enterprise. Neither is it the case that all problems (perhaps not even the majority) benefiting from DMDU analysis necessarily have complex system phenomena at their core. As in other fields, the focus was not so much on understanding the nature of complexity but to deal with its consequences in systems exhibiting complex behaviors—as well as the effects stemming from the other sources of deep uncertainty in real-world application. Rather, the unifying driver was the desire to craft analytically sound systems for addressing the mounting catalog of "wicked" problems confronting decisionmakers in contexts both small and large.[7] This is what technologists do: Develop tools to

---

[3]Prominent among these are Ilya Prigogine in chemical thermodynamics, Benoit Mandelbrot in mathematics, and Norbert Wiener in cybernetics, to name only a few.

[4]Perhaps the best known of these is the Santa Fe Institute, founded in 1984 (https://www.santafe.edu).

[5]Childe (1954).

[6]The most vigorous statement of the linear hypothesis was that of Vannevar Bush in 1945's *Science, the Endless Frontier*. The work of Derek de Solla Price and other historians of science and technology has provided numerous examples refuting unidirectionality from science to technology.

[7]Rittel and Webber (1973) "Dilemmas in a General Theory of Planning." *Policy Sciences*, 4, 155–169.

enhance observation as well as provide the means for maneuver and manipulation within the newly observable space that their tools have illuminated. But in doing so, they have also allowed us to provide analysis of policy in regimes exhibiting the path dependence, nonlinear response, and emergence characteristics of complexity.

The ability to navigate along deeply uncertain pathways has become a pressing need and a hallmark of our times. But it might also be one we have long been familiar with.

16.3 DMDU and Policy in an Era of Perpetual Change

The Third Ape's Problem

A fruit-laden tree grows on a prehistoric savanna. A bipedal ape–human comes to the edge of the brush and ponders the risk of crossing the grasslands to gather a meal. Is a lion lurking in the high cover? She notices the swish of a tail: swishing tail → active lion → danger. She withdraws. Later, a second ape long familiar with this patch comes by. No swishing tail, yet she knows that every so often a troop mate has been lost in gathering fruit here. How often? She considers the risks, balances her need, and determines whether to cross. Now, a third ape approaches this stretch of savanna for the first time. She can draw upon neither concrete information nor familiarity with this patch and the probability of danger as a basis for logical deduction. She is, however, quite hungry, and her survival hangs in the balance. She searches for familiar patterns and weighs them against her experience. She takes a few steps forward and then looks for changes in the patterns. How far can she proceed and still scramble back to safety? Where might a lion hide in this brush? Is that movement over there solely due to the wind? She ventures forth step by step, updating information, planning for contingencies, perhaps tossing a rock or two to probe for any lurking predators.

In this manner, she proceeds into a potentially terrifying unknown.[8]

Technology arises in application. A branch fallen from a tree remains only that until it is picked up by a human hand and transformed into a spear, javelin, awl, hoe, flail, walking stick, roasting spit, cudgel, or lever. This book presents many candidate technologies for creating tools to solve the third ape's problem in its modern manifestations. Hers was the problem of the individual. Ours is in many respects the problem of societies and of their constituent organizations and institutions. And more than a problem of analysis, it is a problem for governance as well.

Surrounding many of the issues confronting us today, from the large issues of global warming, wealth and income inequalities, and the future of work, to local

---

[8] Abstracted from Popper (2003), Lempert and Popper (2004).

decisions regarding infrastructure, regulation, and mobility, we may perceive a more general issue rooted in democratic institutions of governance. From the USA to the UK, from Israel to India, from the states emerging from the shadow of the Cold War to the superstructure of the European Union, democratic institutions (and ultimately the principles upon which they are predicated) are faced with challenges of several kinds that are proving difficult to resolve. True, it has never been easy to deal with clashes of values, conflicting interests, or manifestations of uncertainty, nor with the ensuing policy disagreements. But today, there appears to be a generalized perception in many instances of an inability on the part of the democratic governmental process to either meet these challenges fully or to respond in a timely manner. Despite the varying degree to which this statement might be true for particular issues, what appears to have become widespread is the growing and often implicit perception on the part of the publics whom governments are established to serve that their leaders are incapable of addressing these concerns.

Resolving such issues through means designed to minimize or entirely obviate the role for democratic institutional decisionmaking may undermine values we prize. Markets and market mechanisms are magnificent human artifacts for collating information leading to the efficient allocation of resources. But despite their capacity for fostering allocative efficiency, they were not designed nor are they capable of producing decisions on policy with which we might feel comfortable. While we do care about cost and efficiencies, we also care about equity and balancing other interests, not only among social groupings but across generations. Mechanisms for doing so have often been suggested, but few have addressed the problem of how to avoid further incentivizing the forces seeking to alter market structure in such a way that the outcomes are skewed in a direction best suited to the interests of the economically powerful, who may then translate such market power into political power. On the other hand, authoritarian forms of governance, aided by the technocratic efforts of analysts applying the most scientific of principles, may well achieve many specific priorities, but good governance has never been shown to be among them.

Many countries have looked to democratically elected (and democratically operated) governments to fulfill the role of sovereign policy authority. The role of the analyst who addresses the category of wicked problems or the technologist of complexity who applies his or her skills to the DMDU problem space within the democratic ecosystem, is to enhance and shore up the ability of democratic governance to perform the crucial role of policymaking. They fulfill this function by providing the means to perform two important tasks: (1) make visible and comprehensible those phenomena that are relevant to the decisions being considered but may, by their sheer number and complex interactions, be difficult to observe directly or share widely, and (2) illuminate the limits of the practical trade space within which the political apparatus may find and select short-term policies that may still be reasonably assured not to compromise the ability to achieve longer-term policy goals.

This may hold even more sway in an era in which the roles of hierarchies and authorities may themselves be subject not only to question but the emergence of other forms of informal but nonetheless powerful authority. Previously, governments enjoyed an advantage in having had the means to speak with authoritative expertise on many issues. Now, however, the ease of information gathering means barriers are lower for others to also gain the knowledge necessary to establish strongly held, well-argued positions. Further, the concomitant ease of communications means that such centers of expertise may actively coalesce communities of interest into tangible constituencies, acquire information enabling them to put forward positions and proposals, and communicate these positions widely to gain further adherents. To some extent, this comes at the expense of previous centers of authority as their former concentration (if not monopoly) of expertise, information, and unity of agency ebbs. For it is not only authority that has become dispersed, it is responsibility as well. One of the characteristics of canonically wicked problems is that their solution does not lie clearly with only one, or even a few, public agencies. It is difficult to tell who is responsible, for example, for eliminating roadway fatalities from the US transportation system.[9] At the very least, such responsibility is shared among federal, state, and local agencies, to say nothing of the myriad professional and civic society groups dedicated to this goal.

To the extent that the preceding diagnosis is accurate, business as usual in the analysis and debate of public policy does not appear to be a prescription for success going forward. We would be better advised to consider how best to conduct effective public policy planning, choice, and decisionmaking under conditions that appear now to be more likely characteristic of our future than of our collective past. If, for example, the Dutch decision model of shared responsibility is a potential solution and could be made more ubiquitous, then (as in Chap. 14) it requires a platform providing an integrated system of analysis and decision that is capable of handling not only the full range of real-world complication but its attendant uncertainties as well—a role for DMDU-inspired applications.

16.4 Integrating Analytical and Policy Processes

A basic model of individual decisionmaking, originally developed to explain why pilots of one nation can come to dominate the trained pilots of another, sets decisions within a recursive process that is calibrated by updating knowledge about existing conditions and new developments (Boyd 1996). Figure 16.1 shows a cycle of *observing* the environment and situation, *orienting* within that frame of reference, reviewing a set of possible actions before *deciding* upon one course, and then *acting* (at which point the loop begins once more with results from the preceding action being measured and assessed)—the OODA loop.[10]

---

[9]See Ecola et al. (2018).

[10]Given the phenomena the original OODA loop was designed to explain, it is also a convenient structure for amalgamating into one framework both the fast- and slow-thinking aspects of human cognition discussed by Kahneman (2011); see also Klein (1999) on recognition-primed decisionmaking.

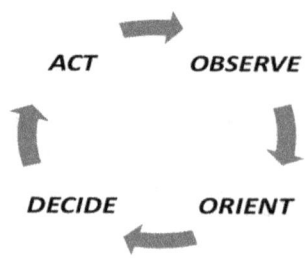

Fig. 16.1 Observe, orient, decide, and act (OODA) loop

Some have seen the OODA loop as a model for decisionmaking within organizations, as well as at the level of the individual.[11] For our purposes, it marries the active with the analytical. An organization wishing to operate strategically must have an ability to scan the horizon (observe) and understand not only the factors that give rise to these observations but also what implications they have for its missions (orient).

The acts of observation and orientation are central to most DMDU approaches in application. Generally referred to as "monitoring," perhaps not surprisingly this theme comes out most strongly in this book's chapters from contributors who work in the Netherlands[12]: Climate and hydraulics interact with slow-moving dynamics that may take decades to show observably certain outcomes but may also result in sudden, catastrophic loss. In such cases, those seeking to perform observation and orientation may not be afforded the luxury of examining actual outcomes. Much of their work must be done by having previously identified early warning metrics and signals for action before resolutions are achieved or through the extensive use of simulations and other modeling. This DMDU work especially considers the role of observation and orientation as precursor to, and trigger of, decision and action.

While this all appears intuitive, the difficulty lies in how to make an OODA approach to public policy operational. This synthesis is not readily observable in most public agency decisionmaking processes (see Chap. 13). Instead, most policy processes are based upon a series of tacitly agreed fictions that in the past either served as useful first approximations to reality or were concessions to the limitations existing in the milieu within which decisionmaking needed to occur. Thus, we see an emphasis on the production of plans rather than on the process of planning itself as a principal tool for effective policy formulation. We also often take an implicit "once-and-done" stance toward policy. That is, having arrived at a policy decision selecting a set of preferred actions, there is a tendency to regard the problem as having been solved.[13] Of course, all involved in such decisions recognize that reality

---

[11] See, for example, Angerman (2004).

[12] See especially the matched Chaps. 3, 4, 8, 9, as well as the second example in Chap. 11, and the history laid out in Chap. 14. Monitoring and triggering are at the core of the planned adaptation case studies provided in Chap. 13.

[13] It is useful to contrast the intrinsic approach to policy implicit in the Kyoto round of climate accords to that which informed the most recent Paris accords. The former was imbued with the

may all too often diverge from plan, but we put off as long as possible having once more to address the inefficiency and ineffectuality that builds up over time until a later round of policymaking is seen to be required.[14]

A policy process that comes closer to the ideal represented by the OODA paradigm would recognize the close relationship between plans and planning, between product and process. It would recognize the value of and enhance the potential for more adaptive approaches to policy formulation. (It would create a more widely shared realization that any policy implemented to address a sufficiently complex issue could also be viewed as a policy *experiment* in the sense that observation of initial outcomes provides opportunity for learning and adaptation.) Instead of policy practices that tend to be confrontational and exclusionary, the means for engagement and inclusion would be incorporated in a manner that still permitted the process to move toward formulation and implementation. And the inherent uncertainties, rather than being at best implicit and at worst disregarded, would instead be explicitly addressed by decisionmakers and policy officials within their deliberative processes.

Analytical support to policy may bring in a further subtle dysfunction. This takes form in a self-reinforcing feedback loop between the way policy processes are currently construed and the requests and demands they place upon the analysts tasked with supporting them. The policy planners and decisionmakers usually ask only of their analysts what they have traditionally provided. Similarly, if the analytical modes themselves reinforce an implicit predisposition toward decision mechanisms predicated on prediction, optimization, and requiring an upfront winnowing of acknowledged assumptions to a small, analytically tractable set, then there is little hope that the decision processes in their turn can pass far beyond the traditional approaches so long in practice and increasingly less capable of satisfying the demands placed upon them.

In this sense, there is an intertwined process. New approaches to policy analysis may provide the means to enable different approaches to policy. And the recognition by policymakers that there exist (and that they should demand) new means for analyses that comport better to the emerging needs of policy would, in turn, allow more rapid diffusion of technique into realms not yet exposed to tools that support decisionmaking under deep uncertainty. The chapters of this volume more than hint at the possibility of a co-evolutionary relationship between changing policy processes and enabling new analytical approaches and tools.[15]

---

sense that the signatories were setting policy for decades to come, while the latter was built around a more sophisticated and recursive policy decision process (Popper 2015). Within this book, the second example in Chap. 7 provides an illustration of how DMDU applications could assist in framing and navigating the type of analysis and policy cycle envisioned in the Paris Agreement.

[14]Classic instances of this phenomenon include the 30-year cycles in the USA between reforms to the tax code or retirement pension policies (i.e., Social Security).

[15]Chapter 13 presents four examples that serve as case studies for a planned adaptation within a nexus of analysis and policy. Chapter 14 gives an exceedingly valuable, detailed history of one such system that stemmed from the cross product of a political disposition toward consensus decisionmaking and the support provided by DMDU analysis.

This theme applies to all applications presented in this volume and certainly is one that each of the represented approaches has been used to address. One of the goals of the book, certainly in the chapters illustrating applications, is to go beyond a narrow intent to enhance training of DMDU methodologists alone. Wider use of DMDU techniques and a greater realization of their latent potential for better supporting policymaking during trying and troubling times also require educating the potential recipients of DMDU outputs and analyses.

16.5 The Two Cultures of Policy and Analysis

If analysts and policymakers are to become partners in devising governance processes in better accord with new conditions while adhering to democratic principles, it would be helpful if they could speak a common language. If their activities are to become more integrated, and therefore less sequential and mutually isolated, there needs to be basis for synchronization. The difficulties of both challenges should not be minimized; nevertheless, the potential for DMDU approaches to provide necessary common ground is worth examining.

To one familiar with the relationship between policymakers and their planning staffs on one hand and scholars and analysts on the other, the obstacles to achieving closer integration between policymaking and analysis would appear to be considerable. Between the two communities of analysis and policy lies a gap in cultures. The traditional culture of analysis is numerate, reductionist in character, systematic in approach, and ultimately rooted in deductive logic. The culture of policy, however, is more narrative based, focused on details, nuanced, and situation-specific in process—and its debates are often framed in the logic of abductive reasoning: "How will we be affected if present trends continue? What could go wrong if we follow this course or that? If the circumstances we most fear come to pass, how will we cope?"

This cultural gulf explains, in part, the frustration felt by analysts who present significant findings with potentially large implications for policy and are thanked for their efforts but next find the door to the chamber in which policy deliberations occur closed in their face. The translation of analytical output into policy failed to occur.

Ultimately, the responsibility for doing so rests on the shoulders of the analysts (principally because their relationship to policy authorities is not that of equals) and will require of them greater sensitivity to culture. Presenting findings couched in a language and structure consistent with scientific norms suitable for academic publication often presents the policymakers with a problem. Such findings may be perceived by them as being less a boon and more an addition of yet more confusion to the welter of information through which they must wade to frame the knowledge base required for policy. If left with the burden of themselves translating research outputs into the terms in which policy is deliberated, meaningful analytical insights may fail to gain entry into the conversation on the other side of the door—or wall—that lies between analysis and policy.

Whether the authors of DMDU applications conceive their role in these terms, the effect of their collective effort has been to create a body of theory and practice with the potential for providing commonality to the work of analysts and of those charged with policy design and implementation during uncertain times. To a greater or lesser degree, most if not all the techniques presented in this book have implicit in them several postures toward analytical practice and the role of analysis in the support of decisionmaking that collectively hold promise for supporting new approaches to both analysis, policy, and the relationship of each toward the other.

16.6 DMDU Principles Enhance the Integration of Analysis with Policy

This book's introduction laid out some commonalities of process among the DMDU approaches presented in subsequent chapters. If we now look to both the predispositions that inform those steps as well as what ensues as the result of having followed them, we can derive several attributes that would themselves appear to be necessary to change the way both analysts and policymakers view their roles given the challenges we now face. While there exists a tight relationship among many of these principles, we list them individually to make them explicit.

Begin from the End. Good analytical practice always seeks to understand the problem being addressed and to frame it appropriately. In many cases, however, there is a tendency after the initial framing to focus primarily on the challenging task of model building that can subtly shift focus. This results from a too narrow view of the role of models, principally seeing them as necessarily representing the most accurate representations of the systems being modeled, and of what constitutes a valid model.[16] To be sure, a DMDU study may begin with an existing model as a legacy and build from there. Often, however, no such model exists and so must be created or an existing one transformed. DMDU approaches construct such an analytical framework backwards: building up what is necessary to answer the questions at the heart of the policy decision (which itself is brought directly into the modeling framework) rather than the more frequent practice of first creating an artifact that attempts to capture all details of the phenomenology involved.[17] The reason for this is simple: a good deal of the detail that would be required for the most accurate system simulator may be shown through rigorous analysis not to be required to provide a full exploration of the decision to be informed in situations involving deep uncertainty. This characteristic of looking through the "wrong" end of the traditional analytical telescope not only affirms the policy question as the lodestone for all that follows, it also improves the

---

[16]Davis, Paul K. and Steven W. Popper (in review). "The Uncertainty Analysis That Policymakers Should Demand."

[17]This approach favors fast simple models (FSM) rather than the more detailed modeling necessary for techniques with less capacity for and less emphasis on iteration. See the Glossary for more detail.

likelihood that results stemming from analysis will fit easily into the policy debate through use of the same language and in addressing the same core issues. (Most of the Part I chapters of this book, especially Chaps. 2 and 6, call out this aspect of DMDU analysis.)

Characterization of Uncertainty. Operating under conditions of deep uncertainty deprives us of high confidence in probability density functions. Instead, the unknowns are characterized by the degree to which they may affect decisions over the choices we face in view of our objectives. (Both examples in Chap. 7 illustrate this point by their use of Scenario Discovery.) That is, DMDU analysis looks across uncertainties and seeks to understand what we would need to believe was true to advocate one course of action over another. The objective then becomes to look for alternatives within the potential solutions set (or to add to that set through modification, hybridization, conditional statements, or novelty) for those short-term courses of action that would appear least vulnerable to the uncertainties present and yet still perform sufficiently well according to our criteria for good outcomes.

This posture also shifts the entire thrust of the analysis from seeking agreement on assumptions to one of agreeing on solutions (Kalra et al. 2014). This is a powerful technique for getting beyond incendiary debates over whose assessment of presently unknowable facts is the most accurate—debates that cannot be resolved to the satisfaction of all parties. The lack of agreement on a set of most valid assumptions greatly inhibits progress toward finding solutions. Instead, the DMDU approach will carry all sets of assumptions forward into the analysis stage and explore courses of action that may satisfy all parties.[18]

Multi-objective Analysis. A DMDU analysis will typically range over several classes of factors to explore. Among these are assumptions about external uncertainties, the differential efficacy of policy and potential policy instruments, alternative specifications of the underlying system model or causal relationships, and the metrics by which outcomes are to be evaluated.[19] While all are crucial to enabling a more comprehensive form of analysis given deep uncertainty and contentious policy debates, the last of these is worth highlighting as an important aid to achieving not only greater correspondence between analytic and policy processes, but also in addressing the phenomena of more widely distributed authoritative expertise and more diffuse government responsibility. This aspect of DMDU analyses allows their authors to confront the multi-attribute character of wicked problems explicitly.

When there are multiple parties to a complex policy issue, there will rarely be widely shared agreement on a single bottom-line metric sufficient to evaluate outcomes. Usually, there are several assessment criteria that may even be in opposition to one another. Even if all may agree that we have an interest, for example, in both

---

[18]This satisfaction is often bought by embedding conditionality at key decision points into the policies that emerge from the process. Resolution of differences therefore does not rest upon agreement among parties. Rather, it stems from the all-too-human principle that, of course, I am willing to concede that should my opponents prove correct we will shift the implementation of the policy to a different course, because I believe that they are almost certain to be wrong.

[19]These are laid out in the introduction to this book (Chap. 1) and discussed in operational terms elsewhere, especially in the paired chapters on RDM (Chaps. 2 and 7) and DAP (Chaps. 3 and 8).

economic growth and environmental sustainability, we may argue about priorities among the outcomes we value. The wider scope for utilizing multiple measures in a DMDU analysis naturally accommodates a wider perspective on normative outcomes. In most cases, this leads to a concomitant shift from searches for optimal outcomes (if exploration of the range of uncertainties has not already disabused too great faith being placed in optimality as a criterion) to an alternative posture of satisficing.[20] In this case, policy sets minimum or maximum threshold levels for selected measures by which to evaluate how well different proposed solutions may meet policy goals across a wide range of plausible futures. (This also more closely resembles the hedging approach taken by senior decisionmakers confronting deep uncertainty than do modes of maximization.) This will be a key to achieving consensus on plans for confronting the major policy issues of today and the future.

The practical routines of DMDU applications will by their nature support a multi-attribute focus. Such studies regularly seek engagement with stakeholder groups, planners, and decisionmakers through the use of real-time analytical support (the water planning example in Chap. 7), workshops (Chap. 8), games (Chap. 9), and Assumption-Based Planning-inspired pre-plan exercises (Ecola et al. 2018).

Iteration. Preceding chapters have demonstrated that most DMDU analyses are inherently iterative. They are applied as a recursive process, similar to the OODA loop discussed above, refining not only the analytical product but also the analytical infrastructure for conducting the analysis. This is the characteristic that enables the "backward" approach to analysis that allows testing and refining of underlying models to bring out greater detail when required and to eliminate the need for more detailed modeling of those aspects of the system that appear to have little or no effect on our normative decisions or the characterization of uncertainties.

In fact, we might coin the dictum that for a problem requiring treatment by DMDU approaches, *"any job worth doing is worth doing superficially."* That is, an initial fast and simple approach to modeling[21] as well as in formulating potential solutions often makes clear many of the interactions between system and policy that may then be fleshed out as warranted in later revisions.

The resource conservation approach to modeling taken by DMDU approaches is more than an efficiency measure. It has the potential for enfranchising categories of knowledge that might otherwise not be codified in a form to fit easily within more traditional analytical modes. This can take several forms. On one end of the spectrum, science generates more knowledge than appears in academic papers. Scientists gain considerable experience and develop hypotheses well before they are positioned to report the conclusive, null-hypothesis-falsifying experiment that will yield a publishable formal result. For sufficiently complex phenomena, such as those involving climate change or potential consequences for human societies of creating genetically modified organisms, it may be nearly impossible to perform a single experiment or even a connected series that will lay all questions to rest. This circumstance places

---

[20] Simon (1956).

[21] See Davis and Bigelow (1998) on the values of multiresolution modeling, Davis and Bigelow (2003) on the use of metamodels, and van Grol et al. (2006) on the use of fast simple models.

scientists in an uncomfortable position. They may possess considerable knowledge with relevance to policy, but potentially risk their academic standing and possibly transgressing professional norms if they lack confidence that their knowledge, if publicly shared, will be viewed correctly as being informed by experience but not itself representing proven scientific truth. But in an iterative process in which this knowledge becomes one of the means used to define potential ranges of values for subsequent exploration, much of this concern is reduced. Their direct experience and any hypotheses may be shared, weighed, and considered appropriately with considerably less risk than earlier, more definitive statements would generate.

At the other end of the spectrum, a good deal of knowledge is not easily codified or formalized. One of the characteristics of many DMDU techniques is to allow multi-dimensional phenomena to be rendered in lower dimensional spaces graphically. This "dimensional collapse"[22] allows observation of connections, relationships, and systemic phenomena that had not been previously visible within the bewildering forest of uncertainties and assumptions. It also makes them shareable. Rather than constructing a *deus ex machina*, a technocratic analysis designed to dictate policy prescriptions, DMDU instead most often provides output that may serve as a "digital campfire"—an occasion for those with different perspectives as well as knowledge that may not previously have found enfranchisement in formal analysis to view DMDU results and share among themselves the significance they find and the insight such results elicit.

The advances in computer technology that make possible the drive toward artificial intelligence also enable the tools of DMDU. The crucial distinction is that in the latter case, the resulting tools seek to preserve and enhance the human-in-the-loop and not eliminate the need for human decision and intervention by making predictions.

Accessibility and Transparency. The enhanced ability to share analytical output and insights provides one further attribute for a more effective engagement of analysis with policy under deep uncertainty. An aspect of uncertainty that particularly bedevils hierarchic organizations as well as cross-organizational decisionmaking is that of uncertainty absorption. As noted by March and Simon (1958), the condition exists when lower levels within an organization possess much information about the quality of their information (how trustworthy may be its source, how well validated, etc.), but this nuance is difficult to convey to the higher levels without having the information channels seize up from sheer overload. The nuance is lost, and uncertainties and indeterminacies are either dropped or presented as solid fact. This can lead to senior-level decisions being reached based on misunderstood premises.

It is in the nature of most DMDU approaches, however, that while they may perform analytically the dimensional collapses that yield observations regarding systemic regularities among actions, assumptions, and outcomes, they also retain a "drill down" capacity. That is, it is possible to easily call up the individual cases from which these massive ensembles are composed to make them available for detailed examination. It is possible, therefore, to provide not only transparency into what cases and instances yielded the observed results, but also to make these underlying

---

[22] See the discussion of Scenario Discovery in Chap. 2.

data available for further questioning if the audience receiving the DMDU output deems it valuable. The nature of DMDU analysis and the framework it creates make possible ongoing and more effective "red teaming"[23] of the analytical output from these techniques.

In a different vein, the fact that the apparatus for DMDU analysis has been constructed to address complicated policy issues, and to be carried out iteratively in evaluating alternative assumptions and evolving potential courses of action, means that the analytical tooling remains useful even after the initial results have been obtained. If we are truly entering an era of testing the implications of assumptions, seeking robust solutions, and approaching policy with greater flexibility and adaptability partly due to a recognition of greater potential for change and surprise, then it stands to reason that the machinery used to conduct policy analysis should continue to be employed, updated with new knowledge, and engaged in the process of navigating the policy adaptations almost certain to be necessary. This is an extension of the deliberation with analysis approach (Chap. 2) beyond the period of policy planning. The nature of DMDU methodology creates analytical artifacts that may continue to provide value within a more integral connection between analysis and policy brought about through a desire to maintain policy effectiveness across the fluctuations that the future may bring.

16.7 Conclusion

This chapter suggests that if one looks at the DMDU applications in Part II of this book, as well as the presentation of approaches in Part I, one may perceive an academic field that has consciously dedicated itself to being relevant to the world of public policy. It has produced a growing library of real-world applications. Any such application will have been intended to serve limited purposes. But viewed in aggregate, the DMDU field is pointed toward supporting governance and decisionmaking in an environment characterized by multiplying centers of authoritative expertise and interests, diffusion of government responsibility beyond the control of any one agency due to the inherent complication of the problems being faced, and creating tools to operate under conditions of deep uncertainty, complexity or both. As such, the field may have relevance more broadly in helping provide means for evolving an approach to governance in which policy and analysis become more integrated than has previously been the norm.

The perceived inability of democratic governance to address the most vexing challenges we face has had its own effect on the public policy sphere. One of the themes that is heard most clearly in the heated debates over policy and in the political

---

[23] A red team is traditionally a specially constituted group, different in membership from the group that developed a plan, report, or conclusion, that is charged with trying to find the holes or flaws in the original product. DMDU techniques may be thought of as a way to make red-teaming part of the analytical process itself.

upheavals manifested in much of the democratic world in recent years is a cry for justice. Indeed, it is this that makes so many issues, from trade to immigration, and on through health and access to opportunity, contentious. Many are now participating in what they perceive as a quest for justice for themselves and for those with whom they identify. Lempert et al. (2013) elicit from Amartya Sen's *The Idea of Justice* (2009) key points that would appear basic to any new synthesis of analysis and policy in support of democratic governance that may flourish in an era of accelerating change. Among these, ethical [policy] reasoning should recognize as fundamental attributes the existence of both diversity of priorities, goals, and values, and an irreducible uncertainty regarding the consequences of our actions. This suggests the value of seeking agreement on which non-ideal options could be viewed as the most just from a range of perspectives and circumstances. The only way to achieve this goal in a participatory democracy is to conduct public deliberation. And this deliberation will work best when it enables re-examination and iterative assessment and permits clear explication of reasoning, logic, and underlying values. This keystone is an "open impartiality" that accepts the legitimacy and importance of the views of others. And these properties are inherent in not only the applications of DMDU but in the approaches themselves.

In many respects, the practice of DMDU analysis is just entering its adolescence. Many questions remain regarding the role of models, how to harness sufficient computational power to fully exploit the latent possibilities for analysis, how to apply DMDU concepts when formal models don't exist, and how the formalisms presented in this book may evolve and be tested in wider application. But the fundamental approach that motivates them all is an intention to move the practice of analysis, as well as the decision processes, policy deliberations, and governance applications DMDU analysts seek to support, in directions that would appear most needful to confront the challenges we will face in years to come.

References

Angerman, W. S. (2004). *Coming full circle with Boyd's OODA loop ideas: An analysis of innovation diffusion and evolution*, Wright-Patterson Air Force Base, Ohio: Air Force Institute of Technology, Air University, Department of the Air Force, AFIT/GIR/ENV/04 M-01.

Boyd, J. R. (1996). "The Essence of Winning and Losing," briefing, Chet Richards and Chuck Spinney, eds.

Bush, V. (1945). *Science, the endless frontier: A report to the President by Vannevar Bush, director of the Office of scientific research and development*. Washington: US Government Printing Office.

Childe, V. G. (1954). Early forms of society. In S. Charles, E. J. Holmyard, & A. R. Hall (Eds.), *The Oxford history of technology*. London: Oxford University Press.

Davis, P. K., Bigelow, J. H. (1998). *Experiments in multiresolution modeling (MRM)*. RAND Corporation, MR-1004-DARPA. https://www.rand.org/pubs/monograph_reports/MR1004.html.

Davis, P. K., Bigelow, J. H. (2003). *Motivated metamodels: Synthesis of cause-effect reasoning and statistical metamodeling*, RAND Corporation, MR-1570-AF.

Ecola, L., Popper, S. W. Silberglitt, R., Fraade-Blanar, L. (2018) The road to zero: A vision for achieving zero roadway deaths by 2050. National Safety Council and the RAND Corporation,

RR-2333-NSC. As of August 31, 2018. https://www.rand.org/pubs/research_reports/RR2333.html.

Kahneman, D. (2011). *Thinking, fast and slow*. New York: Farrar, Straus and Giroux.

Kalra, N., Hallegatte, S., Lempert, R., Brown, C., Fozzard, A., Gill, S., Shah, S. (2014). *Agreeing on robust decisions: A new process for decision making under deep uncertainty*. WPS-6906, World Bank.

Klein, G. A. (1999). *Sources of power: How people make decisions*. Massachusetts: MIT Press.

Lempert, R. J., Groves, D. G., Fischbach, J. R. (2013). *Is it ethical to use a single probability density function?* RAND Corporation WR-992-NSF.

March, J. G., & Simon, H. A. (1958). *Organizations*. Oxford, England: Wiley.

Mitchell, S. D. (2009). *Unsimple truths: Science, complexity, and policy*. Chicago: University of Chicago Press.

Popper, S. W. (2003). The third ape's problem: a parable of reasoning under deep uncertainty, RAND P-8080. This may also be found in abbreviated form in Robert J. Lempert and Steven W. Popper, "High performance government for an uncertain world", in Robert Klitgaard and Paul C. Light (2004), *High Performance Government: Structure, Leadership, Incentives*, (RAND: Santa Monica), pp. 113–135.

Popper, S. W. (2015). Paris gets the (decision) science right. December. https://www.rand.org/blog/2015/12/paris-gets-the-decision-science-right.html.

Rittel, H. W. J., & Webber, M. W. (1973). Dilemmas in a general theory of planning. *Policy Sciences, 4*, 155–169.

Sen, A. (2009). *The idea of justice*. Cambridge: Massachusetts, Belknap Press.

Simon, H. A. (1956). Rational choice and the structure of the environment. *Psychological Review, 63*(2), 129–138.

van Grol, R., Walker W., de Jong, G., Rahman, A. (2006). Using a meta-model to analyse sustainable transport policies for Europe: The SUMMA project's fast simple model. In *21st European Conference on Operational Research*, Reykjavik, Iceland, 2–5 July 2006.

Dr. Steven W. Popper (RAND Corporation) is a RAND Senior Economist and Professor of science and technology policy in the Pardee RAND Graduate School. His work on macrotransitions led to an invitation by President Vaclav Havel to advise the government of Czechoslovakia, participation in an OECD delegation on the first foreign visit to one of the secret cities of the former Soviet Union, and consultation to the World Bank on issues of industrial restructuring in Hungary and in Mexico. His work on microlevel transition focuses on innovation. From 1996 to 2001, he was the Associate Director of the Science and Technology Policy Institute, providing analytic support to the White House Office of Science and Technology Policy and other executive branch agencies. He has taught planning under deep uncertainty at the Pardee RAND Graduate School, the India School of Business, and the Shanghai Climate Institute. He is co-developer of the Robust Decision Making (RDM) approach. He is an elected Fellow of the American Association for the Advancement of Science, served as chair of the AAAS Industrial Science and Technology section, and is the founding Chair for Education and Training of the Society for Decision Making under Deep Uncertainty.

Open Access This chapter is licensed under the terms of the Creative Commons Attribution 4.0 International License (http://creativecommons.org/licenses/by/4.0/), which permits use, sharing, adaptation, distribution and reproduction in any medium or format, as long as you give appropriate credit to the original author(s) and the source, provide a link to the Creative Commons licence and indicate if changes were made.

The images or other third party material in this chapter are included in the chapter's Creative Commons licence, unless indicated otherwise in a credit line to the material. If material is not included in the chapter's Creative Commons licence and your intended use is not permitted by statutory regulation or exceeds the permitted use, you will need to obtain permission directly from the copyright holder.

Chapter 17
Conclusions and Outlook

Vincent A. W. J. Marchau, Warren E. Walker, Pieter J. T. M. Bloemen and Steven W. Popper

Abstract There is increasing interest in policies that can effectively address deeply uncertain conditions or developments.

17.1 Introduction

There is increasing interest in policies that can effectively address deeply uncertain conditions or developments. Researchers, practitioners, and decisionmakers are increasingly confronted with challenges—vulnerabilities or opportunities—for which:

- not all uncertainties can be eliminated;
- ignoring uncertainty could mean that we limit the ability to take corrective action in the future and end up in situations that could have been avoided; and
- ignoring uncertainty can result in missed chances and opportunities.

In Parts I and II of the book, a variety of analytical approaches (and tools) are proposed to handle situations of decisionmaking under deep uncertainty (DMDU). These DMDU approaches all have their roots in Assumption-Based Planning (ABP) which, given an initial policy, tries to protect this policy from failing, by (see Fig. 17.1b): (1) examining each of the underlying assumptions, and seeing what would happen

---

V. A. W. J. Marchau (✉)
Radboud University (RU), Nijmegen, The Netherlands
e-mail: v.marchau@fm.ru.nl

W. E. Walker
Delft University of Technology, Delft, The Netherlands

P. J. T. M. Bloemen
Staff Delta Programme Commissioner, Ministry of Infrastructure and Water Management,
The Hague, The Netherlands
e-mail: pieter.bloemen@deltacommissaris.nl

S. W. Popper
Pardee RAND Graduate School, RAND Corporation, Santa Monica, CA, USA

© The Author(s) 2019
V. A. W. J. Marchau et al. (eds.), *Decision Making under Deep Uncertainty*,
https://doi.org/10.1007/978-3-030-05252-2_17

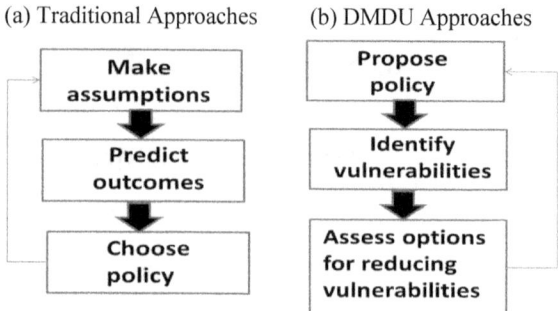

Fig. 17.1 Traditional approaches versus DMDU approaches

to the initial policy if that assumption were not to be true, and (2) developing contingent actions to protect the initial policy from failing and adjusting the policy in the future as needed. In this way, all the analytical approaches in this book try to improve the adaptivity and robustness of a policy—i.e., to make a policy more resistant to significant change, and to help a decisionmaker to identify when to adapt the policy. In this sense, the DMDU approaches turn the traditional approach to policy analysis (Fig. 17.1a) upside down.

In addition to describing DMDU approaches in theory and practice, the book (Part III) looks into the design of institutions and processes to facilitate decisionmaking under deep uncertainty. This requires the specification of procedures and legislation to: (a) enable policies to respond to events and information as they arise, (b) undertake data collection (monitoring), and (c) repeatedly review goals.

Finally, the two chapters constituting Part IV draw upon the material presented in the earlier chapters—not as recapitulation but rather as a synthesis. In a field that has grown so rapidly in both methods and applications, these chapters propose a framing of the field as it has developed to date, as well as a trajectory for application moving forward. Chapter 15 presents a synthesis and guide to the DMDU field with emphasis on its key ideas, and an attempt to frame a taxonomy for its burgeoning set of methods. Chapter 16 derives from the applications presented in the book a theory of where, why, and to what ends DMDU methods may be applied to a growing class of policy conundrums that have confounded the application of more traditional analytical approaches in the past.

17.2 DMDU Approaches—Commonalities and Differences

As stated above, all of the approaches in this book have their roots conceptually in ABP. Comparing the approaches, both commonalities and differences can be identified.

All of the approaches represent uncertainties with a broad range of futures instead of probabilities of specific future states of the world. The way exploratory modeling, a central element in most approaches, is used in combination with adaptive policies explicitly includes uncertainties arising from simulating the real world in a system model. The range of futures may be limited by choosing to prepare "only" for plausible futures. What is considered a "plausible" future is subject to different interpretations (which may be a function of personal beliefs, political setting, etc.), and depends on one's expectations about the future and understanding of the system. Moreover, what is considered acceptable performance of a policy depends on people's values.

All of the approaches aim at enhancing the robustness of a plan by keeping it from failing, or by enhancing its adaptivity by organizing a decisionmaking process that is prepared to adjust the plan to fit new conditions, and to do so in a timely manner. That is, all approaches include (some more explicitly than others) the question: what could make the plan (in its original form or after adjustments) fail, in the sense that the policy goals are not achieved? IG first builds an uncertainty model and then uses this model to evaluate the performance of a set of alternatives over the range of uncertainties. In contrast, RDM begins with the set of alternatives and aims to characterize the uncertainties in a way that usefully highlights the trade-offs among them. DAP explicitly distinguishes different types of actions to keep an initial policy from failing. Such actions can result from the last three steps in RDM. Both DAP and RDM have similar first steps, and go back one step in policymaking compared to the Assumption-Based Planning approach. They do not assume that an initial plan exists, but begin by designing an initial plan to be examined. IG, RDM, DAP, and DAPP have similar mindsets, but have subtle differences in operationalization:

- IG examines a set of predefined alternatives to identify the one that is most robust.
- RDM mainly focuses on increasing the robustness of the initial policy using Scenario Discovery.
- DAP adds triggers to change the initial policy, and specifies the contingent actions.
- DAPP adds the sequencing of changes to the initial policy by adding contingent actions.

The focus of EOA is on exploring the performance of diverse alternative system designs in more detail, each with varying degrees of flexibility in the form of options.

Looking at the DMDU approaches, we note several recurring principles. The essential idea is that decisionmakers facing deep uncertainty create a shared strategic vision of future ambitions, explore possible adaptive strategies, commit to short-term actions, while keeping long-term options open, and prepare a framework (including in some cases a monitoring system, triggers, and contingency actions) that guides future actions. Implicit in this is that decisionmakers accept the irreducible character of the uncertainties about future conditions and aim to reduce uncertainty about the *performance* of their policies. In short:

- Explore a wide variety of relevant uncertainties in a dynamic way.
- Connect short-term targets and long-term goals.
- Commit to short-term actions while keeping options open for the future.

RDM focuses on designing a robust policy by embedding contingencies in its implementation; adaptation can be added on to the policy to respond to external developments. DAP makes explicit the importance of monitoring and adapting to changes over time to prevent the initial policy from failing. DAPP expands the contingency planning concept from DAP by specifying the conditions and time frame under which adaptation of the initial policy is continued into the far future. As such, these three approaches can be seen as extensions of each other. IG can be categorized as an approach for robust decisionmaking. In contrast to the other approaches considered in this paper, unforeseen events (Black Swans) are not incorporated: IG addresses modeled uncertainty, not unexpected uncertainty. EOA, with its detailed analysis of different possible (technical) alternatives provides unique insights, distinct from and complementary to those obtained from the other DMDU approaches.

17.3 DMDU Implementation

DMDU scholars often implicitly assume that their approaches will automatically be welcomed—that this way of reasoning will be embraced politically and accepted institutionally—suggesting that implementing a DMDU approach is mainly a technical and intellectual challenge. Experience, however, shows that in real-life decisionmaking, organizational aspects play a major role in determining the willingness and ultimate success in applying approaches for dealing with deep uncertainty in practice.

First of all, an appealing and convincing narrative is required to mobilize political interest for addressing an issue characterized by deep uncertainty—notably when it involves starting up a costly long-term program. As illustrated in Part III of the book, an adaptive approach can do the job, especially when informative illustrations such as adaptation pathway maps are used.

Following the decision to actually start a long-term program, it is important to create the conditions that can keep politics involved "at arm's length"—well-informed on the main results of the program, but not encouraged or compelled to actively intervene in everyday operational issues. In the case of the Dutch Delta Programme, the Parliament and Senate agreed that the water-related challenges of climate proofing the Netherlands needed to be addressed in a long-term program, secured by the underlying Delta Act, the position of a relatively independent Delta Programme Commissioner assigned to manage the program, and a guaranteed budget of €1 billion per year for decades to come.

A robust, stable basis constitutes not only a precondition for consistency in reacting to new challenges, but also allows for the agility in decisionmaking that is required for adaptive policies. Alertness and willingness at both the political and administrative levels, to adjust strategies and plans to changing conditions, are key in an adaptive approach. Organizational complexes are needed that can effectively plan to adapt. Part III presents a conceptual model of planned adaptation.

Much of the work on adaptive policymaking is about establishing *primary rules* that describe how to sustain system function and integrity—for example, formal standards for minimal heights and for construction requirements of levees that guarantee a certain level of flood safety. These are the rules that determine either how a system should be monitored and sustained, or how the system itself is to function. The DMDU approaches and tools described in Parts I and II of this book are means by which an analyst may systematically evaluate and develop the knowledge informing primary rules for policymakers to manage a given system.

As illustrated in Part III, *secondary rules* should also be considered. They are about how to recognize, create, maintain, and adjudicate primary rules. Secondary rules are "all about [primary rules]; in the sense that while primary rules are concerned with the actions that individuals must or must not do, these secondary rules are all concerned with the primary rules themselves. They specify the ways in which the primary rules may be conclusively ascertained, introduced, eliminated, varied, and the fact of their violation conclusively determined" (Hart 1994). It follows that adaptive processes must strike a balance between the canonical objective of regulation (creating stability) and the potential for chaos if feedback loops trigger changes too frequently.

Implementing adaptive strategies requires organizational arrangements for systematically accommodating adjustments of policies, strategies, and plans, a monitoring system for timely detection of signals, and a decisionmaking process that links directly to its output.

Challenges for further development of adaptive approaches comprise, among others, determining Adaptation Tipping Points in situations of low signal-to-noise ratio (e.g. large natural variability in river discharge) and preparing a switch from incremental to transformational interventions. Among more political challenges are formulating precise policy goals and keeping long-term options open.

On a more general level, it can be concluded that DMDU approaches, such as those described in the scientific literature, can profit from feedback—feedback from other researchers, and feedback from practitioners. Organizing the latter (an instrumental element of "coproduction of knowledge") might be more time-consuming, but is likely to be very effective.

17.4 Concluding Remarks

The main objective of DMDU approaches is to facilitate the development of policies that are robust and/or adaptive, meaning that they perform satisfactorily under a wide variety of futures and can be adapted over time to (unforeseen) future conditions. Agreeing in advance about what future conditions require what contingent actions increases the chances that the policy goals will be reached—with the original strategy or with an adapted version. Doing this might be more time-consuming, but is likely to be very effective.

The breakthroughs in analytical technique over the past two decades have occurred in the presence of increased general awareness of the reality of irreducible uncertainty in practical applications. This has led to a proliferation of methods—and in the description of them. All are subject to interpretation, offer quite some leeway to the user in how to use them, and are changing over time. The present book is a testament to the need to first focus on the building blocks that make up the various approaches. This may lead to a future synthesis where DMDU as a field comes more to be viewed as an approach in itself with particular applications being constructed by swapping one building block with another one. This task is made easier because of the centrality of fundamental concepts in DMDU thinking: exploratory modeling, adaptation, and decision support. Chapter 15 provides the first taxonomic step on the road to exploring the possibility of such a synthesis as well as a guide to its design.

Chapter 16 argues that the time for doing so is ripe. If one may observe a general dissatisfaction with, or general incapacity of, the institutions of democratic governance to meet the challenges now confronting all societies, the origins stem in large part from those circumstances for which DMDU methods were designed: a bewildering number of interacting elements; path dependence, unpredictability, emergence, and other phenomena associated with complex systems; irreducible uncertainty; proliferation of information sources and stakeholder interests; and the inability of single government agencies to deal with all aspects of problems that touch upon our lives in many ways.

DMDU methods collectively represent an evolving capacity to support policy processes better suited to deep uncertainty and dynamic change. DMDU applications collectively present a body of theory and practice with the potential for providing a common vocabulary to the work of both analysts and of those charged with policy design and implementation during uncertain times. To realize this full potential, two things must occur. Analysts who apply DMDU methods to policy problems must think through measures, both individual and collective, to make the possibilities for providing meaningful support to policy more widely known. And the policy community should take more explicit cognizance of the shortcomings in the more traditional use of analysis in the policy process in the presence of the policy challenges they currently face, and so provide the opening that DMDU practitioners require to demonstrate the value of their approaches and tools. The process of education needs to be mutual.

There are still overarching challenges for the DMDU community to address:

- Improvement of the existing DMDU approaches, especially the tools (Part I), in terms of faster, simpler models and their more transparent use. In particular, there is a need to consider how the power of DMDU may be brought to bear on those problems for which there are few formal models.
- Further guidance on when and how to apply a specific DMDU approach and tools is needed (Part II). Dittrich et al. (2016) propose a framework for how different DMDU approaches can work in different circumstances, depending on the characteristics of the adaptation options being considered, the data available, and the time and skills available to the decisionmaker.
- The scope of DMDU applications (Part II) should be broadened from their current focus on climate change and water management to other policy issues that are faced with deep uncertainty. Examples include transportation, energy, security, health care, and spatial planning. How can the current DMDU knowledge be applied to these and other domains?

- "Monitor and adapt" is gradually becoming preferred by policy analysts to "predict then act" as the strategy for long-term decisionmaking in the face of deep uncertainty. i.e., as a means to design policies that are able to achieve economic, environmental, and social objectives for a long-term uncertain future. More needs to be done to bridge the gap between DMDU researchers and policymakers to improve mutual understanding (Part III).

The authors of the chapters in this book are already hard at work in developing solutions to these challenges. They and their students will carry DMDU concepts, tools, and approaches to the next level to meet the challenges in the coming years.

References

Dittrich, R., Wreford, A., & Moran, D. (2016). A survey of decision-making approaches for climate change adaptation: Are robust methods the way forward? *Ecological Economics, 122*, 79–89.

Hart, H. L. A. (1994). *The concept of law* (2nd ed.). Clarendon Law Series. Oxford: Oxford University Press.

Prof. Vincent A. W. J. Marchau (Radboud University (RU), Nijmegen School of Management) holds a chair on Uncertainty and Adaptivity of Societal Systems. This chair is supported by The Netherlands Study Centre for Technology Trends (STT). His research focuses on long-term planning under uncertainty in transportation, logistics, spatial planning, energy, water, and security. Marchau is also Managing Director of the Dutch Research School for Transport, Infrastructure and Logistics (TRAIL) at Delft University of Technology (DUT), with 100 Ph.D. students and 50 staff members across 6 Dutch universities.

Prof. Warren E. Walker (Emeritus Professor of Policy Analysis, Delft University of Technology). He has a Ph.D. in Operations Research from Cornell University, and more than 40 years of experience as an analyst and project leader at the RAND Corporation, applying quantitative analysis to public policy problems. His recent research has focused on methods for dealing with deep uncertainty in making public policies (especially with respect to climate change), improving the freight transport system in the Netherlands, and the design of decision support systems for airport strategic planning. He is the recipient of the 1997 President's Award from the Institute for Operations Research and the Management Sciences (INFORMS) for his 'contributions to the welfare of society through quantitative analysis of governmental policy problems.'

Drs. Pieter J. T. M. Bloemen (Ministry of Infrastructure and Water management—Staff Delta Programme Commissioner) is the Chief Strategic Officer of the Dutch Delta Programme. He is responsible for the development and application of Adaptive Delta Management. He led the Strategic Environmental Assessment of the Delta Decisions and preferred regional strategies published in 2014. He presently works on the development of a monitoring and evaluation system that matches the adaptive approach of the Delta Programme and is responsible for the first six-yearly review of the Delta Decisions and preferred regional strategies, planned for 2020. Bloemen is Visiting Researcher at IHE Delft Institute of Water Education (Chair Group Flood Resilience) since January 2015 and works on a Ph.D. thesis on the governance of the adaptive approach.

Dr. Steven W. Popper (RAND Corporation) is a RAND Senior Economist and Professor of science and technology policy in the Pardee RAND Graduate School. His work on macrotransitions led to an invitation by President Vaclav Havel to advise the government of Czechoslovakia, participation in an OECD delegation on the first foreign visit to one of the secret cities of the former Soviet Union, and consultation to the World Bank on issues of industrial restructuring in Hungary and in Mexico. His work on microlevel transition focuses on innovation. From 1996 to 2001, he was the Associate Director of the Science and Technology Policy Institute, providing analytic support to the White House Office of Science and Technology Policy and other executive branch agencies. He has taught planning under deep uncertainty at the Pardee RAND Graduate School, the India School of Business, and the Shanghai Climate Institute. He is co-developer of the Robust Decision-Making (RDM) approach. He is an elected Fellow of the American Association for the Advancement of Science, served as chair of the AAAS Industrial Science and Technology section, and is the founding chair for education and training of the Society for Decision Making under Deep Uncertainty.

Open Access This chapter is licensed under the terms of the Creative Commons Attribution 4.0 International License (http://creativecommons.org/licenses/by/4.0/), which permits use, sharing, adaptation, distribution and reproduction in any medium or format, as long as you give appropriate credit to the original author(s) and the source, provide a link to the Creative Commons licence and indicate if changes were made.

The images or other third party material in this chapter are included in the chapter's Creative Commons licence, unless indicated otherwise in a credit line to the material. If material is not included in the chapter's Creative Commons licence and your intended use is not permitted by statutory regulation or exceeds the permitted use, you will need to obtain permission directly from the copyright holder.

Glossary

This document provides a glossary of terms commonly used in DMDU studies. Providing such a glossary is challenging for two reasons. First, many of these terms (e.g., uncertainty, scenario, policy) have different meanings across the different scholarly and practice communities with which DMDU analysts interact. Second, the terms in the glossary describe both aspects of the analysis and corresponding attributes of the real world. Nevertheless, we offer the following key to frequently used terms as a rough guide to better assist the reader in understanding the language used in DMDU studies, and its mapping to the real world facing decisionmakers.[1]

Adaptation Tipping Point (ATP) An ATP is reached when a policy is no longer able to achieve its objectives; i.e., the conditions under which the current policy starts to perform unacceptably (Kwadijk et al. 2010). (This is sometimes called a '**threshold**'.). An ATP is related to a '**trigger point**,' in that the adaptive policy should actually be changed (triggered) at a time sufficiently before the ATP that is greater than the lead time required to implement the change.

Adaptive policy A **policy** (see) that is designed *ex ante* to be changed over time as new information becomes available or as the situation changes (see also **flexible policy**).

Analytic model A model that is derived, often mathematically, from a theoretical statement of relationships that hold in the real world. The relationships are used to project how a policy will perform in a given **future** (see), evaluated according to the specified **outcome indicators** (see). The analytic model is

---

[1] We are aware that there is discussion on whether to write policy-making, policy making, or policymaking. (The same goes for decisionmaker and policymaker.) This is a question of style. There are a variety of style manuals available (e.g. Harvard, University of Chicago, RAND). However, there is one style manual that focuses on policy analysis—the RAND Style Manual. We, therefore, selected this style, and so we use policymaking, decisionmaking, policymaker, etc., as single, unhyphenated words.

© The Editor(s) (if applicable) and The Author(s) 2019
V. A. W. J. Marchau et al. (eds.), *Decision Making under Deep Uncertainty*,
https://doi.org/10.1007/978-3-030-05252-2

typically embodied in computer code. Bankes (1993) distinguishes between 'consolidative' and 'exploratory' models. The former are validated and predictive. The latter provide a mapping of assumptions to consequences without any judgment regarding the validity of alternative assumptions. DMDU analyses typically regard models as exploratory and use them in performing **Exploratory Modeling** (see).

Approach A process to design a plan or strategy to react to a **problem or opportunity** (see). It focuses on the steps to be taken in performing the study.

Case study This book includes several 'case studies,' which are applications of a DMDU approach to a specific real-world situation.

Case In DMDU analyses, the term 'case' has a very specific meaning. It is a run of the system model for one future (scenario) and one policy (strategy, plan). DMDU analyses typically generate a database of many model runs. Each entry in such a database is a case. Each database entry typically includes numbers describing the future, the policy, and the values of the outcome indicators that result from pursuing the policy in that future.

Deep uncertainty 'The condition in which analysts do not know or the parties to a decision cannot agree upon (1) the appropriate models to describe interactions among a system's variables, (2) the probability distributions to represent uncertainty about key parameters in the models, and/or (3) how to value the desirability of alternative outcomes' (Lempert et al. 2003). In this book, deep uncertainty is seen as the highest of four defined levels of uncertainty.

Exploratory Modeling (EM) EM is a **tool** (see) that uses computational experiments to analyze complex and uncertain systems. In each experiment, a large number of runs are made over the input space in order to generate an ensemble of runs. Because it requires making a large number of runs, EM usually makes use of a '**Fast Simple Model**' (Kwakkel et al. 2009) (see).

Fast Simple Model (FSM) An FSM is 'a relatively small, simple model intended to mimic the behavior of a large complex model' (Davis and Bigelow 2003, p. 8; van Grol et al. 2006). In DMDU studies, FSMs are used to assess large numbers of alternative actions under a range of plausible futures. The main purpose of FSMs is not to provide 'the solution,' but to provide information on a (future) problem situation on which decisionmakers can base their decisions. FSMs are also sometimes called metamodels (Davis and Bigelow 2003), compact models (Gildenblat 2010), repro models (Meisel and Collins 1973), surrogate models (Razavi et al. 2012), emulation models (Machac et al. 2016), computationally efficient models, low-fidelity models, or screening models.

Flexible policy A flexible policy can be employed differently (or easily modified) to keep on meeting the desired objectives as new information becomes available or as the situation changes (see also **adaptive policy**).

Future A future, context, future state of the world (all synonymous terms) is a specific set of assumptions about the future (usually about the external environment of the system, but sometimes also about the internal structure of the system and stakeholder preferences among the outcome indicators). DMDU studies typically use sets of multiple futures to represent uncertainty. A DMDU analysis typically represents each future with a vector of specific values for each of the uncertain parameters of the model (see also **scenario**).

Implementation The entire process by which a change in policy is brought into practice. Implementation is mostly outside the scope of this book. However, the 'theory' part of the book (Part I) deals with the design of a policy, which might include some consideration of how the policy might be implemented (e.g., setting up a monitoring system); the 'practice' part (Part II) sometimes addresses implementation issues; and Part III of the book deals directly with policy implementation issues.

Objectives The set of outcomes (in general terms) that the problem owner and other stakeholders desire to achieve (sometimes called 'goals'). Sometimes, the objectives are related to preferences (weights) among the outcomes (which help to determine how the policy alternatives are to be ranked in order of desirability).

Outcome indicators The quantifiable (measurable) outcomes related to the objectives (which are generally the outcomes of the system models used in the analysis). These are the (measurable) results that the system models suggest would ensue were the policy to be adopted and implemented under the assumed conditions (they are also called 'metrics' or 'performance measures').

Policy analysis process: Most of the DMDU approaches apply the policy analysis process (Walker 2000), which is a methodical, iterative process for designing a policy, which involves choosing among alternative actions (policies) in a complex system. It is characterized by a series of steps that include identification of the problem and the objectives of the analysis, choosing policy evaluation criteria (**outcome indicators**), using a model of the system, and comparing the alternative policies.

Policy A policy (also often called a strategy or plan) represents a distinct choice facing a planner or decisionmaker. It is often defined by the amount, location, and timing of different interventions, programs, or regulations under the control of the actors in the policy domain whose decision is being analyzed. Different policies represent different alternative conceptions for a course of future actions (also called policy options or alternatives). A DMDU policy is often designed to be adaptive. In this case, it is composed of initial actions (implemented when the policy is first implemented) and contingent or long-term actions (implemented if an **Adaptation Tipping Point** (see) is reached).

Problem (or opportunity or question) This is both the catalyst and the guiding principle for a DMDU study. For such a study to be undertaken, someone or organization (the 'problem owner') must be dissatisfied with the current or anticipated outcomes of a system (or must see opportunities for the future) and want help in discovering how to bring about more preferred outcomes. DMDU analyses can be used to present arguments and information to help win acceptance for a course of action (policy). Thus, DMDU analyses serve as decision support.

Resilient policy A resilient policy can resume meeting the desired objectives (or can 'bounce back') reasonably quickly after a shock to the system or as the situation changes (de Haan et al. 2011). Resilience can also be a property of the system itself.

Robust Policy A robust policy can keep on meeting the desired objectives as new information becomes available or as the situation changes.

Scenario Discovery (SD) SD is a **tool** (see) to distinguish futures in which proposed policies meet or miss their objectives. It begins with a large database of model runs (e.g., from **EM**, see) in which each model run represents the performance of a strategy in one future. The SD algorithms identify those combinations of future conditions that best distinguish the cases in which the policy does or does not meet its objectives.

Scenario A scenario (or decision-relevant scenario) may be used synonymously with **case** (see above) or even **future** (see above; but this usage is more usually found in the scenario planning literature.) Alternatively, in some DMDU studies the term is used to refer to a set of cases that share some decision-relevant attribute. For instance, a region in a plot of case outcomes where a strategy performs poorly might be considered such a scenario (this is the meaning of scenario in SD).

Threshold See Adaptation Tipping Point.

Tool A tool refers to the computational support, models, statistical analyses, etc., that can be used to carry out one or more steps of an **approach** (see).

Trigger point See Adaptation Tipping Point.

Uncertainty An uncertainty is a single factor affecting the future context (external to the system), model structure, model parameters, or stakeholder preferences among the outcome indicators that lie outside the control of the actors in the policy domain and about which the policy actors possess insufficient knowledge when the policy is being planned or implemented.

Vulnerability A vulnerability is a characteristic of a policy (strategy, plan) indicating that some plausible future situation might cause the policy to fail to meet its objectives. It is closely related to an **Adaptation Tipping Point** (see), which is reached when a policy is no longer able to achieve its objectives.

References

Bankes, S. C. (1993). Exploratory modeling for policy analysis. *Operations Research, 1*(3), 435–449.

Davis, P. K., Bigelow, J. H. (2003). *Motivated metamodels: Synthesis of cause-effect reasoning and statistical metamodeling*, RAND Corporation, MR-1570-AF, Santa Monica, CA.

de Haan, J., Kwakkel, J. H., Walker, W. E., Spirco, J., Thissen, W. A. H. (2011). Framing flexibility: Theorising and data mining to develop a useful definition of flexibility and related concepts. *Futures, 43*, 923–933. http://dx.doi.org/10.1016/j.futures.2011.06.002.

Gildenblat, G. (ed.) (2010). *Compact Modeling: Principles, Techniques and Applications*. Springer.

Kwadijk, J. C. J., Haasnoot, M., Mulder, J. P. M., Hoogvliet, M., Jeuken, A., van der Krogt, R., et al. (2010). Using adaptation tipping points to prepare for climate change and sea level rise: A case study in the Netherlands. *Wiley Interdisciplinary Review Climate Change, 1*(5), 729–740.

Kwakkel, J. H., Wijnen, R. A. A., Walker, W. E., Marchau, V. A. W. J. (2009). A fast and simple model for a quick scan of airport performance. In P. Hooper (Ed.), *Proceedings of the 13th Air Transport Research Society World Conference, Air Transport Research Society*, Abu Dhabi, United Arab Emirates.

Lempert, R. J., Popper, S. W., & Bankes, S. C. (2003). Shaping the next one hundred years: New methods for quantitative, long-term policy analysis. MR-1626-RPC, Santa Monica, CA: RAND.

Machac, D., Reichert, P., Rieckermann, J., Albert, C. (2016). Fast mechanism-based emulator of a slow urban hydrodynamic drainage simulator. *Environmental Modelling & Software, 78*, 54–67. http://dx.doi.org/10.1016/j.envsoft.2015.12.007.

Meisel, W. S., Collins, D. C. (1973). Repro-modeling: An approach to efficient model utilization and interpretation. *IEEE Transactions on Systems, Man, and Cybernetics, SMC-3*(4), 349–358. http://dx.doi.org/10.1109/TSMC.1973.4309245.

van Grol, R., Walker, W., de Jong, G., Rahman, A. (2006). Using a meta-model to analyse sustainable transport policies for Europe: The SUMMA project's fast simple model. In *21st European Conference on Operational Research*, Reykjavik, Iceland, 2–5 July 2006.

Walker, W. E. (2000). Policy analysis: A systematic approach to supporting policymaking in the public sector. *Journal of Multicriteria Decision Analysis, 9*(1–3), 11–27.

CPSIA information can be obtained
at www.ICGtesting.com
Printed in the USA
LVHW080154071022
730184LV00003B/15